D1690293

Microfauna Marina
Vol. 11

Akademie der Wissenschaften und der Literatur
Mathematisch-naturwissenschaftliche Klasse

Kommission für Zoologie

Akademie der Wissenschaften und der Literatur · Mainz · 1997

Microfauna Marina

Editor: Peter Ax

Vol. 11

Gustav Fischer Verlag · Stuttgart · Jena · Lübeck · Ulm · 1997

Gefördert mit Mitteln des
Bundesministeriums für Bildung, Wissenschaft, Forschung und Technologie, Bonn,
und des Niedersächsischen Ministeriums für Wissenschaft und Kultur, Hannover

Anschrift des Herausgebers
Prof. Dr. *Peter Ax*, II. Zoologisches Institut und Museum
der Universität Göttingen, Berliner Straße 28, D-37073 Göttingen

Abbildung auf dem Einband:

Convolutriloba longifissura (Acoela) – erstes Beispiel einer Längsteilung bei Plathelminthen

---

Die Deutsche Bibliothek – CIP-Einheitsaufnahme

Microfauna marina / Akad. d. Wiss. u. d. Literatur,
Math.-Naturwiss. Klasse, Komm. für Zoologie;
Akad. d. Wiss. u. d. Literatur, Mainz –
Stuttgart; Jena; New York: Fischer
ISSN 0176–3296
Erscheint unregelmäßig
Forts. von: Mikrofauna des Meeresbodens
Vol. I (1984) –

© 1997 by Akademie der Wissenschaften und der Literatur, Mainz.
Alle Rechte einschließlich des Rechts zur Vervielfältigung, zur Einspeisung in elektronische Systeme sowie Übersetzung vorbehalten. Jede Verwertung außerhalb der engen Grenzen des Urheberrechtsgesetzes ist ohne ausdrückliche Genehmigung der Akademie und des Verlages unzulässig und strafbar.
Gesamtherstellung: Röhm GmbH, Sindelfingen. Printed in Germany.
Gedruckt auf säurefreiem, chlorfrei gebleichtem Papier.

ISBN 3-437-25696-3
ISSN 0176-3296

Inhalt/Contents

Thomas Bartolomaeus and Ivonne Balzer: *Convolutriloba longifissura* nov. spec. (Acoela) – the first case of longitudinal fission in Plathelminthes 7

Peter Ax: *Beklemischeviella angustior* Luther and *Vejdovskya parapellucida* n. sp. (Rhabdocoela, Plathelminthes) from brackish water of the Winyah Bay, USA 19

Beate Sopott-Ehlers: First report on the fine structure of unpigmented rhabdomeric photoreceptors in a free-living species of the "Dalyellioida" (Plathelminthes, Rhabdocoela) 27

Sybille Seifried: Three new species of *Ectinosoma* Boeck, 1865 (Harpacticoida, Ectinosomatidae) from Papua New Guinea and the Fiji Islands 35

Beate Sopott-Ehlers: Ultrastructural observations on the "eye spot" of *Halammovortex nigrifrons* (Plathelminthes, Rhabdocoela, "Dalyellioida") 59

Kai Horst George: *Mielkiella spinulosa* gen. n. sp. n., a new taxon of the Laophontidae (Copeopda, Harpacticoida) from Porvenir, (Tierra del Fuego, Chile) 71

Rudolf Meyer und Thomas Bartolomaeus: Ultrastruktur und Morphogenese der Hakenborsten bei *Psammodrilus balanoglossoides* – Bedeutung für die Stellung der Psammodrilida (Annelida) 87

Wolfgang Mielke: Interstitial Fauna of Galapagos. XXXIX. Copepoda, part 7 115

Wolfgang Mielke: Interstitial Fauna of Galapagos. XL. Copepoda, part 8 153

BEATE SOPOTT-EHLERS and ULRICH EHLERS: Electronmicroscopical investigations of male gametes in *Ptychopera westbladi* (Plathelminthes, Rhabdocoela, "Typhloplanoida") 193

BEATE SOPOTT-EHLERS: Submicroscopic anatomy of female gonads in *Ciliopharyngiella intermedia* (Plathelminthes, Rhabdocoela, "Typhloplanoida") 209

WOLFGANG MIELKE: On a small collection of Laophontidae (Copepoda) from Sulawesi, Indonesia 223

BEATE SOPOTT-EHLERS: Fine-structural features of male and female gonads in *Jensenia angulata* (Plathelminthes, Rhabdocoela, "Dalyellioida") 251

WOLFGANG MIELKE: New findings of interstitial Copepoda from Punta Morales, Pacific Coast of Costa Rica 271

ULRICH EHLERS and BEATE SOPOTT-EHLERS: Plasma membranes flanked by cisternae of the endoplasmic reticulum: a remarkable organization of polarized cells in small Plathelminthes 281

ULRICH EHLERS and BEATE SOPOTT-EHLERS: Ultrastructure of protonephridial structures within the Prolecithophora (Plathelminthes) 291

PETER AX: Two *Prognathorhynchus* species (Kalyptorhynchia, Plathelminthes) from the North Inlet Saltmarsh of Hobcaw Barony, South Carolina, USA 317

Convolutriloba longifissura, nov. spec. (Acoela) – first case of longitudinal fission in Plathelminthes

Thomas Bartolomaeus & Ivonne Balzer

Abstract

Convolutriloba longifissura nov. spec. was isolated from a sample of coral debris from the Indonesian coast. The animals have a maximum length of 6 mm and belong to the acoel plathelminths. The animals are green with areas of red pigmentation. The green colour is caused by symbiontic algae which belong to the taxon *Tetraselmis*. The anterior region of the animals is rounded and contains two unpigmented areas, which are presumed to be eyefields. Caudally they bear three lobes. The animals were kept alive at 25° C; at temperatures below 20° C they degenerate. The animals also die within four days, if they are kept in darkness. It is concluded from this observation that they essentially depend on their symbionts. *Convolutriloba longifissura* reproduces asexually by longitudinal fission. Fisson always starts anteriorly and one eyefield is adopted by each daughter individual. Although fisson occurs along the median longitudinal axis of the animal, the caudal lobes are inequally divided, so that one of the daughter individuals retains one, the other one two lobes. The internal organs are completed during fission (architomy) and the new individuals soon acquire the same size as the mother individual. *Convolutriloba longifissura* is the first record of a plathelminth with longitudinal fission and even within the Bilateria this mode of fission has thus far been unknown for vagile organisms.

A. Introduction

Asexual reproduction occurs in several Plathelminthes either as paratomy or as architomy and is regarded as a plesiomorphic characteristic of the Plathelminthes (RIEGER 1986). Architomy is characterized by a desintegration of the mother animal into pieces, before new organs are formed, whereas paratomy is characterized by the formation of new organs before the animals separate. Both kinds of asexual reproduction occur as transversal fissions, i.e. the new individuals initially lie one behind the other. Besides this, experimental fragmentation has repeatedly been reported to result in generation of new individuals from each fragment, but fragmentation does not seem to be the normal mode of asexual reproduction. Paratomy (AX & SCHULZ 1959, DÖRJES 1966, 1968) and architomy (HANSON 1960) have been reported from the Acoela. Recently, HENDELBERG and ÅKESSON (1988) described a new representative of the Acoela, *Convolutriloba retrogemma* Hendelberg et Åkesson 1988, which asexually reproduces by budding and which is regarded as a special mode of architomy. The polarity of the daughter individudals, which are generated at the caudal edge of the mother animal, is reversed 180° relative to the mother individual (HENDELBERG & ÅKESSON 1991). *Convolutriloba retrogemma* is characterized by three caudal lobes and contains algal symbionts which lead to the green colour of the animals. Additional red spots are scattered over the entire surface, sometimes forming large clusters.

This paper describes a new species of the taxon *Convolutriloba*, which contains symbiontic algae and shares the same external characteristics like *Convolutriloba retrogemma*, but reproduces asexually in a completely different manner. Individuals of *Convolutriloba longifissura* separate into two daughter animals by longitudinal fission, which thus far has never been described for plathelminths.

B. Material and Methods

Material examined: The material was obtained from coral pieces bought at a local supplier for aquaristic supplements. The coral pieces were collected in Indonesia. The animals were kept alive in small glass dishes with autoclaved sea water at 25° C for several years and were fed with nauplia of *Artemia salina* (Linné, 1758). More than 100 animals were observed alive and 6 animals were sectioned for light microscopy. Five series of sections (5 µm thickness) were stained with Weigert's hämatoxiline and eosin and examined under a

ZEISS Axioskop. Additionally a few animals were fixed for electron microscopy according to the method described in MEYER & BARTOLOMAEUS (1997) and examined under a ZEISS EM 10. The animals were cooled down to relax them, but as they are very sensible to low temperatures, the fixation was extremely poor.

Holotypus of *Convolutriloba longifissura*: Cross sections of an unfertile individuum. Holotypus and four paratypes are deposited at the Zoological Museum of the University of Göttingen (Holotypus = P2271, Paratypes = P2272 - P2275).

The symbiontic algae are new to science and belong to the taxon *Tetraselmis* (Chlorodendrales, Chlorophyta). They were isolated and kept in culture at the SAG collection of algae at the University of Göttingen (SAG B 35.93).

C. Description

External feature

The animals measure up to 6 mm in length. A lateral constriction at the first quarter divides the body. The anterior tip of the animal is rounded, and the caudal end bears three lobes. The animals have a green colouration with varying areas of red pigmentation (Fig. 1A-B). Under the microscope these appear as several yellow spots. A brighter area at the anterior egde of the posterior region marks the pharynx. Occasionally the animals possess a single dark spot in front of the pharynx. Two anterior-lateral, elipsoid areas lack any pigmentation (Fig. 1C). They will be termed eye fields (according to HENDELBERG & ÅKESSON (1988)) and presumably mark the position of photoreceptive organs in deeper tissue layers close to the cerebral ganglia (Fig. 3). The green colour of the animals results from symbiontic algae, which were found between the bases of epidermal cells, but occured also in deeper tissues layers. There was no polarity in the frequence of the algae.

Internal organization

The epidermis is multiciliated and shows that arrangement of ciliary rootlets characteristic to acoelans. Gland cells with a rod to lancet shaped, electron dense material inside a large vacuole are frequently found (Fig. 2C). In the light microscope these structures are yellow and in reflecting light they are red and, thus, turned out to be responsible for the red pigmentation of the animals. When the animals were squeezed for light microscopy these gland cell products soon dissolved and became coalescent, but still remained inside the va-

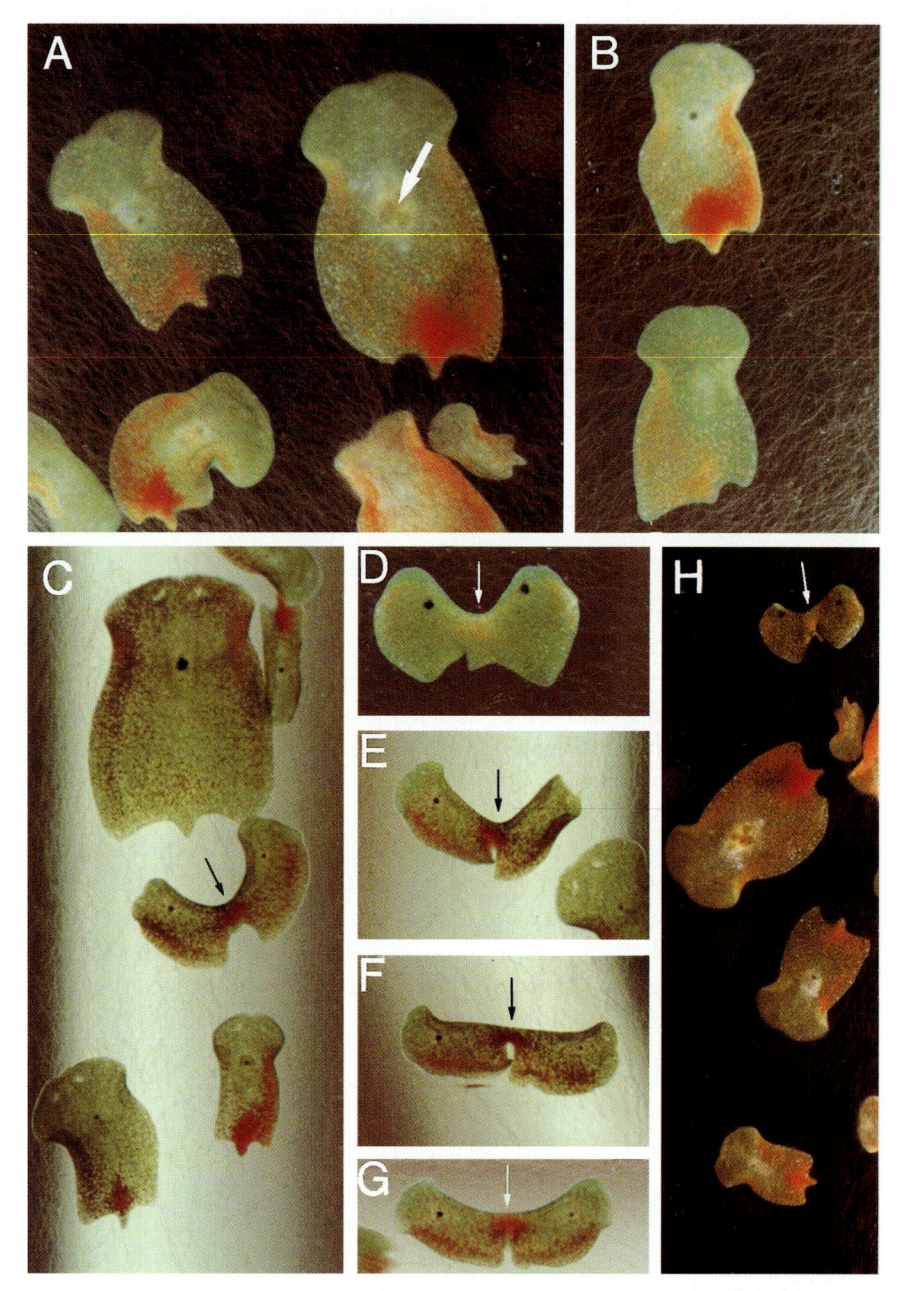

Fig. 1: Life photographs from *Convolutriloba longifissura* (Acoela) at different life cycle stages. A. Larger animals (arrow indicates fed nauplius of *Artemia salina*). B. Black vesicle anterior to the pharynx. C. Bright spots in the anterior represent eye fields. Arrow indicates advanced fission. D.-G. Different stages of longitudinal fission. Arrows indicate longitudinal axis. H. Animals of various sizes, fission is indicated by an arrow.

Convolutriloba longifissura, nov. spec.

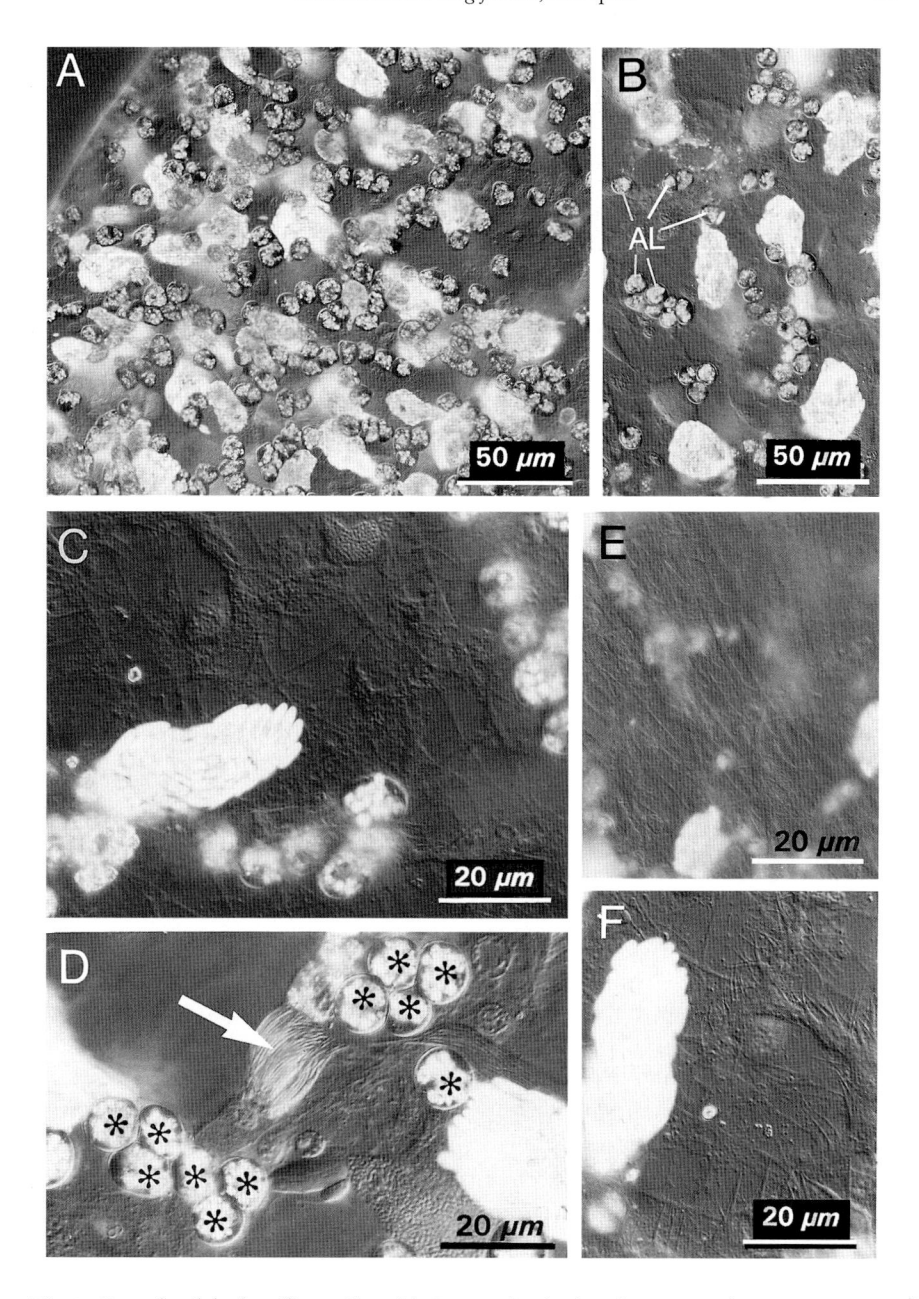

Fig. 2: *Convolutriloba longifissura* (Acoela). Squeezed animal under Normarsky contrast. A., B. Epidermis with symbiontic algae (*AL*) and large gland cells. C. Rod-shaped granules inside a gland cell. D. Spermatozoa (arrow) in the seminal vesicle and algae (asterisks) in deeper tissue layers. E., F. Muscular system appears as fibers.

cuole. The animals have no complete layers of musculature; instead, circular, longitudinal and transversal muscle strands form a net-like arrangement which can be observed under Normarsky contrast (Fig. 2E, F). The muscle cells do not show any striation. More densely stained cells are located medially to the muscle cells and between them. They are followed by highly vacuolized cells, which entirely occupy the centre of the animals. They are regarded as resorptive cells and are lacking in those animals which were fixed soon after feeding. The pharyngeal opening appears as a simple whole within the ventral epidermis. However, a gut lumen is lacking as well as pharyngeal opening, in those animals which were fixed several hours after feeding. In these animals a small invagination of the ventral epidermis indicates the position of the pharynx and algal symbionts are lacking here. A black spot anterior to the pharynx could be observed in some animals.

Sexual reproduction was never observed in our culture. Nevertheless, sperm cells were observed in one squeezed animal. The spermatozoa are slender and elongated cells, with an almost constant diameter, except for the narrow terminal section which comprises almost one fifth of the entire length. The cells were motile and moved by propagating waves. In a few animals the spermatozoa were observed inside a pair elongated, pear-shaped vesicles, which are located on each side on the longitudinal axis close to the medio-caudal lobe (Fig. 2D, 3). None of the animals sectioned were in a reproductive state.

Behaviour

The animals move by ciliary gliding on the surface of the glass bowls they were kept in. The anterior end is extremely flexible and is normally wider than the remaining body. This is essential for capturing prey, but during rapid locomotion the anterior end has the same diameter as the remaining body. During rapid locomotion they bulge their lateral margins towards the ventral side and it seems likely that they only glide on these infolded lateral margins. When the animals moved slower their anterior end is extended in the same manner as in those animals which calmly rest on the surface of the culture bowls (Fig. 3). Some animals were observed to glide on the water surface with their upside down and sometimes they were seen to float in the water column.

When they were fed with nauplia of *Artemia salina* it could be observed that the animals can recognize their prey, if it is in close vicinity. If they catch a nauplius larva in the water column or on the surface of the water, they completely enclose it and swallow it. On the surface of the bowls they rise the anterior end to form a cage-like structure and to arrest the prey while pressing the lateral

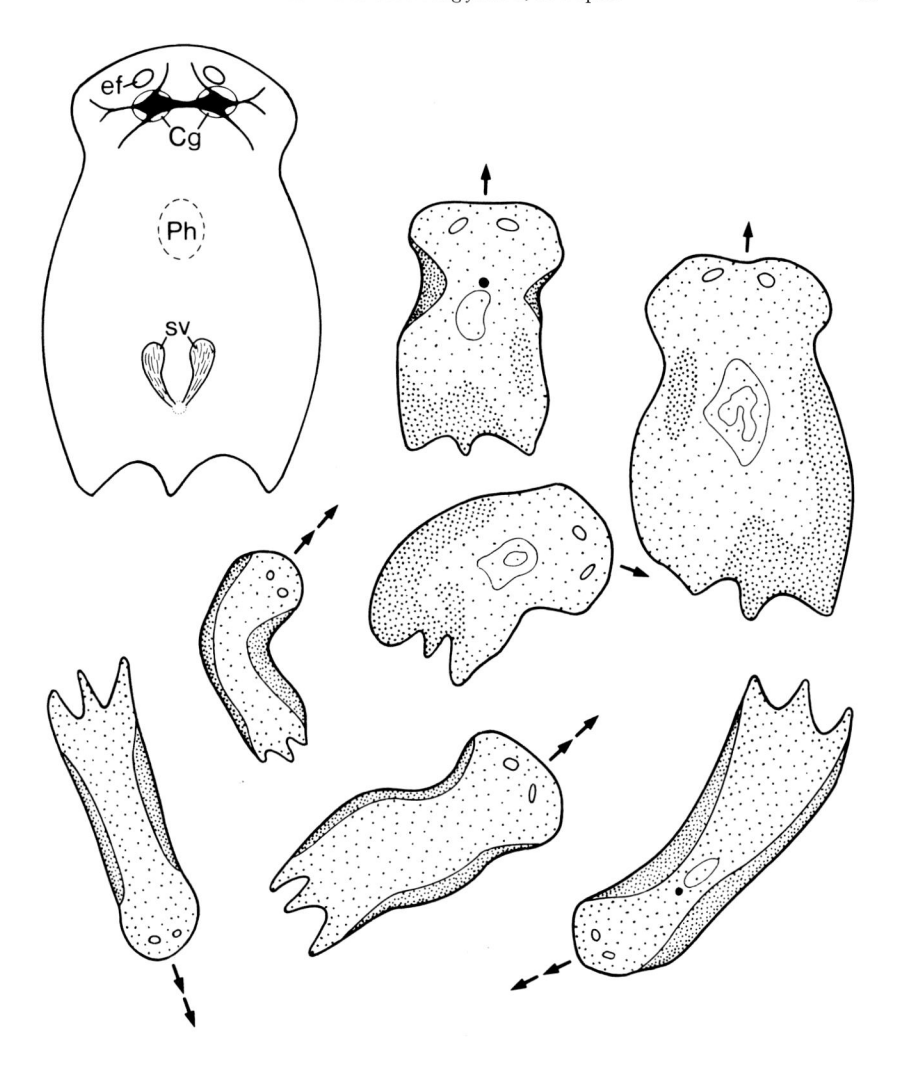

Fig. 3:. Organization of *Convolutriloba longifissura* (Acoela) and position of the relative widely separated cerebral ganglia (*Cg*), eye fields (*ef*), pharynx (*Ph*) and seminal vesicles (*Sv*). Drawing of different kinds of locomotion (arrow indicate slow (one arrow) and fast (two arrows) gliding).

margins of the anterior end onto the surface of the bowl. Then the central resorptive tissue form a cavity and the whole larva is ingested by the pharynx (Fig. 1A).

The regenerative potential of these animals is enormous. When we tried to isolate the symbiontic algae, the animals were squeezed with a paint-brush and

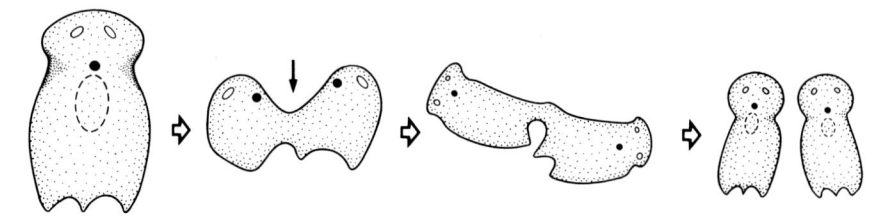

Fig. 4: Longitudinal fission in *Convolutriloba longifissura* (Acoela).

desintegrated into numerous pieces. These soon developed into entire animals which grew up to their final size.

Asexual reproduction

Asexual reproduction seems to occur in large animals only. It always starts at the anterior end and propagates along the longitudinal axis of the animal (Fig. 1D-E, 4). Thus, one eye field is transferred to each daughter individual. Towards the caudal end the fission occurs somewhat laterally to the longitudinal axis and the medial lobe of the anterior end is adopted by only one of the daughter organisms, which usually seems to be the right one. The daughter organisms form additional lobes and, thus, aquire the habitus characteristic for the species. One eye field is already newly formed during fission and initially appears to be smaller than that one adopted from the mother individual. Some stages of asexual reproduction were observed to contain a dark spot anterior to the pharynx which is present in each daughter individual (Fig. 1D-H). As this spot apparently marks the anterior tip of the pharyngeal pouch, it is concluded from this observation that the entire pharyngeal region is divided into two equal halfs during fission. The daughter organisms, which initially are half as large as the mother individual, aquire the same size the mother organisms had within a few days.

Algal symbionts

The algal symbionts of *Convolutriloba longifissura* belong to the taxon *Tetraselmis* (Chlorodendrales, Chlorophyta). They were isolated from the animals and survived in culture at 25° C, so that they seemingly do not depend on any substance produced by the animals. *Convolutriloba longifissura*, on the other hand, dies within four days when kept in darkness. *Tetraselmis* spec. thus presumably produces a photosynthetical substance, the animals essentially depend on.

D. Discussion

External shape with a rounded anterior end with two bright eye fields and three caudal lobes, green colour with clusters of red pigments and behavioural characterisitic like locomotion and feeding mechanisms clearly indicate that *Convolutriloba longifissura* belong to the taxon *Convolutriloba*, erected by HENDELBERG and ÅKESSON (1988). Like *Convolutriloba retrogemma*, *Convolutriloba longifissura* has no statocyst. The algal symbionts and the mode of asexual reproduction will be discussed in more detail in the following.

Symbiontic algae are described from several species of the Acoela (DROOP 1963, TAYLOR 1971, s. also SMITH & DOUGLAS 1987). The algal symbionts of *Convolutriloba longifissura* belong to the taxon *Tetraselmis* and are an still undescribed species. Another representative of *Tetraselmis*, *Tetraselmis convolutae* is the symbiontic alga of *Convoluta roscoffensis*, a small acoel from the french coast. Other representative of the taxon *Tetraselmis* are freeliving algae (MARTIN et al. 1996) and *Tetraselmis* individuals isolated from *Convoluta roscoffensis* and *Convolutrilona longifissura* survive in culture. Moreover, young individuals of *Convoluta roscoffensis*, which were sexually produced, have to be infected by their symbiontic algae. After this, they change from predators into biotrophic animals, which totally depend on their algal symbionts for nutrition. Utilizing their symbionts directly as a source for nutrition is seldom (MUSCATINE et al. 1974).

Convolutriloba retrogemma survived 21-24 days in darkness, when they were fed with copepods twice a week, but 34-36 days of total darkness were letal for this species (ÅKESSON & HENDELBERG 1989). In contrast to this *Convolutriloba longifissura* seems to depend more essentially on certain products of their symbionts, because they degenerate within four days when kept in darkness. Nevertheless, both species use their symbionts for nutrition. Experiments on the asexual reproduction in *Convolutriloba retrogemma* clearly indicate that feeding on animals (holozoic feeding) increases the reproductive potential, although the animals are able to reproduce asexually when they purely use their algae for nutrition (ÅKESSON & HENDELBERG 1989). In addition to products from algal photosynthesis which are presumed to be utilized for nutrition, entire algae are transported into the central resorptive tissue. The remnants of the algae, i.e. shell material, accumulate in a dark spot at the tip of the pharynx, before it is released from the body. Such a nutrition from the symbiontic algae lacks in *Convoluta roscoffensis* and other *Convoluta* and *Amphiscolops* species (see ÅKESSON & HENDELBERG 1989). Because some dividing *Convolutriloba longifissura* were observed with a dark spot anterior to the pharynx, it is assumed that algae are also utilized during fission. The same has been described for *Convolutriloba retrogemma*. Therefore, both *Convolutriloba* species are thus far the only known plathelminths which regularly feed on their symbionts and where digestion of them already starts during the formation of daughter individuals during asexual reproduction.

Convolutriloba longifissura is the first plathelminth described which reproduces asexually by longitudinal fission. According to the terms mentioned intially, this mode of asexual reproduction is a modified architomy, because one half of each animal formed one functional half of the mother individual and those parts of the daughter organism, which are lacking after fission, are replaced by regeneration. However, architomy and paratomy have generally been reported to occur as transversal fission, so that the daughter individuals lie one behind the other. Such a transversal fission has been reported from Acoela, Catenulida, and Macrostomida (see Ax & Schulz 1959, Rieger 1986, Palmberg 1991) and has also been reported from certain annelids (see Clark 1977, Dozsa-Farks 1996, Korn 1981, Radashevsky 1996). In other bilaterians budding also seems to be a common mode of asexual reproduction (Kamptozoa: see Emschermann (1892), Phoronida: see Emig (1982)). Longitudinal fission has up to date never been described for vagile bilaterians. The mode of asexual reproduction in *Convolutriloba longifissura* thus is not only the first case of longitudinal fission in Plathelminthes, but also in the Bilateria at all.

Acknowledgements

This investigation was financially supported by the Akademie der Wissenschaften und der Literatur, Mainz. We would like to thank Prof. Dr. P. Ax for helpful comments on the manuscript and Prof. Dr. U. Schlösser for determing the algal symbionts.

Zusammenfassung

Convolutriloba longifissura nov. spec. wurde aus einer Lieferung Korallen von der indonesischen Küste isoliert. Es handelt sich um einen maximal 6 mm großen Vertreter der acoelen Plathelminthen. Der Habitus der Tiere entspricht dem von *Convolutriloba retrogemma* Hendelberg et Åkesson, 1988. Die Organismen besitzen ein globuläres Vorderende mit zwei frontal gelegenen, pigmentfreien Zonen, die als Augenfelder gedeutet werden. Das Vorderende ist durch eine laterale Konstriktion vom übrigen Körper abgesetzt. Das

Hinterende der Tiere ist in drei caudale Lappen ausgezogen. Die Tiere enthalten Symbionten (*Tetraselmis* spec.) und sind daher olivgrün. Zusätzlich zeigen sie caudal und lateral Anhäufungen rot pigmentierter Flächen sowie leuchtend gelb gefärbte Vakuolen. Manche Individuen zeigen einen fronto-median gelegenen, dunklen Fleck. Unter Bedingungen der Dauerdunkelheit starben die Tiere bereits nach 4 Tagen. Das deutet darauf hin, daß die Tiere neben der Nahrungsaufnahme essentiell auf ihre Symbionten bzw. deren Stoffwechselprodukte angewiesen sind. *Convolutriloba longifissura* wurde bei 25° C in autoklaviertem Meereswasser aus Helgoland gehalten und mit Nauplien von *Artemia salina* ernährt. Die Nahrungsaufnahme erfolgt, indem die Tiere ihre Beute umwölben und dann ingestieren. Während der Fortbewegung zeigen sie eine charakteristische Wölbung der lateralen Ränder nach ventral; im Ruhezustand liegen sie abgeflacht an der Wand der Kulturgefäße. In Kultur vermehren sich die Tiere ausschließlich durch Längsteilung, die stets am Vorderende beginnt und sich nach caudal fortsetzt. Dabei wird jeweils ein Augenfeld auf ein Tochterindividuum übertragen, wogegen der caudo-mediane Lappen nur an eines der Tochtertiere weitergegeben wird. Befand sich vor der Teilung ein medianer, dunkler Fleck aus degenerierten Algen im Muttertier, so teilt sich dieser und wird beiden Tochtertieren mitgegeben. Die fehlende Organe werden noch während der Teilung ergänzt (Architomie). Die so entstandenen Individuen sind anfangs nur halb so groß wie die Muttertiere, gleichen sich aber bald an deren ursprüngliche Größe an. Bei Plathelminthen sind bisher nur Knospungen oder Querteilungen beobachtet worden. Mit *Convolutribloba longifissura* wird nun erstmalig ein Vertreter der Plathelminthen beschrieben, der sich durch Längsteilung asexuell vermehrt. Gleichzeitig ist dieses der erste Nachweis asexueller Vermehrung für einen vagilen Vertreter der Bilateria.

References

ÅKESSON, B. & J. HENDELBERG (1989): Nutrition and asexual reproduction in *Convolutriloba retrogemma*, an acoelous turbellarian in obligate symbiosis with algal cells. In: RYLAND, J.S. & A.A. TYLER (eds.): Reproduction, genetics and distributions of marine organisms, 23rd European Marine Biology Symposium. Olsen & Olsen, Fredensborg. pp. 13-21.

AX, P. & E. SCHULZ (1959): Ungeschlechtliche Fortpflanzung durch Paratomie bei acoelen Turbellarien. Biol. Zentralbl. **78**, 613-621

CLARK, R.B. (1977): Reproduction, speciation and polychaete taxonomy. In: REISH, D.J. & K. FAUCHALD (eds.): Essays on polychaetous annelids in memory of Dr. Olga Hartman. Allan Hancock Found., Los Angeles. pp. 477-501.

DÖRJES, J. (1966): *Paratomella unichaeta* nov. gen. nov. spec., Vertreter einer neuen Familie der Turbellaria Acoela mit asexueller Fortpflanzung durch Paratomie. Veröff. Inst. Meeresforsch. Bremerhaven, Sonderheft II, 187-200.

DÖRJES, J. (1968): Die Acoela (Turbellaria) der Deutschen Nordseeküste und ein neues System der Ordnung. Z. zool. Syst. Evolut.-Forsch. **6**, 56-452.

DOZSA-FARKAS, K. (1996): An interesting reproduction type in enchytraeids. Acta Zool. Acad. Sci. Hungaricae **42**: 3-10.

DROOP, M.R. (1963): Algae and invertebrates in symbiosis. In: NUTMAN, P.S. & B. MOSSE (eds.): Symbiotic associations. Cambridge Univ. Press, Cambridge. pp. 147-172

EMIG, C.C. (1982): The biology of the Phoronida. Adv. Mar. Biol. **19**: 1-89.

EMSCHERMANN, P. (1982): Les Kamptozoaires. État actuel de nos connaissances sur leur anatomie, leur dévelopement, leur biologie et leur position phylogénétique. Bull. Soc. Zool. France **107**, 317-344.

HANSON, E.D. (1960) Asexual reproduction in acoelous Turbellaria. Yale J. Biol. Med. **33**, 107-111.

HENDELBERG, J. & B. ÅESSON (1988): *Convolutriloba retrogemma* gen. et sp. n., a new turbellarian (Acoela, Plathyhelminthes) with reversed polarity of reproductive buds. Fortschr. Zool **36**, 321-327.

HENDELBERG, J. & B. ÅKESSON (1991): Studies of the budding process in *Convolutriloba retrogemma* (Acoela, Plathyhelminthes). Hydrobiologia **227**: 11-17.

KORN, H. (1982): Annelida. In: SEIDEL, F.(ed.): Morphogenese der Tiere, Erste Reihe, Vol. **5**: H-I. G. Fischer, Jena. pp. 1-599.

MARTIN, B., K. HOEF-EMDEN & M. MELKONIAN (1996): Light and electron microscope observations on *Tetraselmis desikacharyia* sp.nov. (Chlorodendrales, Chlorophyta). Nova Hedwigia **112**, 461-475.

MEYER, R. & T. BARTOLOMAEUS (1997): Ultrastruktur und Morphogenese der Hakenborsten bei *Psammodrilus balanoglossoides* – Bedeutung für die Stellung der Psammodrilida (Annelida). Microfauna Mar. **11**, 87-113.

MUSCATINE, L., J.E. BOYLE & D.C. SMITH (1974): Symbiosis of the acoel flatworm *Convoluta roscoffensis* with the alga *Platymonas convolutae*. Proc. R. Soc. London Ser. B. **187**, 221-234.

PALMBERG, I. (1991): Differentiation during asexual reproduction and regeneration in a microturbellarian. Hydrobiologia **227**, 1-10.

RADASHEVSKY, V.I. (1996): Morphology, ecology and asexual reproduction of new *Polydorella* species (Polychaeta, Spionidae) from the South China Sea. Bull. Mar. Sci. **58**, 684-693.

RIEGER, R.M. (1986): Asexual reproduction and the turbellarian archetype. Hydrobiologia **132**, 35-45.

SMITH, D.C. & A.E. DOUGLAS (1987): The biology of symbiosis. Edward Arnold, London.

TAYLOR, D.L. (1971): On the symbiosis between *Amphidinium klebsii* (Dinophyceae) and *Amphiscolops langerhansi* (Turbellaria: Acoela). J. mar. biol. Ass. U.K. **51**, 301-313.

Priv.-Doz. Dr. Thomas Bartolomaeus
II. Zoologisches Institut und Museum, Universität Göttingen
Berliner Str. 28, D-37073 Göttingen, Germany
Priv.-Doz. Dr. Ivonne Balzer
I. Zoologisches Institut, Universität Göttingen
Berliner Str. 28, D-37073 Göttingen, Germany

Beklemischeviella angustior Luther and *Vejdovskya parapellucida* n. sp. (Rhabdocoela, Plathelminthes) from brackish water of the Winyah Bay, South Carolina, USA.

Peter Ax

Abstract

This paper deals with two rhabdocoel plathelminths found in the brackish water of the Winyah Bay in South Carolina. The population of one species is interpreted as a member of *Beklemischeviella angustior* Luther, known from brackish water biotopes in Finland and Canada. The other population will be described as a new species *Vejdovskya parapellucida* n. sp. It is very similar to the brackish water species *Vejdovskya pellucida* (M. Schultze) from Europe and Canada.

A. Introduction

For numerous brackish water Plathelminthes we could demonstrate a circumpolar distribution in the boreal climate zone (Ax 1991, 1995).

The Winyah Bay on the Atlantic coast of South Carolina proved to be a suitable place for comparative research in a warm temperate climate. The result was unexpected. In a sandy beach of the bay at the border between marine and limnic conditions, more than a dozen of free-living plathelminths exist that are identical with or extremely similar to brackish water species known up to now only from comparable biotopes in the boreal and subarctic region. In advance of a more complete elaboration of my material, I will treat one example of each of the two possibilities in this paper.

The investigations were carried out at the Marine Field Laboratory, Georgetown, of the Belle W. Baruch Institute for Marine Biology and Coastal Research of the University of South Carolina, Columbia.

I thank Dr. Dennis Allen and Dr. David Bushek for their hospitality in April 1995 and October 1996. Furthermore I am very grateful for the help of Paul Kenny, including his measurements of abiotic factors.

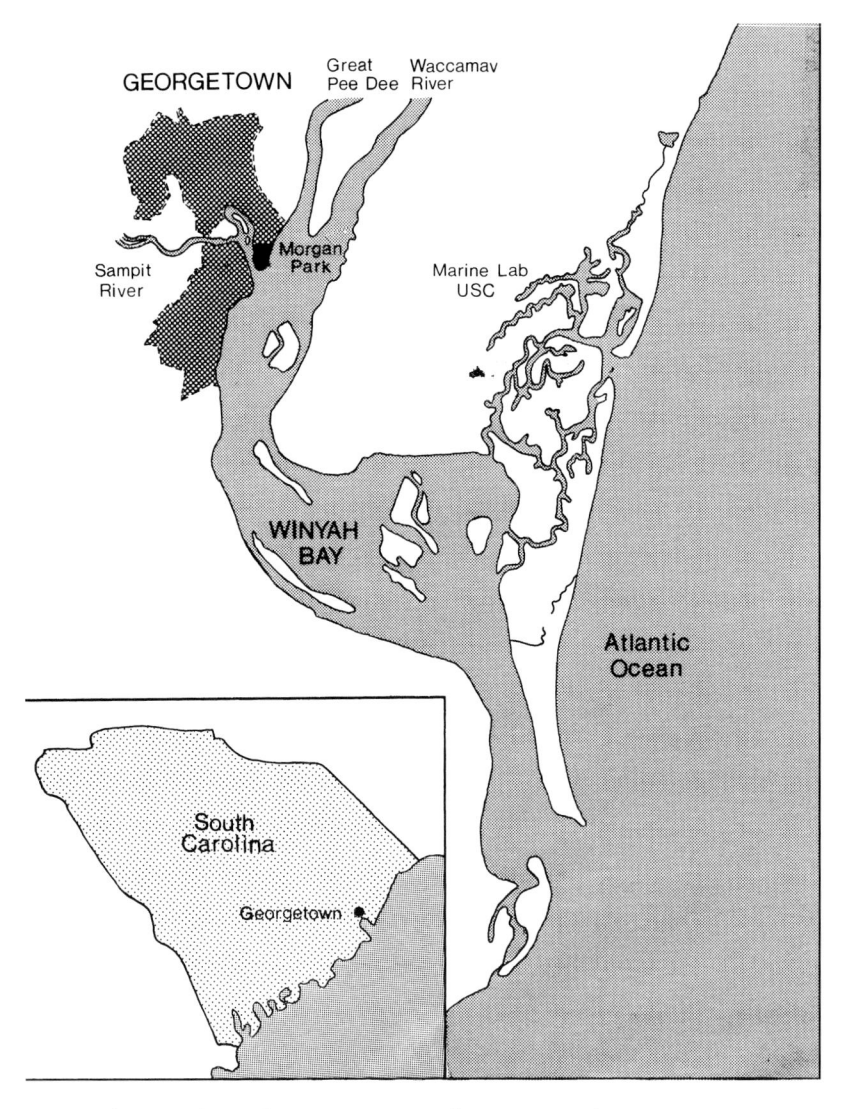

Fig. 1. Winyah Bay with sample site at Morgan Park (East Bay Park), Georgetown, S. C. Insert: South Carolina, USA.

B. Biotope
(Fig. 1, 2)

The Morgan Park or East Bay Park of the city of Georgetown is located in the northern part of the Winyah Bay, where the Sampit River, the Pee Dee and the Waccamav River empty. The park ends at the bay front in a small sandy beach between large populations of *Phragmites*. Samples were collected from the intertidal – partly from clean medium sand without vegetation and partly from more or less muddy sand between *Phragmites*. Cushions of cyanobacterians occasionally grow at the surface in the latter case.

The salinity of the locality is very low. It does not exceed 5–6 ‰, and it is independent of the tide stage.

Fig. 2. Sandy beach at the bay front of Morgan Park (East Bay Park), Georgetown.

DATE	TIME	TIDE STAGE	WATER TEMP °C	SALINITY ‰
7/17/95	13.30	high	28.5	0.1
9/18/95	12.50	1/4.ebb	26.9	2.1
11/02/95	13.00	low	21.3	0.2
12/20/95	12.30	low	10.5	3.6
2/09/96	13.00	mid.ebb	8.5	0.1
3/21/96	13.00	mid.ebb	13.5	0.1
5/08/96	13.00	high	22.9	1.3
8/13/96	13.00	mid.ebb	29.1	5.1
10/07/96	11.00	low	18.6	5.5

Morgan Park (East Bay Park), Georgetown.
Measurements of water temperature and salinity just in front of the sandy beach (Paul Kenny).

C. Results

Beklemischeviella angustior Luther, 1943

(Fig. 3A, 4A, B)

Material
A few individuals from the beach of Morgan Park. Muddy sand with cushions of cyanobacterians (April 1995).

Up to now, *Beklemischeviella angustior* was known only from SW Finland in the Baltic Proper (LUTHER 1943, 1948, 1962; KARLING 1974) and from New Brunswick along the Atlantic Coast of Canada (AX & ARMONIES 1987).
Animals of South Carolina are ~0,7 mm long, uncolored, with eyes. A quick movement is striking. My sketches and photos of the hard copulatory organ of individuals from South Carolina are exchangeable with the figures of animals from SW Finland (LUTHER 1943, fig. 81, 82; 1948, fig. 120; 1962 fig. 23; KARLING 1974, fig. 105, 106) and from Canada (AX & ARMONIES 1987, fig. 26A).
The stylet begins with a funnel and tapers distally into a tube with a descending and an ascending part. The characteristic spiral striping is identical with that of animals found in Finland. The length of the stylet up to the inflexion of the tube is 50 µm in South Carolina, as against 61 µm in the Canadian specimen and 54 µm in Finland (Lappvik, 1997). Aside from this small variation in size, there is complete congruence in the construction and shape of the

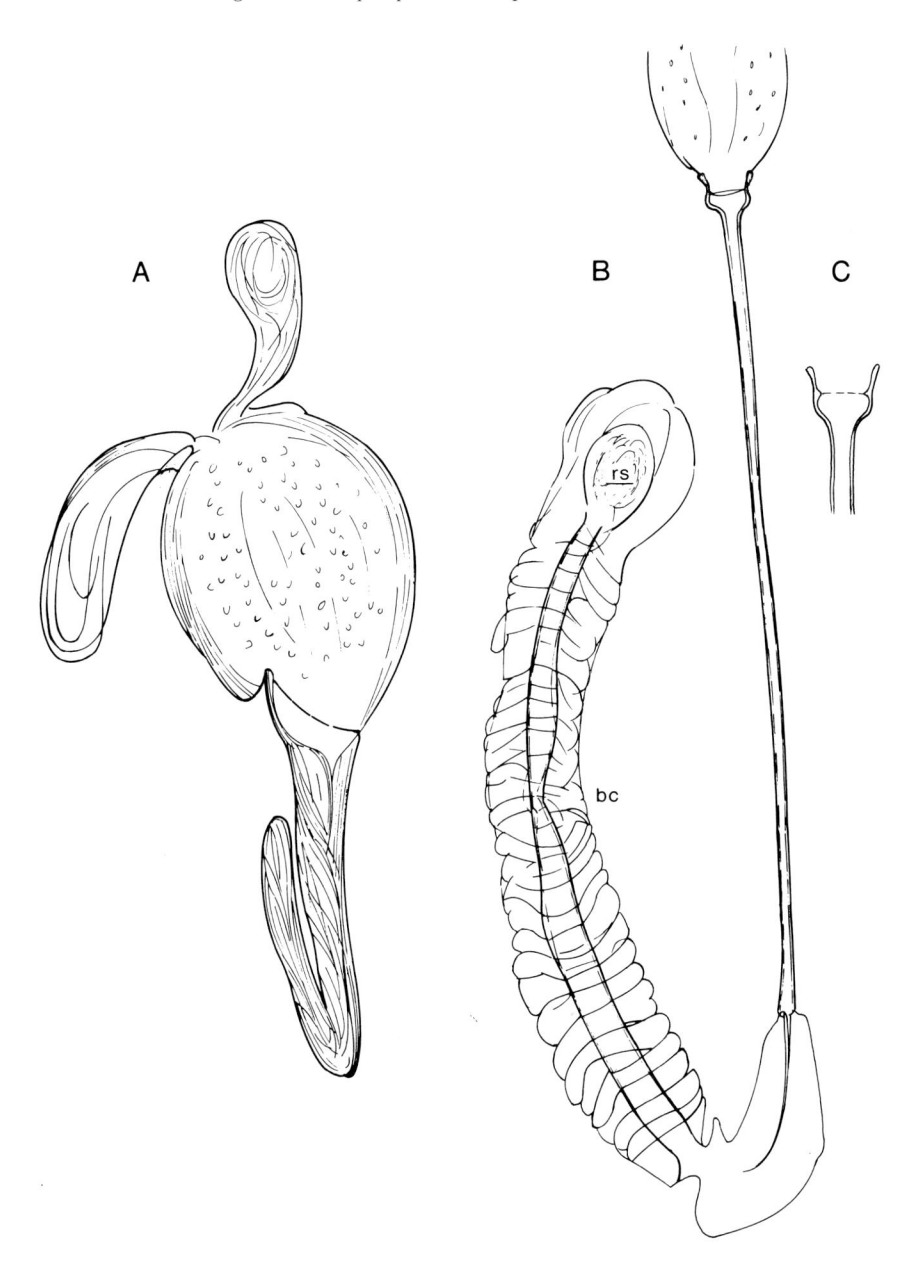

Fig. 3. A. *Beklemischeviella angustior*. Copulatory organ with paired vesiculae seminalis (vs), muscular bulb (mb) and stylet (st). Winyah Bay, South Carolina. B–C. *Vejdovskya parapellucida*. B. Stylet and bursa copulatrix (bc) with receptaculum seminis (rs). C. Higher magnification of proximal part of the stylet. Winyah Bay, South Carolina.

copulatory organ in the populations of Finland, Canada and South Carolina. We treat them as members of one species.

Vejdovskya parapellucida n. sp.

(Fig. 3B, C, 4C–F)

Material
 Two adults and one juvenile organism from the beach of Morgan Park. Clean and muddy sand (October 1996). Locus typicus.

Description

Tiny animals with a length of only 0,4 mm. Uncolored. Without pigmented eyes.

The stylet of the copulatory organ has a length of ~105 µm. It is differentiated in a proximal funnel (diameter 6 µm), a long tube (diameter ~2 µm) and a minute distal flagellum. The latter is curved only slighty and measures 18 µm.

The adults have a voluminous bursa copulatrix terminating in a receptaculum seminis. A regular system of hard lamellae sourrounds the central canal at right angles to the longitudinal axis of the bursa.

Discussion

Extensive congruence exist between the specimens of South Carolina and populations of the brackish water species *Vejdovskya pellucida* (M. Schultze, 1851) from Europe and Canada. There are however, distinct differences in two points which force the discrimination of the two species *V. parapellucida* and *V. pellucida*.

(1) The stylet of *V. pellucida* is twice as long as that of *V. parapellucida* – Finland: 222, 230 µm (LUTHER 1962), Kiel (Baltic Sea): 190–200 µm (AX 1951), New Brunswick (Canada): 220 µm (AX & ARMONIES 1987).

(2) In all figures of *V. pellucida* the flagellum of the stylet is a long, spirally coiled structure with a terminal hook (KARLING 1957, fig. 6; LUTHER 1962, fig. 6; AX 1951, fig. 18; AX 1956, fig. 19; AX & ARMONIES 1987, fig. 46).

I emphasize the fact that the European and Canadian populations of *Vejdovskya pellucida* are identical with respect to size and shape of the stylet. Differences that indicate considerable evolutionary change exist along the American Atlantic Coast between the population of *V. pellucida* studied in Canada and the individuals of *V. parapellucida* observed in South Carolina.

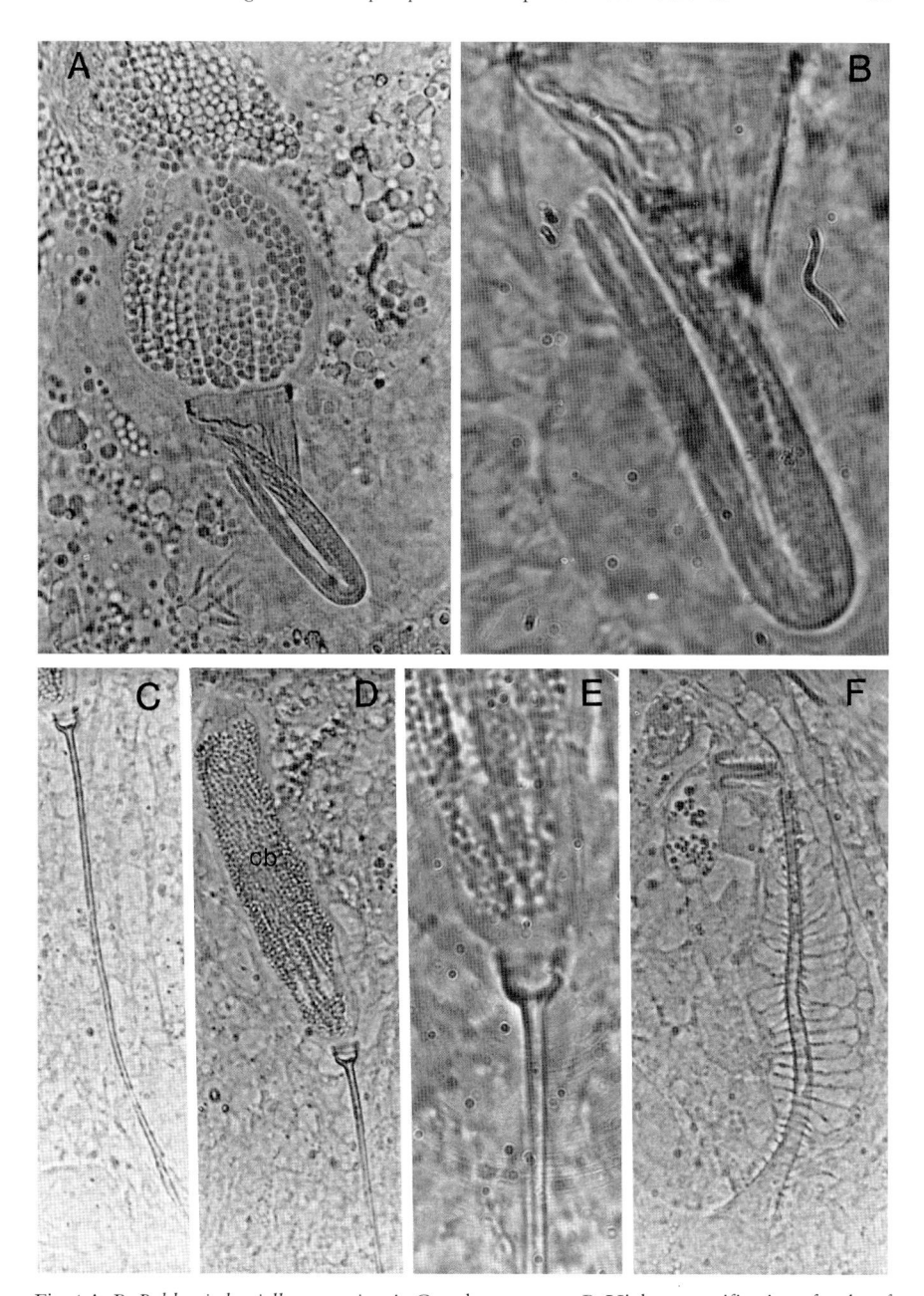

Fig. 4 A–B. *Beklemischeviella angustior*. A. Copulatory organ. B. Higher magnification of stylet of the copulatory organ. Winyah Bay, South Carolina. C–F. *Vejdovskya parapellucida*. C. Stylet. D. Cylindric bulb (cb) and stylet of the copulatory organ. E. Proximal part of the stylet with funnel. F. Bursa copulatrix. Winyah Bay, South Carolina.

Zusammenfassung

Zwei Arten des Plathelminthen-Taxons Rhabdocoela aus dem Brackwasser der Winyah-Bucht, South Carolina werden behandelt. Die Population der einen Art gehört zu *Beklemischeviella angustior* Luther, die bisher aus Brackwasser-Biotopen von Finnland und Kanada bekannt war. Die zweite Population wird als neue Art *Vejdovskya parapellucida* n. sp. beschrieben; sie ist der Brackwasser-Art *Vejdovskya pellucida* (M. Schulze) von Europa und Kanada sehr ähnlich.

References

Ax, P. (1951): Die Turbellarien des Eulitorals der Kieler Bucht. Zool. Jb. Syst. **80**, 277–378.

Ax, P. (1956): Les Turbellariés des étangs côtiers du littoral méditerranéen de la France méridionale. Vie et Milieu Suppl. **5**, 1–215.

Ax, P. (1991): Northern circumpolar distribution of brackish water plathelminths. Hydrobiologia **227**, 365–368.

Ax, P. (1995): Brackish water Plathelminthes from the Faroe Islands. Hydrobiologia **305**, 45–47.

Ax, P. & W. Armonies (1987): Amphiatlantic identities in the composition of the boreal brackish water community of Plathelminthes. A comparison between the Canadian and European Atlantic Coast. Microfauna Marina **3**, 7–80.

Karling, T. G. (1957): Drei neue Turbellaria Neorhabdocoela aus dem Grundwasser der schwedischen Ostseeküste. Kungl. fysiogr. Sällsk. Lund Förh. **27**, 25–33.

Karling, T. G. (1974): Turbellarian fauna of the Baltic Proper. Identification, ecology and biogeography. Fauna Fennica **27**, 1–101.

Luther, A. (1943): Untersuchungen an rhabdocoelen Turbellarien. IV. Über einige Repräsentanten der Familie Proxenetidae. Acta Zool. Fennica **38**, 1–95.

Luther, A. (1948): Untersuchungen an rhabdocoelen Turbellarien. VII. Über einige marine Dalyellioida VIII. Beiträge zur Kenntnis der Typhloplanoida. Acta Zool. Fenn. **55**, 1–122.

Luther, A. (1962): Die Turbellarien Ostfennoskandiens. III. Neorhabdocoela 1. Dalyellioida, Typhloplanoida: Byrsophlebidae und Trigonostomidae. Fauna Fennica **12**, 1–71.

Prof. Dr. Peter Ax
II. Zoologisches Institut und Museum der Universität Göttingen
Berliner Straße 28, D-37073 Göttingen

First report on the fine structure of unpigmented rhabdomeric photoreceptors in a free-living species of the "Dalyellioida" (Plathelminthes, Rhabdocoela)

Beate Sopott-Ehlers

Abstract

The submicroscopical structure of light-sensing organs in *Jensenia angulata* is discribed. These photoreceptors consist of two sensory cells bearing microvilli-like extensions of the surface membrane, so-called rhabdomeres and a single cup or mantle cell devoid of pigment granules. These data confirm the hypothesis that reductions of pigment granules in cup or mantle cells have occurred more than once within the Plathelminthes, since such a type of photorceptors also exists in free-living stages of Neodermata and in other free-living taxa, as well.

A. Introduction

Pigmented rhabdomeric photoreceptors situated in antero-dorsal or dorso-lateral positions are a wide spread feature of many free-living Plathelminthes and many free-living stages of Monogenea and Digenea, as well. Rhabdomeric photoreceptors devoid of shading devices in free-living Plathelminthes were first described by von HOFSTEN (1918) for a representative of the Otoplanidae, a taxon of the Proseriata, at the light-microscopical level. Electron-microscopy however, has revealed that unpigmented photoreceptors are more common than presumed. Owing to the minor size of these light-sensing organs, electron microscopy is necessary to see whether these eyes exist, and to elucidate their precise structure.

To date, photoreceptors of this nature have been found in all members of the Proseriata Lithophora studied so far (for ref. see RIEGER et al.1991; SOPOTT-

EHLERS 1991, 1995a), in *Archimonotresis limophila*, a species of the Prolecithophora (see SOPOTT-EHLERS 1988), in the cercaria of *Cryptocotyle lingua* (see REES 1975) and in the oncomiracidia of *Diplozoon paradoxum* and *Diplozoon nipponicum* (see KEARN 1978; LAMBERT & DENIS 1982).

The present study was prompted to enlarge our knowledge on the distribution of unshielded rhabdomeric photoreceptors and to see, whether this characteristic could contribute to an elucidation of phylogenetic relationships. In this respect there is a special point of view: the search for features suitable to help to understand the systematic position of taxa ascribed to the non-monophyletic "Dalyellioida" and "Typhloplanoida" within the Rhabdocoela.

B. Materials and Methods

Specimens of *Jensenia angulata* derive from sediment samples taken on beaches of the small island Hallig Langeneß (North Sea). EM-preparations followed conventional steps described in detail by SOPOTT-EHLERS (1995b).

Series of sections were examined using a Zeiss electron microscope EM 900.

C. Results

Individuals of *Jensenia angulata* (Jensen, 1978) appear blind when viewed with the light microscope. During electronmicroscopical examinations of consecutive sections through the cerebral region, however, two light-sensing organs were discovered. These photoreceptors lie embedded in processes of the brain in a dorso-lateral position (Fig. 1), close to the body surface. Each of the eyes has the same essential components as pigmented rhabdomeric photoreceptors, namely, retinular or sensory cells and a cup or mantle cell, in this instance devoid of pigment granules.

In the photoreceptors of *Jensenia angulata* two sensory cells exist (Fig. 2A). On entering the eye cup through narrow openings which face the body surface in a dorso-lateral and dorsal direction, the dendritic portion of each sensory cell expands to form a disk. The surface membrane of these sections is evaginated to densely packed unbrached microvilli interdigitating in the center of the eye. Each disk-like cytoplasmic core is crowded with organelles including mitochondria, microtubules, agranular endoplasmic reticulum, clear and dense-cored neurovesicles as well as glygogen. Concentrations of glycogen deposits were also observed in the cytoplasm of the microvilli. The sensory cells expand outside the eye cup to about the half width of the rhabdo-

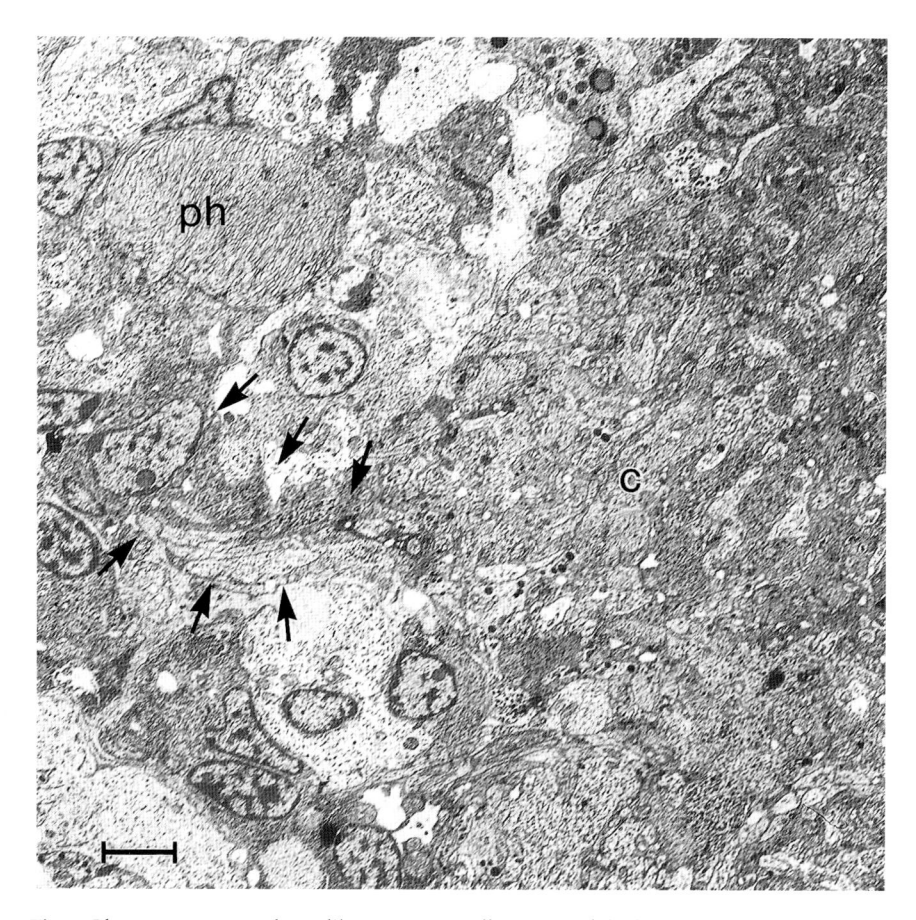

Fig. 1: Photoreceptor on a horn-like process (small arrows) of the brain. Scale = 2 μm.

meric complex. This segment often appears to contain fewer inclusions, albeit some mitochondria, dense-cored vesicles, a small number of profiles of rough endoplasmic reticulum and multivesicular bodies are also present. This portion of the sensory cells also includes the nucleus roundish in shape.

The eye cup consists of a shallow thin walled mantle cell devoid of pigment granules. It is a very delicate extension, difficult to determine in its entirety. Just in the area containig the elongated nucleus the cell lumen widens (Fig. 3A). The nucleus takes a dorso-lateral to median position. The cytoplasm is densely granular with glycogen islets and free ribosomes. Furthermore it contains smooth endoplasmic reticulum, a few small vesicles and multivesicular bodies. Occasional mitochondria were also observed. In one instance a large

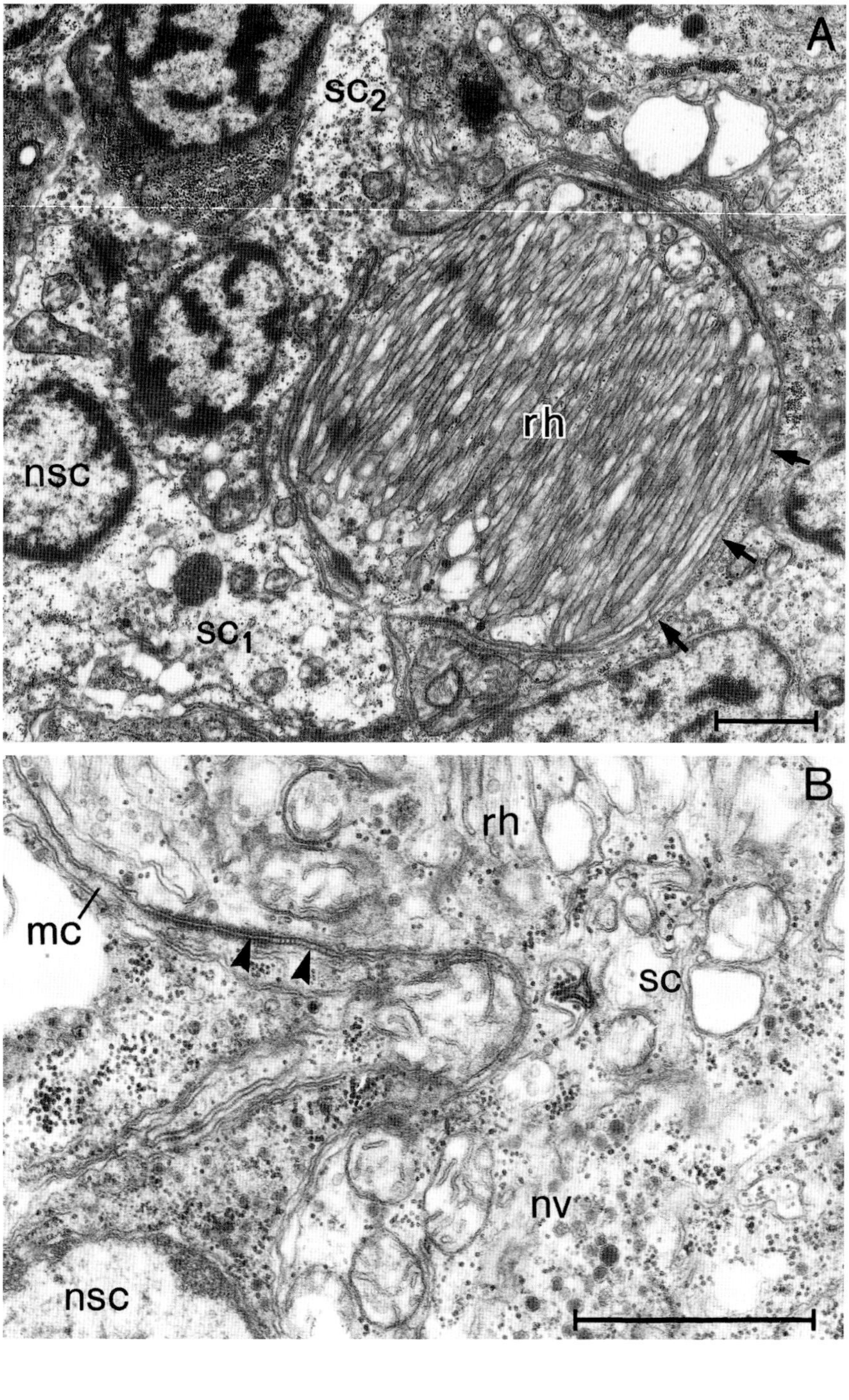
A
sc2
rh
nsc
sc1
B
rh
mc
sc
nv
nsc

mitochondrion containing electron-dark inclusions much larger in diameter than dense-cored vesicles was found near the nucleus (Fig. 3B).

Around the cup opening the adjacent cell membranes of mantle and retinular cells are thickened to form an attachment zone. This area of cell junction is made up by zonulae adhaerentes and septate junctions (Fig. 2B).

D. Discussion

The existence of so-called pigment cup ocelli consisting of one or several sensory cells and one or several cup cells has been hypothesized to be a basic feature of the Neoophora. In regard to the findings of KUHNERT & EHLERS (1987) for *Macrostomum spirale* and data on *Paromalostomum notandum* (SOPOTT-EHLERS unpublished), two species of the Macrostomida, this characteristic has to be postulated to be typical of the stem species of the Rhabditophora.

Despite of intense electronmicroscopical research on light sensing organs during the last decade, the distribution of unpigmented rhabdomeric photoreceptors still appears quite spotty. However, from the findings available so far it could be concluded that 1. pigmented and unpigmented mantle cells are homologous to each other and 2. that reductions of pigment in cup cells have occurred convergently more than once (see EHLERS 1985; SOPOTT-EHLERS 1988). The existence of such a kind of photoreceptor in *Jensenia angulata* corroborates both hypotheses.

Studies on the submicroscopical structure of light-sensing organs have revealed, that species of the Rhabdocoela "Dalyellioida" and "Typhloplanoida" as well, possess dioptric devices made up by mitochondria (for ref. see SOPOTT-EHLERS 1996). From these data it is hypothesized that all species with mitochondrial lenses belong to one monophylum, the taxon Rhabdocoela (SOPOTT-EHLERS 1996, p.100). The lack of mitochondrial lenses in the photoreceptors of *J. angulata* does not imply that this species does not belong to the Rhabdocoela, since the loss of pigment granules in the cup cell makes the existence of dioptrics useless and futile.

◄ Fig. 2: A. Dorsal and dorso-lateral sensory cell entering the eye cup. Small arrows mark the delicate mantle cell. B. Cell junctions (arrow heads) between sensory cell and mantle cell. Scales = 1 µm.

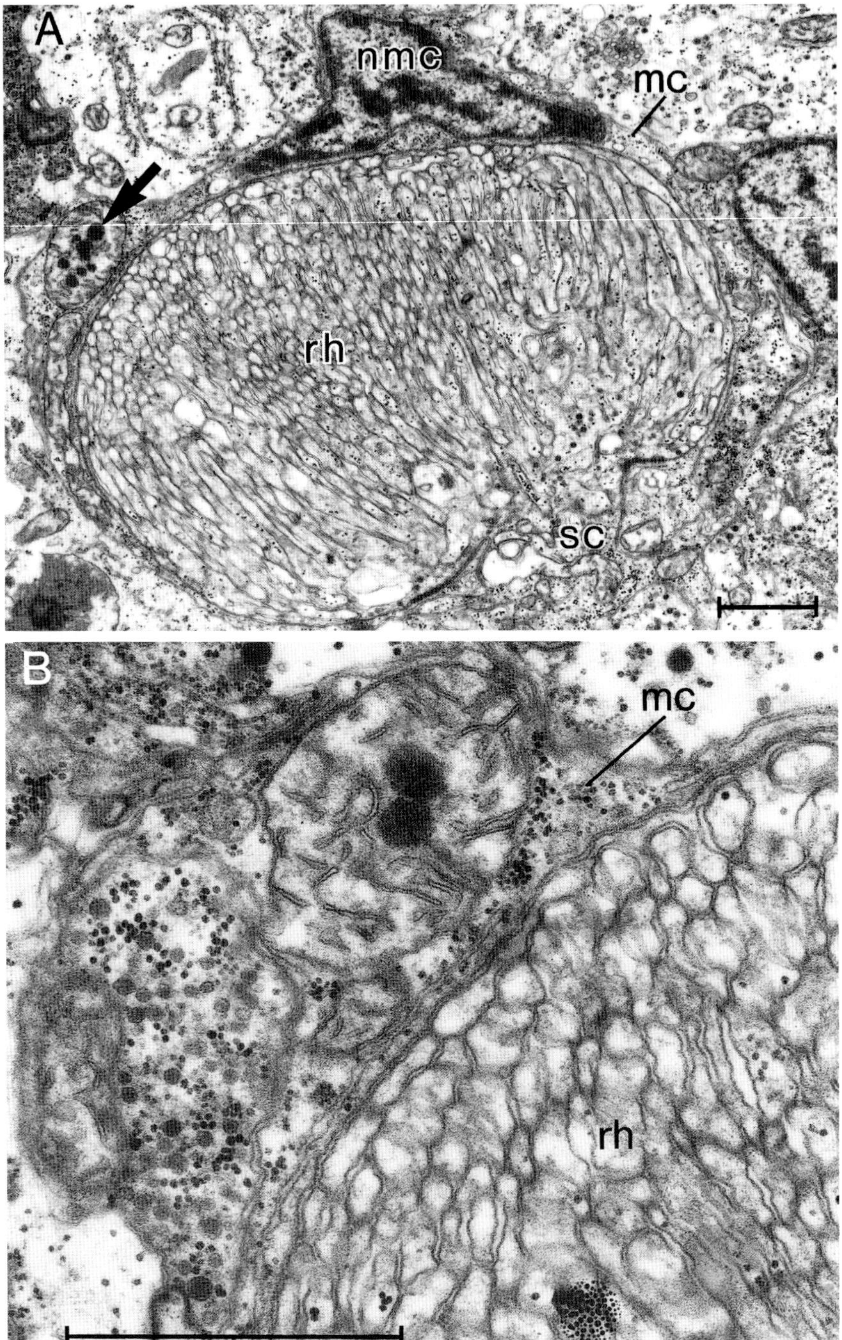
A
nmc
mc
rh
sc
B
mc
rh

Acknowledgements

Financial support was provided by the Akademie der Wissenschaften und der Literatur, Mainz. Mrs. K. Lotz kindly provided the sand samples. The technical assistance of Mrs. S. Gubert is greatly acknowledged. Mr. B. Baumgart helped with the figures.

Zusammenfassung

Die submikroskopische Struktur der Lichtsinnesorgane von *Jensenia angulata* wird dargestellt. Diese Photoreceptoren bestehen aus zwei Sinneszellen, die mikrovilliartige Fortsätze der Oberflächenmembran, sogenannte Rhabdomeren, tragen und einer einzigen Becher- oder Mantelzelle frei von Pigmentgranula. Diese Befunde stützen die Hypothese, daß Reduktionen der Pigmentgranula in Becher- oder Mantelzellen mehrfach innerhalb der Plathelminthen stattgefunden haben, da derartig gestaltete Photoreceptoren auch bei freilebenden Stadien der Neodermata sowie einigen anderen Taxa der freilebenden Plathelminthen existieren.

Abbreviations

c	brain	nv	neurovesicles
mc	mantle cell	ph	photoreceptor
nmc	nucleus of the mantle cell	rh	rhabditic complex
nsc	nucleus of the sensory cell	sc, sc1, sc2	sensory cell

References

EHLERS, U. (1985): Das phylogenetische Sytem der Plathelminthes. G. Fischer, Stuttgart, New York, pp. 1-317.
HOFSTEN, N. v. (1918): Anatomie, Histologie und systematische Stellung von *Otoplana intermedia* du Plessis. Zool. Bidr. Upps. **7**, 1-74.
KEARN, G.C. (1978): Eyes with and without pigment shields in the oncomiracidium of the monogenean parasite *Diplozoon paradoxum*. Z. Parasitenkd. **57**, 35-47.
LAMBERT, A. & A. DENIS (1982): Etude de l'oncomiracidium de *Diplozoon nipponicum* Goto, 1891. Ann. Parasitol. (Paris) **57**, 533-542.

◀ Fig. 3: A. A sensory cell enwrapped by the mantle cell with its nucleus (nmc) and a large mitochondrion containing electron-dense inclusions. B. Large mitochondrion with dark deposits located in a dilatation of the mantle cell. Scales = 1 µm.

KUNERT, T. & U. EHLERS (1987): Ultrastructure of the photoreceptors of *Macrostomum spirale* (Macrostomida, Plathelminthes). Microfauna Marina **3**, 391-409.

REES, G. (1975): Studies on the pigmented and unpigmented photoreceptors of the cercaria of *Cryptocotyle lingua* (Creplin) from *Littorina littorea* (L.). Proc. R. Soc. London B. **188**, 121-138.

RIEGER, R.M., S. TYLER, J.P.S. SMITH III & G. RIEGER (1991): Platyhelminthes. Turbellaria. In: Microscopic anatomy of invertebrates, vol 3. F.W. HARRISON & B. J. BOGITSH (eds.) Wiley Liss, New York, pp. 7-104.

SOPOTT-EHLERS, B. (1988): Fine structure of photoreceptors in two species of the Prolecithophora. Fortschr. Zool./ Prog. Zool. **36**, 221-227.

– (1991): Comparative morphology of photoreceptors in free-living plathelminths - a survey. Hydrobiologia **227**, 231-239.

– (1995a): Electronmicroscopical studies on the photoreceptors of *Provortex tubiferus* (Plathelminthes, Rhabdocoela). Mikrofauna Marina **10**, 79-88.

– (1995b): Ultrastructural features of *Bresslauilla relicta* (Plathelminthes, Rhabdocoela). The eyes. Microfauna Marina **10**, 31-40.

– (1996): First evidence of mitochondrial lensing in two species of the "Typhloplanoida" (Plathelminthes, Rhabodocoela): phylogenetic implications. Zoomorphology **116**, 95-101.

Dr. Beate Sopott-Ehlers
II. Zoologisches Institut und Museum der Universität Göttingen
Berliner Straße 28, D-37073 Göttingen

Three new species of *Ectinosoma* Boeck, 1865 (Harpacticoida; Ectinosomatidae) from Papua New Guinea and the Fiji Islands

Sybille Seifried

Abstract

Males and females of three new species of *Ectinosoma* are described. *E. barbararum* sp. n. and *E. papuarum* sp. n. were found in a beach of Ednago Island of the coast of Kavieng (New Ireland), Papua New Guinea. *E. tegula* sp. n. was collected in sublittoral sands of the Fiji Islands and is closely related to *E. papuarum* sp. n. and *E. nonpectinatum* Mielke, 1979. A diagnosis of the Ectinosomatidae Sars, 1903 is given. The determining character is mainly one or two setae on the distal side of the cutting edge mandible. The genus *Ectinosoma* can be distinguished from the sister group *Halectinosoma* Lang, 1948 by the characteristic u-shaped tubercles on the cuticle of the body and P1–P5. *H. porosum* Wells, 1967 has tubercles on the cuticle and an *Ectinosoma*-like P5 and is to be transferred to the genus *Ectinosoma*.

A. Introduction

Harpacticoida of the Southern Hemisphere and the Pacific region are only incompletely known. First results from samples taken along the coasts of Papua New Guinea and the Fiji Islands in 1984 show a high percentage of new species. Out of 18 species of Ectinosomatidae, for example, 13 are recorded for the first time. Nine of them belong to the genus *Ectinosoma*. Three of these nine new *Ectinosoma* species will be presented here.

The few species of *Ectinosoma* reported from the Southern Hemisphere are: *E. melaniceps* Boeck, 1865, a cosmopolitan species, *E. dentatum* Vervoort, 1964 from the coast of the Karoline Islands, *E. acutorostratum* Vervoort, 1962 from New Caledonia and *E. pectinatum* Mielke, 1979, *E. nonpectinatum* and

E. spec. I–V From Galapagos Island. GEORGE (1996) reports on an undescribed *Ectinosoma* species from Panitao near Puerto Montt, Chile. It is remarkable that despite ist ubiquity in samples from the littoral down to the deep sea this genus is only rarely treated in the literature. This may be explained by the fact that it is almost impossible to determine species of *Ectinosoma* because the descriptions lack the detail necessary to distinguish the species.

B. Material and methods

The material was collected during an expedition to Papua New Guinea and the Fiji Islands by Prof. H. K. Schminke in 1984. The specimens were fixed in 5 % buffered formalin and later transferred into Zeiss W15 embedding medium. Drawings were made with the aid of a camera lucida on a Leitz Dialux 20 interference microscope. The type material is stored in the Copepod Collection of the Arbeitsgruppe Zoomorphologie, University of Oldenburg.

The terminology for the hyaline frill is adopted from MOORE (1976). The setae of the caudal rami are numbered according to HUYS & BOXSHALL (1991).

Abbreviations used in the text: C. R.: Caudal rami, aes.: aesthetasc, P1–P6: swimming legs 1–6.

C. Results

Family **Ectinosomatidae**

Nauplius eye absent. Short female antennule at most 7-segmented; aesthetasc on second or third segment. Antenna with distinct basis; exopod at most 3-segmented. Mandible with 1 or 2 setae on distal side of cutting edge. Allobasis of maxilla enlarged. Maxilliped characteristic, not prehensile. Baseoendopod P5 of both sexes at most with 2 setae. Exopod P5 of both sexes with at most 4 setae, only *Bradya confluens* Lang, 1936 with 5 setae (?); exopod sometimes reduced or fused with baseoendopod. Anal somite divided. Penultimate somite with pseudoperculum. Female with one egg-sac.

Genus *Ectinosoma*

So-called tubercle which is a relative big, chitinuous characteristic u-shaped pore, is to be found on the cephalothorax, all body somites, except the penultimate one, on the distal endopod segment P2–P4 and on exopod P5. All 4 setae exopod P5 are in distal position.

Description

Ectinosoma papuarum sp. n.

(Figs. 1–5)

Material: 13 females and 2 males. Locality: 2°35' S, 150°5' E, Ednago Island, near Kavieng, New Ireland, Papua New Guinea; sublittoral sand from 1 m depth, 11. 11. 1984. The holotype (female) is dissected and mounted on 8 slides (Coll. No. 1996.6/1–1996.6/8). The allotype (male) is dissected and mounted on 6 slides (Coll. No. 1996.7/1–1996.7/6). Of the paratypes 2 females are dissected (Coll. No. 1996.8/1–1996.8/5; Coll. No. 1996.9/1–1996.9/7).

Description of the female (holotype)

Body length (incl. C.R.)	491 µm
Caudal rami	15 µm
Maximum body width	139 µm
Cephalothorax length	147 µm

Rostrum (Fig. 2C) hyaline and tongue-shaped, fused with cephalothorax. **Body** (Figs. 1A–E, 2A) fusiform. Cuticle of body and P1–P5 with minute perforations on its surface; cuticle of cephalothorax and body somites, except the penultimate, with sensilla and tubercles. Hyaline frill of cephalothorax and all somites strengthened with subcuticular stripes of chitin; hyaline frill of cephalothorax and first three of free thoracic somites plain, of the fourth free thoracic and all following somites semi-incised subulate. Broad hyaline palisades over the hyaline frill of the genital double and the following somite. First three free thoracic somites with one row of relatively long slender cuticular hairs. Thoracic and abdominal somites with rows of minute cuticular hairs, all of the same length. Genital and first abdominal somite fused to form a genital double somite, which is subdivided dorsally by chitinous patches and ventrally by a transverse w-shaped chitinous stripe. Pseudoperculum semicircular, semi-incised subulate and with subcuticular stripes of chitin. **Caudal rami** (Figs. 1E–G) about as long as wide and with 7 setae. Posterior edge of rami terminating ventrally as acuminate lappet. **Antennule** (Fig. 2B) 7-segmented. Third and fifth segments elongated. Armature formula (?): 1, 9, 7+aes, 1, 0, 2, 1+2aes. **Antenna** (Fig. 2D). Coxa short and bare. Basis as long as first endopod segment, with 2 very long spinules at inner margin. Exopod 3-segmented, much shorter than endopod; setal formula: 0, 1, 2. Endopod 2-segmented; segment 1 bare, segment 2 with 2 spinulose setae on inner margin and 6 terminal setae, 4 of which are spinulose on one side. **Labrum** (Fig. 3C) prominent and trapezoidal, terminating in a spinous projection. In the middle with 2 rostrate protrusions and 2 semicircular rows of spinules. **Mandible** (Fig. 2E). Coxa

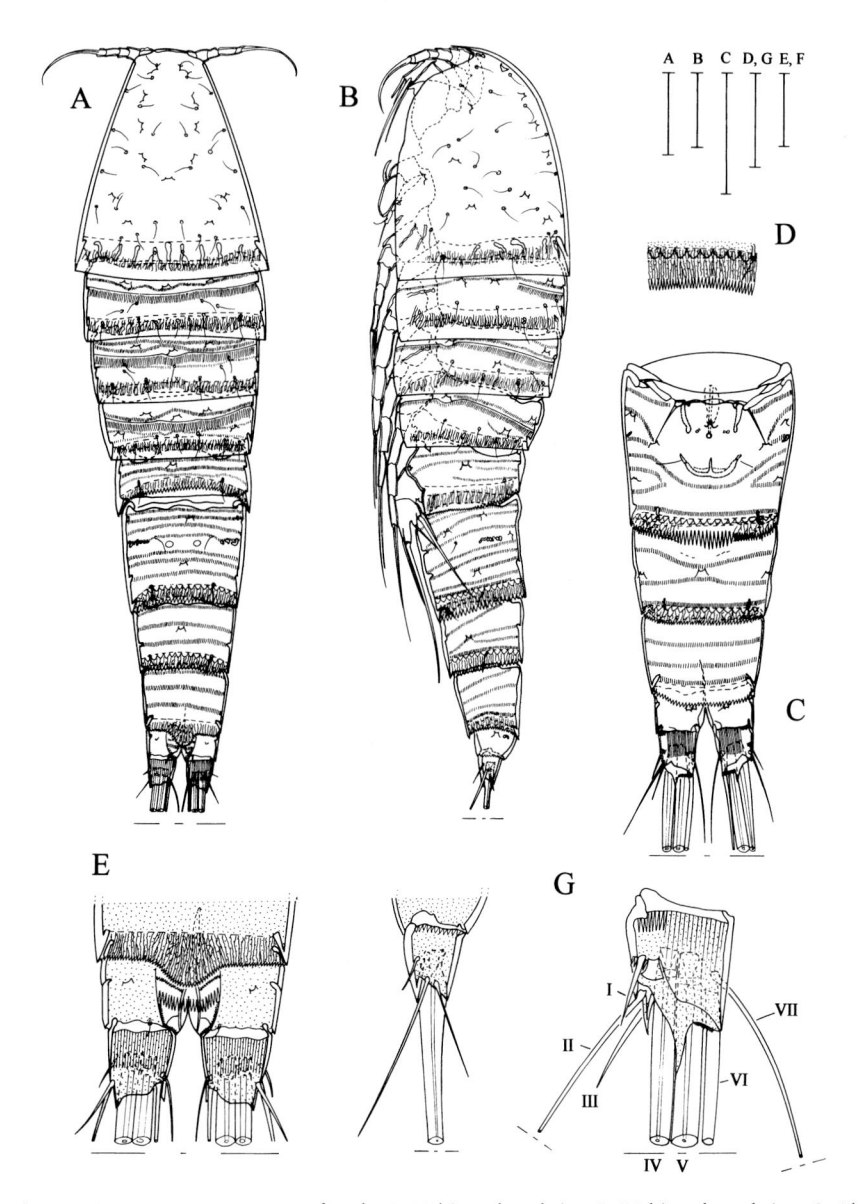

Fig. 1. *Ectinosoma papuarum* sp. n., female. A. Habitus, dorsal view. B. Habitus, lateral view. C. Abdomen, ventral view. D. Hyaline frill of the genital double somite. E. Furca, dorsal view. F. Furca, lateral view. G. Furcal, ventral view. Scale bars A-C: 50 μm; D-F: 20 μm.

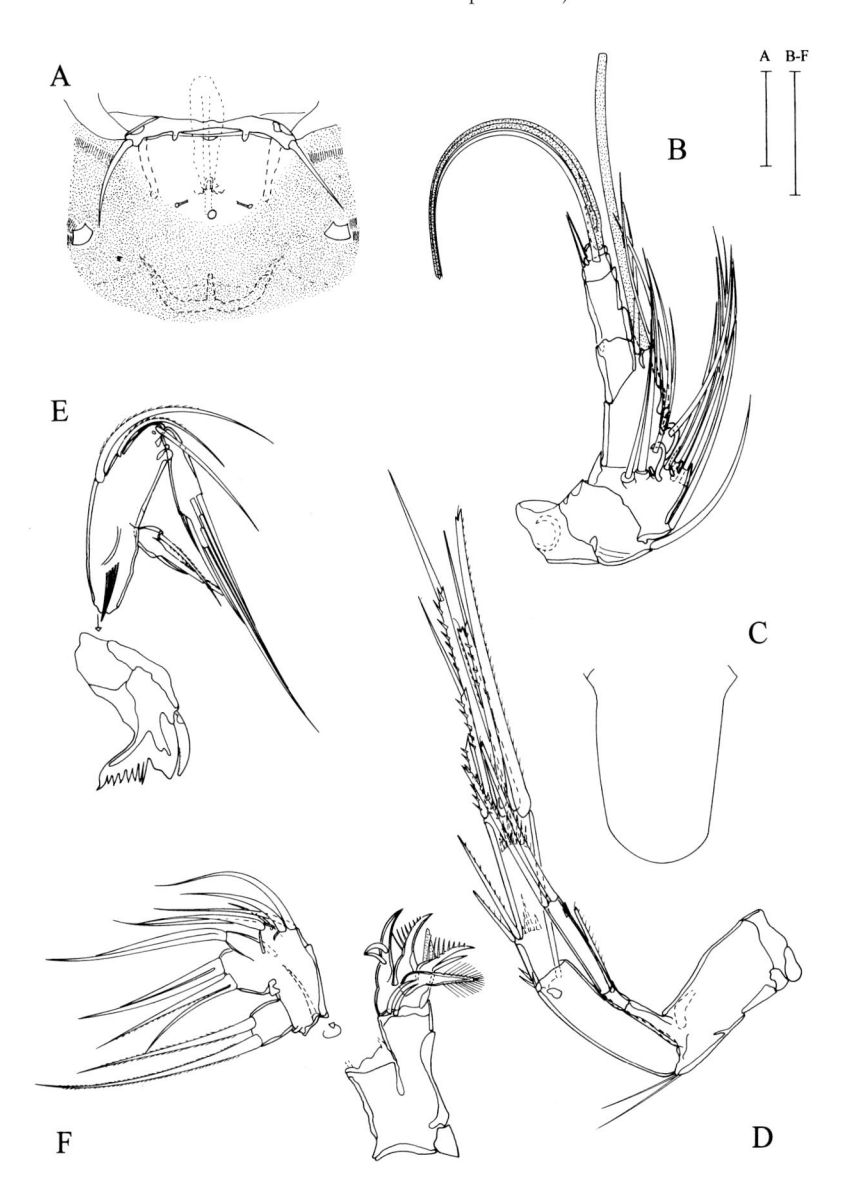

Fig. 2. *Ectinosoma papuarum* sp. n., female. A. Genital field. B. Antennule. C. Rostrum. D. Antenna. E. Mandible. F. Maxillula. Scale bars: 20 μm.

with 1 strong seta on the distal side of cutting edge, 1 big tooth and 8 smaller acute teeth. Big tooth with a small bulge. Basis with a row of long spinules and 3 setae at distal inner corner. Exopod 1-segmented, with 3 setae, the middle one with undefined base. Endopod 1-segmented, with 7 (or more?) setae. **Maxillule** (Fig. 2F). Arthrite of praecoxa apically with 3 unguiform spines, 1 very strong pinnate seta and two bare setae. One of the spines with a small triangular hyaline structure, not easily seen. Coxa fused with basis and armed with 1 seta. Basis with 6 bare setae. Exopod with 2 pinnate setae. Endopod not defined, armed with 6 setae, basally fused to three pairs. **Maxilla** (Fig. 3A). Syncoxa with 3 endites; proximal endite with 3 spinulose spines and 1 bare seta, middle endite with 2 bare setae and distal endite with 1 spinulose spine and 2 bare setae. Allobasis armed with 3 bare setae medially. Endopod 1-segmented, with 2 strong partly pinnate setae and 3 (?) bare setae. **Maxilliped** (Fig. 3B). Syncoxa slightly wider than long, with 1 long bifid seta. Basis about 7 times longer than greatest width, with 2 parallel rows of long spinules; endopod segment with 4 setae, two terminal ones fused at base.

P1–P4 (Figs. 3D–E, 4A–B). Exopod and endopod 3-segmented with setal formula as follows:

	Exopod	Endopod
P1	I–0; I–1; III,I,2	0–1; 0–1; 1,2,2
P2	I–1; I–1; III,I,3	0–1; 0–1; 1,2,2
P3	I–1, I–1; III,I,4	0–1; 0–1; 1,2,2
P4	I–1; I–1; III,I,4	0–1; 0–1; 1,2,2

P1 (Fig. 3D). Exopod extending to end of second endopod segment. Terminal end of third endopod segment strengthened with characteristically formed cuticular thickening. Inner seta of middle exopod segment conspicuously spinulose at tip. **P2–P4** (Figs. 3E, 4A–B). Intercoxal sclerites v-shaped. Inner distal corner of basis with dentiform process. Exopod extending to about the proximal part of last endopod segment. Middle segment of exopod shortest. Inner seta of middle exopod segment conspicuously spinulose at tip. Third segment of endopod with a tubercle subdistally on anterior surface. Terminal end of third segment endopod and exopod strengthened with characteristically formed cuticular thickening. **P5** (Fig. 4C). Inner expansion of baseoendopod extending to middle of exopod; inner seta about twice as long as outer one; outer seta pinnate, not lanceolate. Exopod as long as wide; basal limit of exopod only visible on posterior side of P5; row of spinules on proximal part of exopod; two inner lobules of exopod subequal in length, separated by a deep cleft and reaching far beyond the insertion points of both outer setae; of

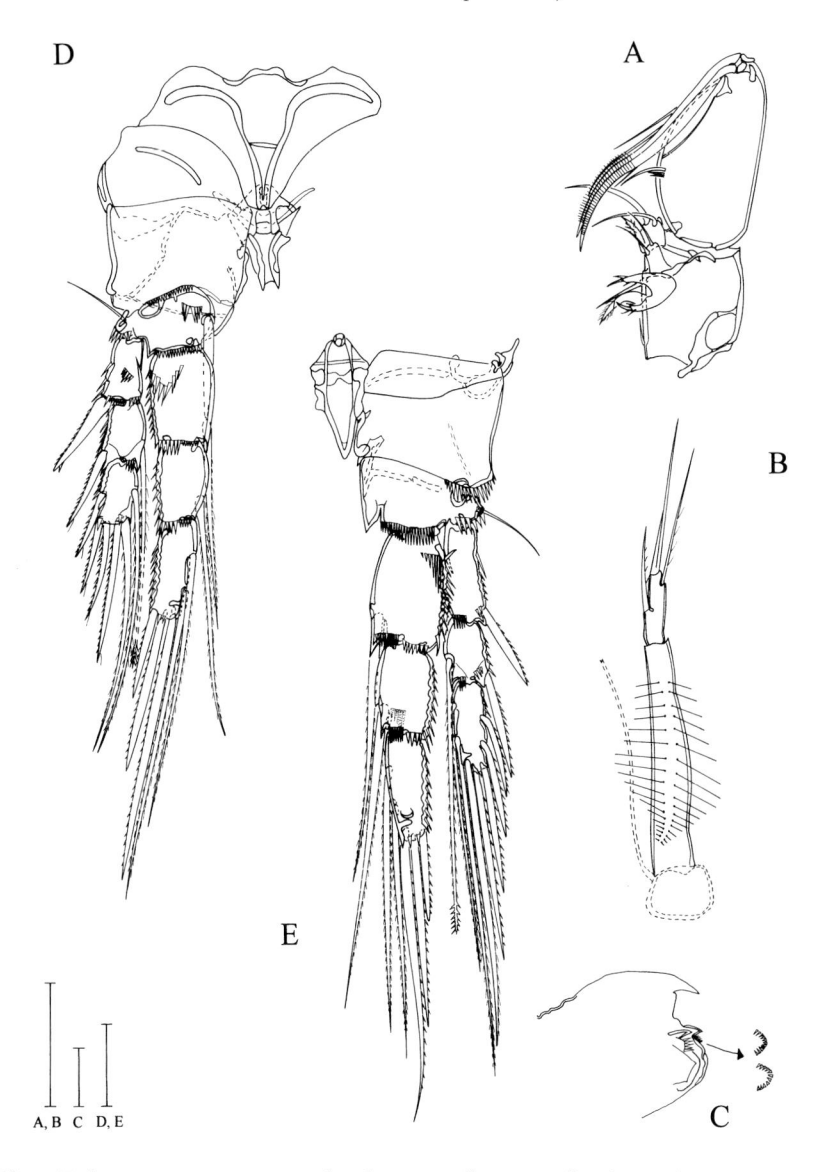

Fig. 3. Ectinosoma papuarum sp. n., female. A. Maxilla. B. Maxilliped. C. Labrum. D. P1. E. P2. Scale bars: 20 μm.

the 4 terminal setae the innermost but one seta longest; inner edge of the exopod with small cuticular spine. Big tubercles in middle of inner edge of exopod and outer part of baseoendopod.

Description of the male (allotype)

Body length (incl. C. R.) 356 µm
Caudal rami 13 µm
Maximum body width 75 µm
Cephalothorax length 116 µm
Spermatophore 31 µm

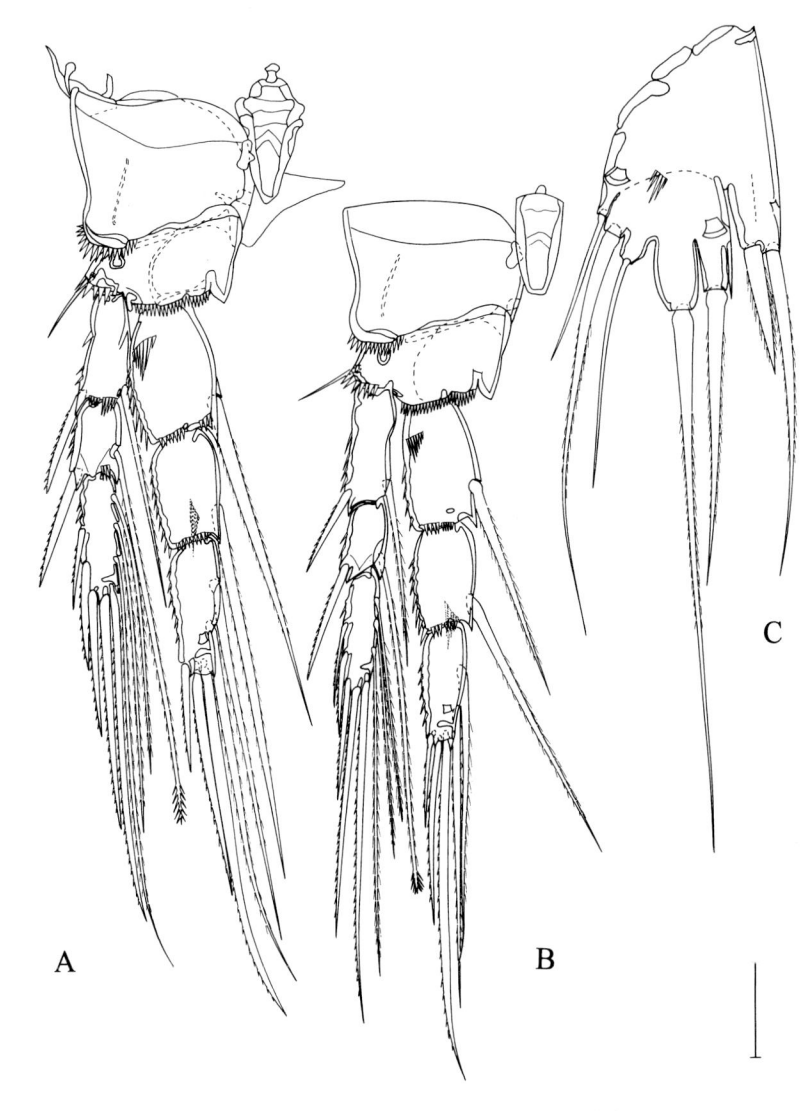

Fig. 4. *Ectinosoma papuarum* sp. n., female. A. P3. B. P4. C. P5. Scale bar: 20 µm.

Ectinosoma from Papua and Fiji   43

Habitus (Figs. 5A–B), caudal rami, antenna, mouthparts and P1–P4 as in female.

Body ornamentation. On the ventral side of second abdominal somite two tubercles. Otherwise as in female. **Antennule** (Fig. 5C) haplocer and 6-seg-

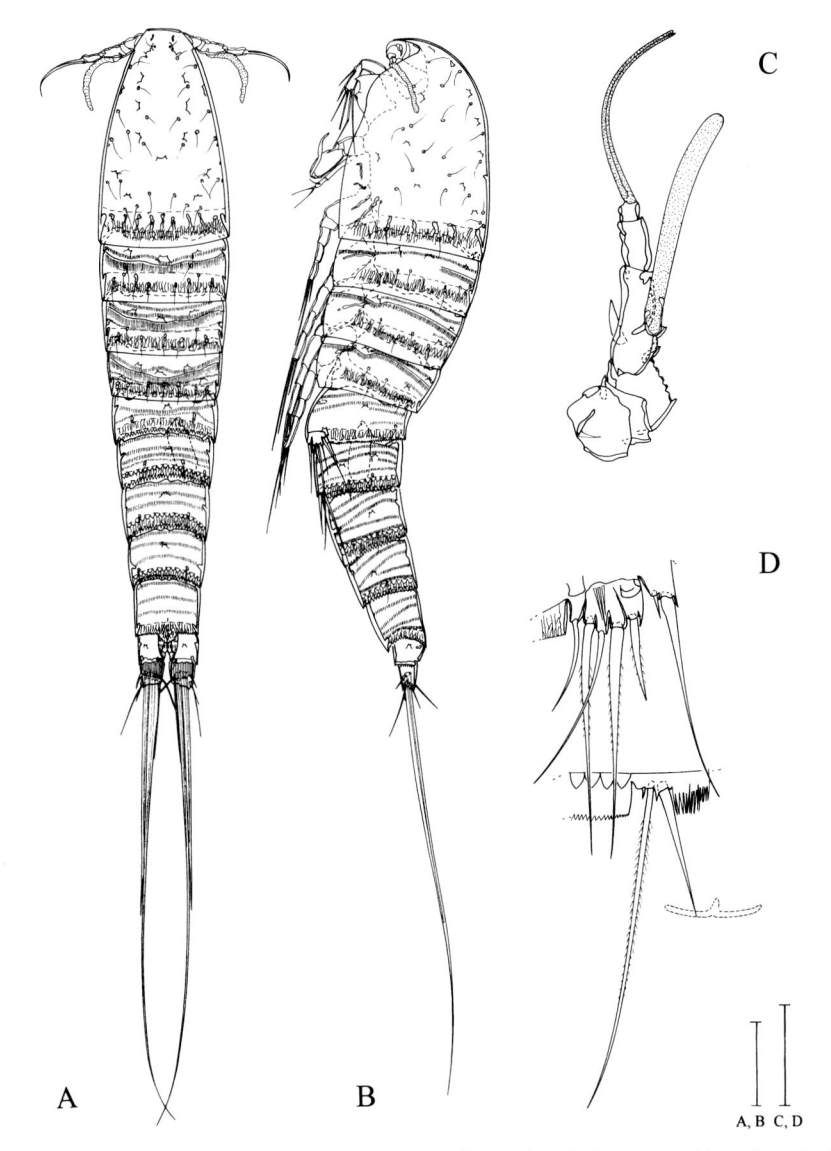

Fig. 5. *Ectinosoma papuarum* sp. n., male. A. Habitus, dorsal view. B. Habitus, lateral view. C. Antennule. D. P5, P6. Scale bars A, B: 50 μm; C, D: 20 μm.

mented. Segment 4 elongate and slightly swollen forming a joint with 5th segment, with one big aesthetasc and a cuticular cone. Armature formula (?): 1, 2, 6, 4+aes, 1, 1+2aes. **P5** (Fig. 5D). Inner expansion of baseoendopod extending to about middle of exopod; outer seta remarkably small. Exopod wider than long, with row of spinules proximally and a deep cleft between the inner and the innermost but one seta. **P6** (Fig. 5D) with 2 setae. Outer seta twice as long.

Variability
Body length of female syntypes varies between 473 µm and 556 µm ($\bar{x}$ =519 µm; n = 10). Body length of males varies between 350 µm and 356 µm. The number of spinules of P1–P4 is different.

Etymology
This species is dedicated to the women of the indigenous people of Papua New Guinea, the Papua.

Ectinosoma tegula **sp. n.**

(Fig. 6)

Material: 13 females and 1 male. Locality: Gridreference 60K XE 4093, Viti Levu, Joske Reef west of Suva, Fiji Islands; coars-grained coral sand from 3 m depth, 20. 08. 1984. The holotype (female) is dissected and mounted on 7 slides (Coll. No. 1996.10/1–1996.10/7). The allotype (male) is dissected and mounted on 5 slides (Coll. No. 1996.11/1–1996.11/5). Of paratypes 1 female is dissected (Coll. No. 1996.12/1–1996.12/5).

Description of the holotype (female)
Body length (incl. C.R.)   487 µm
Caudal rami   16 µm
Maximum body width   121 µm
Cephalothorax length   153 µm

Habitus, caudal rami, antennule, antenna, labrum and mouthparts exactly as in *E. papuarum* sp. n.

The following differences are found:
Ornamentation of the body (Figs. 6C–D). The rows of relatively long slender cuticular hairs of the first three free thoracic segments have the form of smal palisades. The indentation of the hyaline frill of the genital double somite is not stronger ventrally than laterally. **P2–P4** (Fig. 6B). Setal formula as in *E. papuarum* sp. n. Terminal end of third segment exopod not strengthened with cuticular thickening. **P5** (Fig. 6A). The cleft between insertion points of the two inner setae exopod is not so deep. No row of spinules on proximal

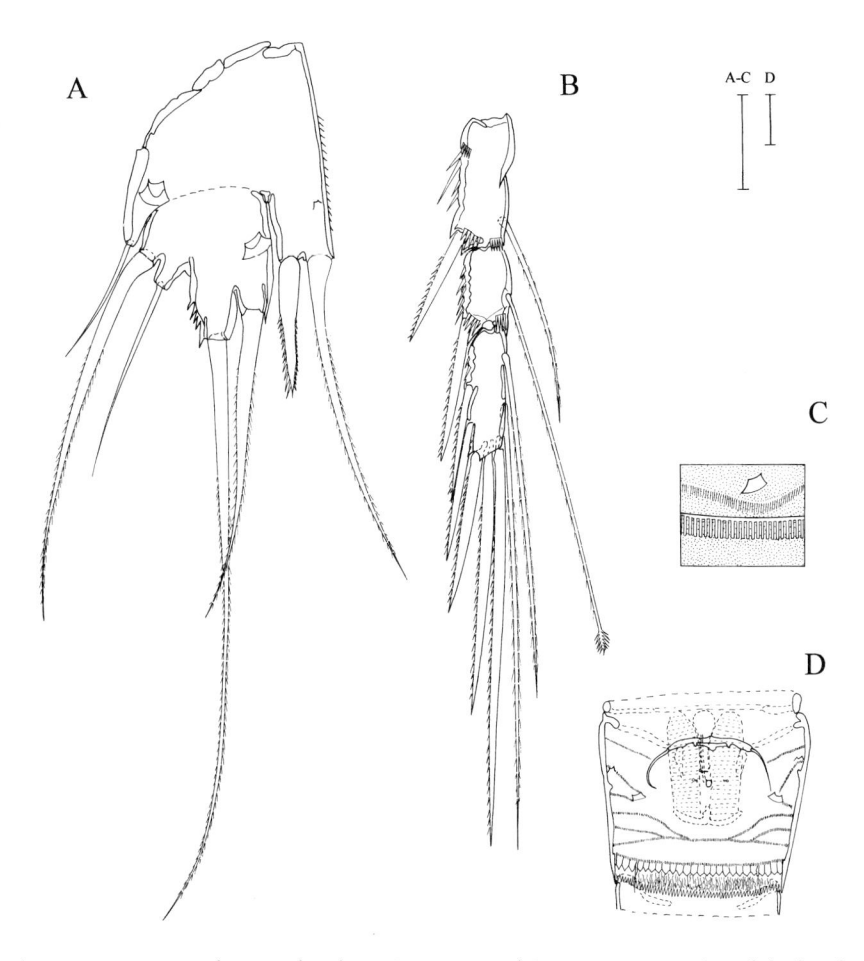

Fig. 6. *Ectinosoma tegula* sp. n., female. A. P5. B. Exopod P2. C. Ornamentation of the first free thoracic somite. D. Genital double somite, ventral view. Scale bars: 20 μm.

part of exopod. Spinules on inner edge of inner expansion baseoendopod and on outer edge of the lobule from innermost but one seta.

Description of allotype (male)
Body length (incl. C.R.)   410 μm
Caudal rami   12 μm
Maximum body width   104 μm
Cephalothorax length   146 μm
Spermatophore   34 μm

Males show the same differences from *E. papuarum* sp. n. as females.

Ecology

The dissected paratype has a complete nauplius in the gut of the abdomen. It probably is a nauplius of the family Diosaccidae. This leads to the conclusion, that this species feeds on harpacticoid nauplii at least as a part of its diet.

Variability

Body length of the 2 female syntypes varies between 455 µm and 533 µm ($\bar{x}$ = 502 µm; n = 2).

Etymology

Derived from the latin *tegula*, which means brick or tile and refers to the presence of small hyaline palisades median on the first three free thoracic segments and broad hyaline palisades over the hyaline frill of the genital double somite and the following segment.

Ectinosoma barbararum sp. n.

(Figs. 7–12)

Material: 14 females and 1 male. Locality: 2°35` S, 150°5` E, Ednago Island, near Kavieng, New Ireland, Papua New Guinea; sublittoral sand from 1 m depth, 11. 11. 1984. The holotype (female) is dissected and mounted on 6 slides (Coll. No. 1996.13/1–1996.13/6). The allotype (male) is dissected and mounted on 7 slides (Coll. No. 1996.14/1–1996.14/7). Of the paratypes 3 females are dissected (Coll. No. 1996.15/1–1996.15/4; Coll. No. 1996.16/1–1996.16/5; Coll. No. 1996.17/1–17/7).

Description of the holotype (female)

Body length (incl. C. R.)   780 µm
Caudal rami   32 µm
Maximum body width   156 µm
Cephalothorax length   224 µm

Rostrum (Fig. 9C) broad and hyaline, fused with cephalothorax. **Body** (Figs. 7A–D, 8A, 8E) fusiform. Cuticle of body and P1–P5 with minute perforations on ist surface; cuticle of cephalothorax and body somites, except the penultimate one, with sensilla and tubercles. Hyaline frill of cephalothorax and all somites strengthened with subcuticular stripes of chitin; hyaline frill of cephalothorax plain; hyaline frill of first and second free thoracic somites denticulate, of the following somites semi-incised subulate. First three thoracic somites with one row of relatively long slender cuticular hairs. Thoracic and abdominal somites with rows of minute cuticular hairs of different length. Genital and first abdominal somite fused to form a genital double somite, which is subdivided dorsally by chitinous patches and ventrally by a transverse w–

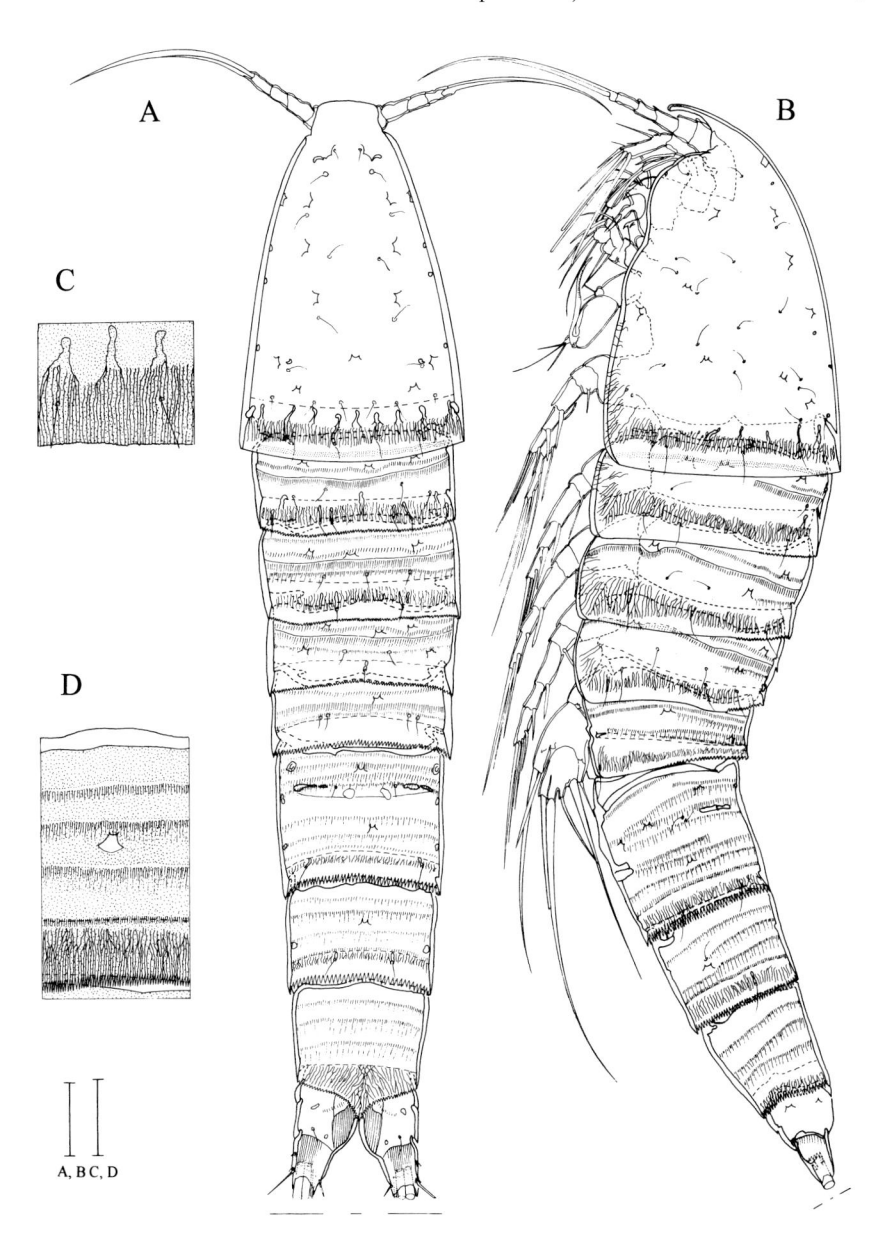

Fig. 7. *Ectinosoma barbararum* sp. n., female. A. Habitus, dorsal view. B. habitus, lateral view. C. Hyaline frill of the cephalothorax. D. Ornamentation of the genital double somite, dorsal view. Scale bars A, B: 50 μm; C, D: 20 μm.

shaped chitinous stripe. Pseudoperculum acute with subcuticular stripes of chitin. **Caudal rami** (Figs. 8C–D) about 1.5 times longer than greatest width, with 7 setae. Posterior edge of rami terminating ventrally as acuminate small lappet. **Antennule** (Fig. 9A) 7-segmented. First segment with row of spinules near outer margin. Armature formula: 1, 10, 8+aes, 1, 1, 3, 4+aes. **Antenna** (Fig. 9B). Coxa short and bare. Basis with 3 very long spinules at inner distal margin. Exopod 3-segmented, somewhat shorter than endopod; setal formula: 1, 1, 2. Endopod 2-segmented; segment 1 bare, segment 2 with 2 spinulose setae on inner margin and 6 terminal setae, 4 of which are spinulose on one side. **Labrum** (Fig. 10C) priminent and trapezoidal, terminating in a spinous projection. In the middle of labrum 4 rostrate protrusions. **Mandible** (Fig. 9D). Coxa with 1 strong spinulose seta on the distal side of the cutting edge, 1 big tooth and 6 smaller acute teeth. Basis with row of long spinules, row of short spinules and 3 setae at distal inner corner. Exopod 1-segmented armed with 3 setae, 2 of them with undefined bases. Endopod 1-segmented, with 7 setae. **Maxillule** (Fig. 9E). Arthrite of praecoxa apically with 3 unguiform spines and 1 pinnate seta. Coxa fused with basis and armed with 1 seta. Basis with 6 bare setae. Exopod with 2 pinnate setae. Endopod not defined, armed with 6 setae, basally fused to three pairs. **Maxilla** (Fig. 10A). Syncoxa with 3 endites; proximal endite with 3 spinulose spines and 1 bare seta on a lobule, middle endite with 2 bare setae and distal endite with 1 spinulose spine and 2 bare setae. Allobasis armed with 3 bare setae medially. Endopod 1-segmented with 2 strong partly pinnate setae and 5 bare setae. **Maxilliped** (Fig. 10B). Syncoxa trapezoidal with 1 long bifid seta. Basis about 6 times longer than greatest width, with 2 parallel rows of short spinules and one row of long spinules. Endopod segment with 4 setae; the shorter seta of the 2 terminal ones with blunt end.

P1–P4 (Figs. 10D–E, 11A–B). Exopod and endopod 3-segmented with setal formula as follows:

	Exopod	Endopod
P1	I–0; I–1; III,I,2	0–1; 0–1; 1,2,2
P2	I–1; I–1; III,I,3	0–1; 0–1; 1,2,2
P3	I–1; I–1; III,I,4	0–1; 0V1; 1,2,2
P4	I–1; I–1; III,I,4	0–1; 0–1; 1,2,2

P1 (Fig. 10D). Exopod extending to end of second endopod segment; inner seta of middle exopod segment conspicuously spinulose at tip. **P2–P4** (Figs. 10E, 11A-B). Intercoxal sclerites u-shaped. Inner distal corner of basis with relatively small dentiform process. Exopod extending to about the proximal part of last endopod segment. Middle segment of exopod shortest. Inner seta

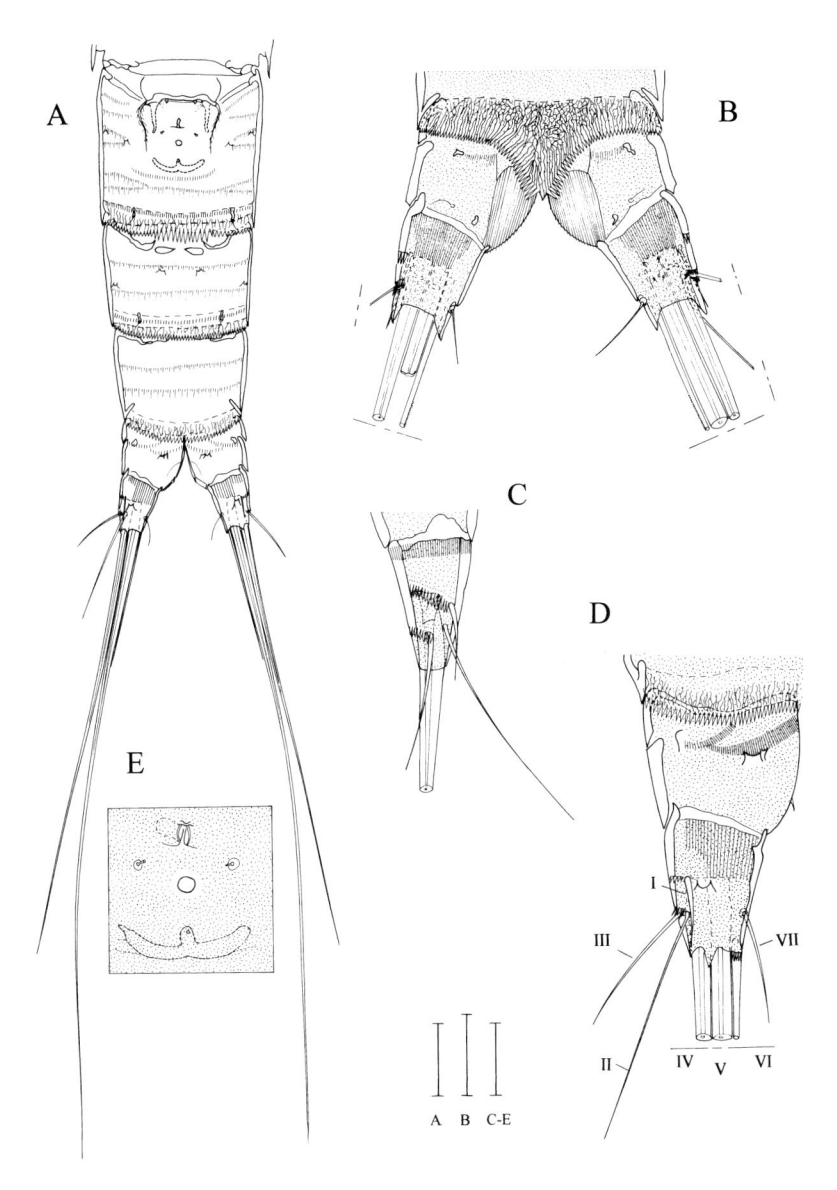

Fig. 8. *Ectinosoma barbararum* sp. n. A. Abdomen, ventral view, female. B. Caudal part, dorsal view, male. C. Furca, lateral view, female. D. Furca, ventral view, female. E. Genital Field. Scale bars A: 50 μm; B-E: 20 μm.

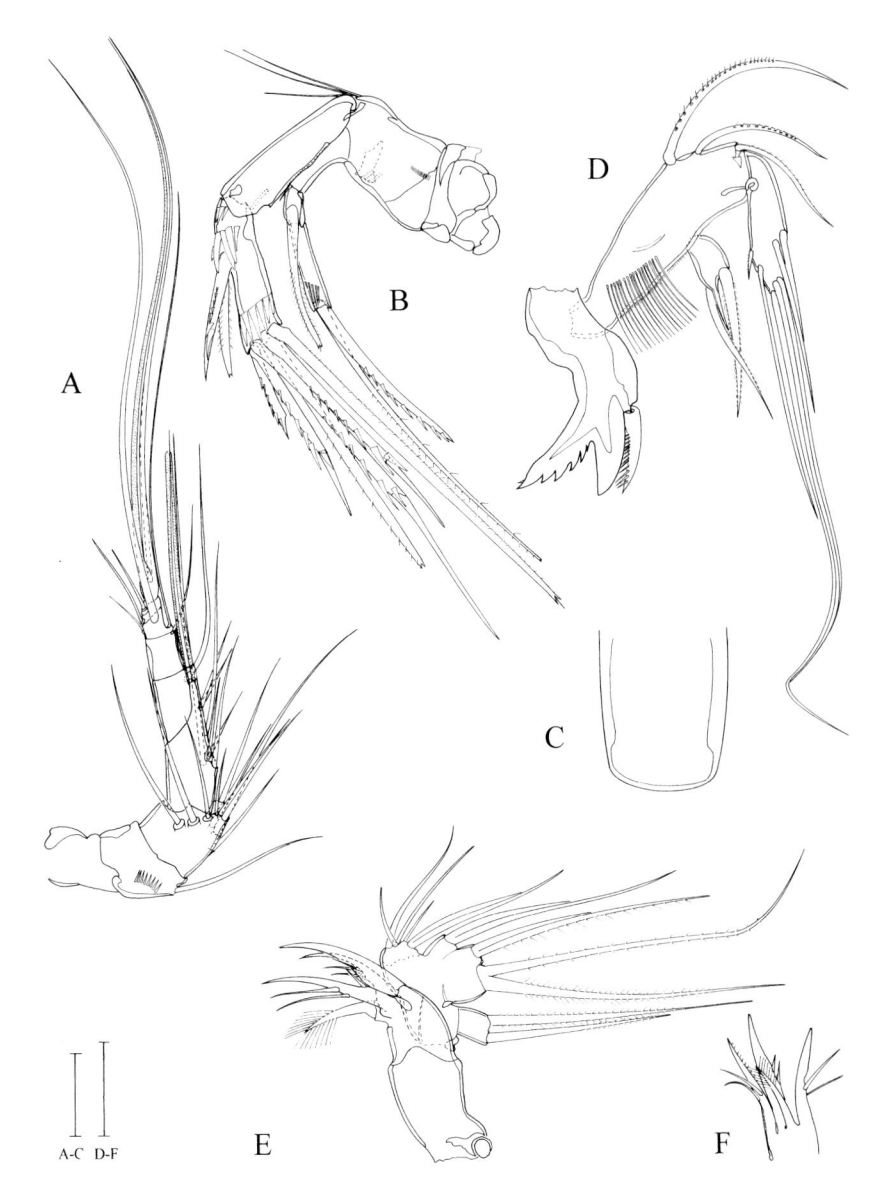

Fig. 9. *Ectinosoma barbararum* sp. n., female. A. Antennule. B. Antenna. C. Rostrum. D. Mandible. E. Maxillula. F. Arthrite of praecoxa, maxillula. Scale bars: 20 μm.

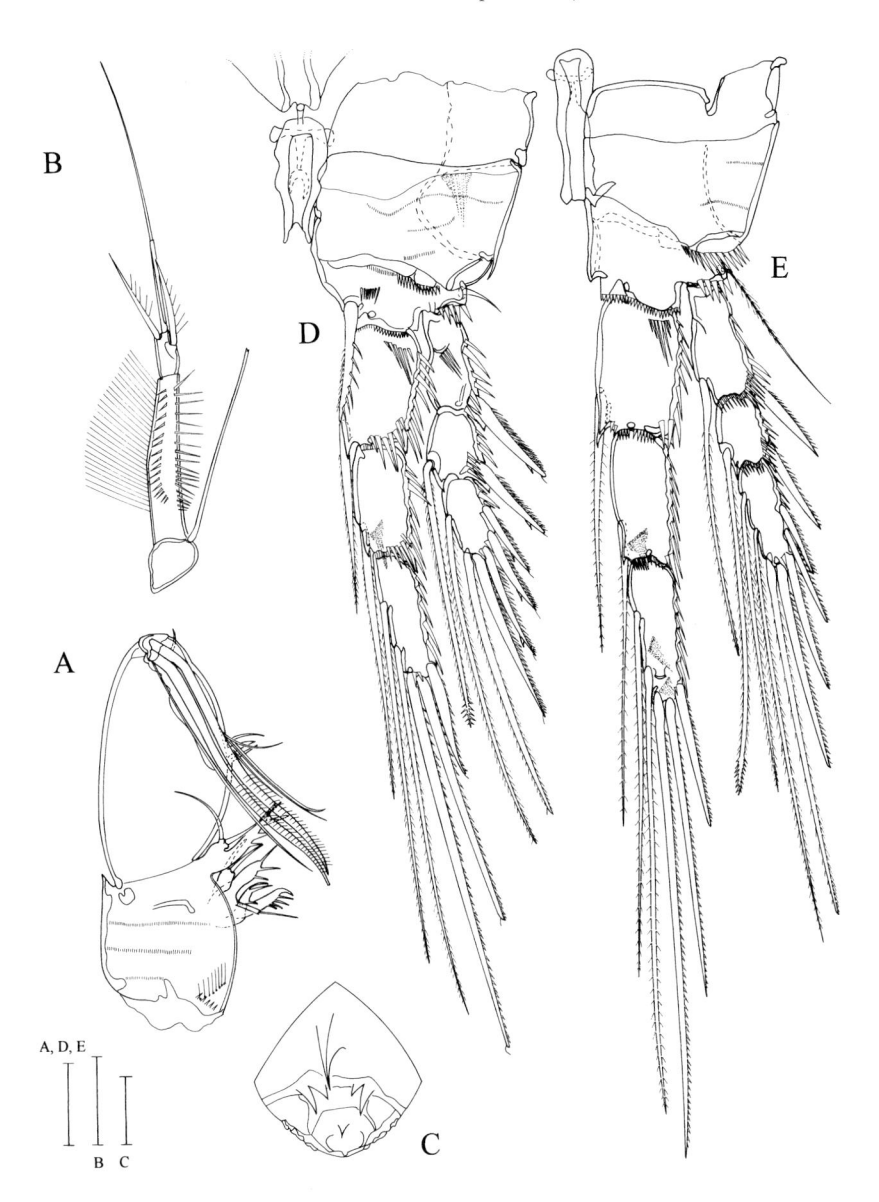

Fig. 10. *Ectinosoma barbararum* sp. n., female. A. Maxilla. B. Maxilliped. C. Labrum. D. P1. E. P2. Scale bars: 20 μm.

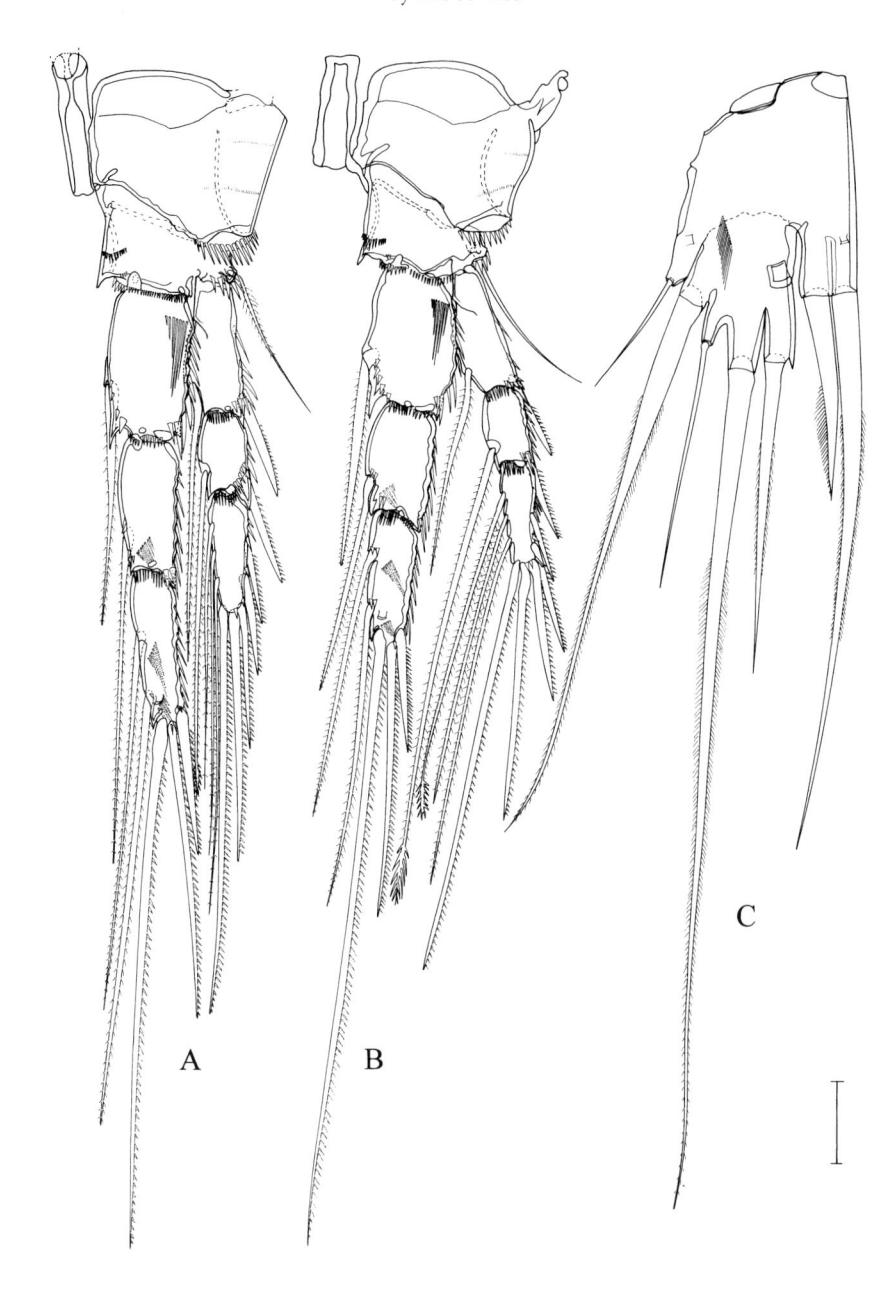

Fig. 11. *Ectinosoma barbararum* sp. n., female. A. P3. B. P4. C. P5. Scale bars: 20 µm.

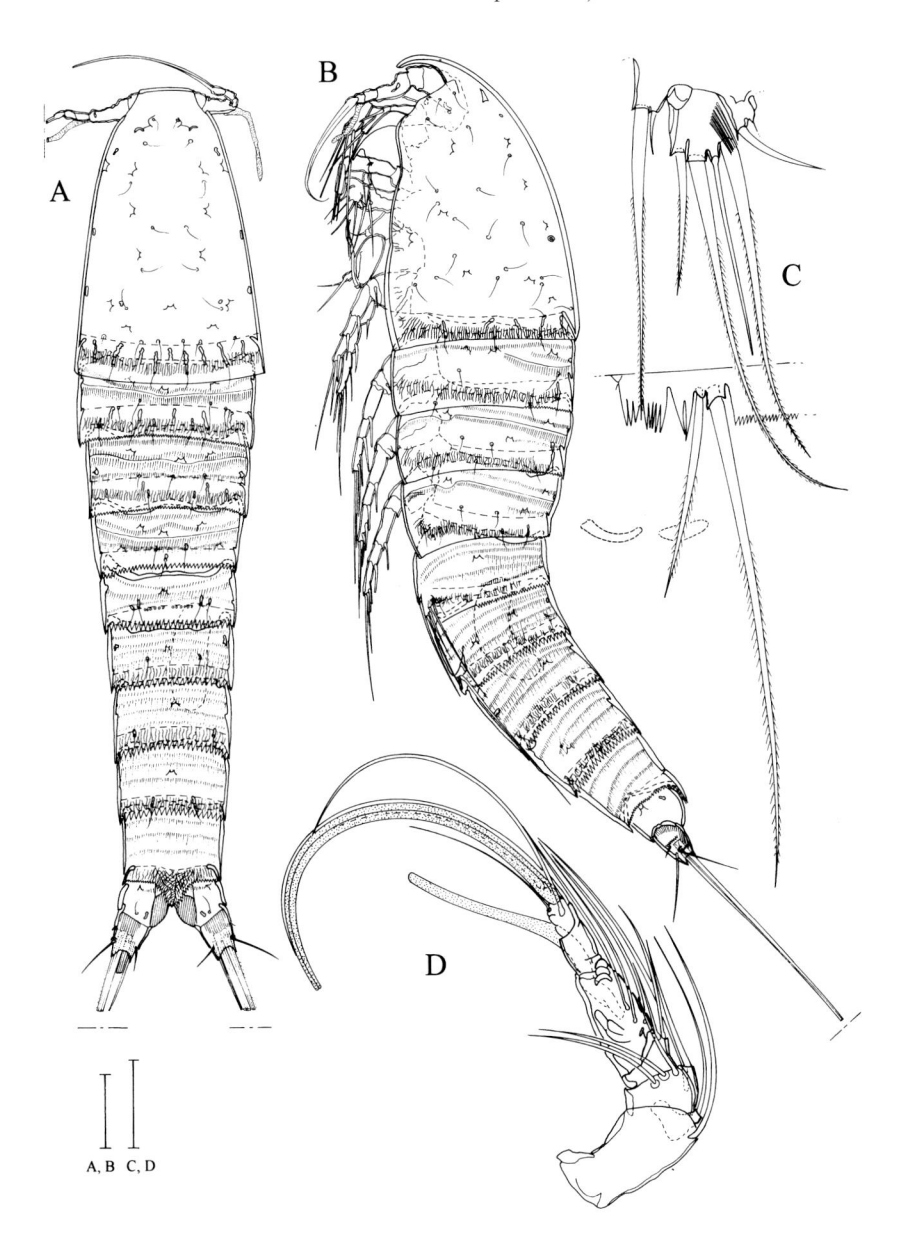

Fig. 12. *Ectinosoma barbararum* sp. n., male. A. habitus, dorsal view. B. Habitus, lateral view. C. P5, P6. D. Antennule. Scale bars A, B: 50 μm; C, D: 20 μm.

of middle exopod segment of P2-P4 and inner proximal seta of third exopod segment P4 conspicuously spinulose at tip. Third segment of endopod with tubercle subdistally on anterior surface. **P5** (Fig. 11C). Inner expansion of baseoendopod extending to middle of exopod and slightly cleft; inner seta about three times as long as outer one; outer seta one-sided pinnate, not lanceolate. Exopod distinctly longer than wide; the basal limit of exopod only visible on posterior side of P5; row of spinules on proximal part of exopod; two inner lobules of exopod subequal in length, separated by a deep cleft and reaching not far beyond the long outermost but one lobule. Of the 4 terminal setae the innermost but one seta is longest; 1 big tubercle in middle of inner edge of exopod.

Description of the allotype (male)
Body length (incl. C.R.)   524 µm
Caudal rami   27 µm
Maximum body width   96 µm
Cephalothorax length   135 µm
Spermatophore   48 µm

Habitus (Figs. 12A-B), body ornamentation, caudal rami (Fig. 8B), antenna, mouthparts and P1-P4 as in female.

Antennule (Fig. 12D) haplocer and 6-segmented. Segments 1 and 4 elongate. Segment 4 slightly swollen forming a joint with 5th segment, with one big aesthetasc and a characteristic three-piece cone. Armature formula (?): 1, 6, 2, 3+aes, 1, 4+2aes. **P5** (Fig. 12C). Inner expansion of baseoendopod extending to 1st third of exopod; outer seta remarkably small. Exopod a little longer than wide, with a row of spinules. **P6** (Fig. 12C) with 2 setae. Outer seta 2.5 times as long as inner one.

Variability
The body length of the female syntypes varies between 768 µm and 832 µm ($\bar{x}$ = 797 µm; n = 10). The number of spinules of P1-P4 is different.

Etymology
This species is dedicated to Barbara Hosfeld and Barbara Seifried.

Discussion

E. papuarum sp. n. and *E. tegula* sp. n. share a number of characters with *E. nonpectinatum* Mielke, 1979 from the Galapagos Islands, mainly some details of body ornamentation and the shape of P5. *E. nonpectinatum, E. papuarum*

sp. n., *E. tegula* sp. n., *E. barbicauda* Bozic, 1979, and *E. californicum* Lang, 1965 have hyaline palisades over the hyaline frill of the genital and following segment. But in *E. barbicauda* and *E. californicum* the deep incision between the two inner setae exopod P5 is absent and the setal formula is different. *E. papuarum* sp. n. and *E. tagula* sp. n. can be distinguished from *E. nonpectinatum* by the following characters:
– first segment exopod antenna longer than second.
– cutting edge of mandible with 1 big tooth and 8 smaller teeth.
– middle seta of the 3 setae exopod mandible not defined at base.
– middle endite maxilla with 2 setae.
– first and second segment endopod P1 without rows of fine cuticular hairs.
– first segment endopod P2-P4 without rows of fine cuticular hairs.
– terminal end of third endopod segment P1-P4 strengthened with characteristically formed cuticular thickening.
– ornamentation of female abdomen ventrally: all cuticular hairs are of the same length; penultimate segment without hyaline palisades over hyaline frill.
– female: projection of innermost seta of exopod P5 shorter than that of innermost but one; inner seta exopod as long as inner seta baseoendopod.
– male: deep incision between innermost and innermost but one seta of exopod P5, outer seta of inner expansion smaller than that of *E. nonpectinatum*.

Further more *E. tegula* sp. n. shows characters that distinguish this species from *E. papuarum* sp. n. and *E. nonpectinatum* as shown in the description of *E. tegula* sp. n.

MIELKE (1979) mentions that besides *E. nonpectinatum* there are other varieties along the coast of the Galapagos Islands, which are very closely related to *E. nonpectinatum*. *E.* spec. V shares the following characters with *E. papuarum* sp. n. and *E. tegula* sp. n.:
– first segment exopod antenna a little longer than second.
– middle endite maxilla with 2 setae.
– penultimate segment without hyaline palisades over hyaline frill.
– male: incision between innermost and innermost but one seta of exopod P5.

Nevertheless, it is impossible to unite *E. papuarum* sp. n. from Papua New Guinea, *E. tegula* sp. n. from the Fiji Islands and *E.* spec. V from the Galapagos Islands in one species because of the mentioned morphological differences.

E. barbararum sp. n. differs from all other *Ectinosoma* species in a suite of characters:
– ornamentation of body.
– cutting edge of mandible with 1 big tooth and 6 smaller acute teeth.

- shorter seta of the 2 terminal setae second endopod segment maxilliped with blunt end.
- inner distal corner of basis P2-P4 relatively small.
- inner expansion of baseoendopod P5 slightly cleft.
- pseudoperculum acute.

At the moment it is impossible to determine the phylogenetic position of *E. barbararum* sp. n. The taxonomy of the genus *Ectinosoma* is made difficult by only minute morphological differences between the species. Only detailed descriptions such as that of *E. nonpectinatum* and *E. pectinatum* and careful drawings of the whole habitus dorsal and lateral, will shed light upon the taxonomy and evolution of this widespread and species rich genus.

In 1865, BOECK established the genus *Ectinosoma* with the type species *E. melaniceps* Boeck, 1865. LANG (1948) split the genus into two subgenera: *Ectinosoma (Halectinosoma)* includes all species of *Ectinosoma* with one surface seta and 3 distal setae exopod P5, *Ectinosoma (Ectinosoma)* embraces all species with 4 distal setae exopod P5. In 1965 LANG "looked upon them as separate genera ...", "since no transitional forms can exist between them ...". But sometimes it is difficult to decide whether there is a surface seta or not, as shown in the case of *Halectinosoma porosum* Wells, 1967. This species has a very peculiar P5, because the exopod is completely fused with the baseoendopod and there is no real surface seta but a seta on a lobule in distal position. This species has pores on the cephalothorax that resemble the characteristic u-shaped tubercles that can only be found in the *Ectinosoma* species. For this reason, *H. porosum* has to be transferred to the genus *Ectinosoma*. LANG (1965) pointed out, that tubercles can probably be found in all species of the genus *Ectinosoma* and that "pores of the very same appearance occur on all somites, the penultimate one excepted." I could find the big characteristic u-shaped tubercles on the cephalothorax, on all body somites, except the penultimate one, and on the distal endopod segment P1-P4 as well as on exopod P5 of several as yet undescribed *Ectinosoma* species, of *E. barbararum* sp. n., *E. papuarum* sp. n., *E. tegula* sp. n., *E. melaniceps*, *E. nonpectinatum*, *E. reductum* Bozic, 1954 and *E. litorale* (noodt, 1958). I was able to see paratypes of *E. litorale* from Noodt's collection. They have the characteristic u-shaped tubercles not mentioned by NOODT (1958). Except for *E. soyeri* Apostolov, 1975, all species of *Ectinosoma* have been shown to possess a tubercle on exopod P5. Most species are also reported to have tubercles on the abdomen or the whole body. Contrary to LANG's (1965) opinion it is possible to count them and to map their distribution. The only problem is, that it is very difficult to see tubercles in specimens that have just moulted.

It is necessary to examine whether *E. porosum* and some more *Ectinosoma* species also have tubercles on all body somites, except the penultimate one and on P1-P4 and whether *E. soyeri* really has tubercles. The species of *Halectinosoma*, the sister group of *Ectinosoma*, have little round inconspicuous pores instead of tubercles.

At the moment it is very difficult to determine *Ectinosoma* species. Very often there are no differences in the number of setae of the appendages and the distinctive features are very small. The drawings frequently are not good enough or lacking. As LANG (1965) pointed out, the mandible carries very good diagnostic characters. The cutting edge is unique for a species in most cases. Unfortunately, the mandible is often not mentioned in the description. MIELKE (1979) shows that characters of the antenna, female P5, male P5-P6, and the ornamentation of the abdominal somite reveal distinctive features. These become apparent when detailed descriptions are available. This also holds for the ornamentation of the body. The presence, number and distribution of rows of fine cuticular hairs, tubercles, sensillae and the shape of hyaline frills are constant in a species and provide many useful characters which can be seen in whole specimens without dissection. The form of the labrum also seems to be useful in distinguishing between the species, as in the case of *E. papuarum* sp. n., and *E. barbararum* sp. n. Provided detailed drawings of the antenna, mandible, labrum, P5-P6, and the ornamentation of the whole body are available it would be possible to revise this genus. Including *E. barbararum* sp. n., *E. papuarum* sp. n., *E. porosum* and *E. tegula* sp. n. *Ectinosoma* holds a total of 25 valid species and 10 species incertae. In my opinion some of the valid species are probably of doubtful position.

All species of Ectinosomatidae carry one or two setae on the distal side of the cutting edge of the mandible. These setae are found in no other copepod species and are only visible in dissected specimens. There is no doubt that the Ectinosomatidae are monophyletic.

Zusammenfassung

Weibchen und Männchen dreier neuer *Ectinosoma*-Arten werden beschrieben. *E. barbararum* sp. n. und *E. papuarum* sp. n. stammen von den Ednago Islands, Papua-Neuguinea. *E. tegula* sp. n., eine mit *E. papuarum* sp. n. und *E. nonpectinatum* Mielke, 1979 verwandte Art, stammt von den Fiji-Inseln. Eine Diagnose der Ectinosomatidae wird erstellt. Die Monophylie der Ectinosomatidae wird vor allem durch ein oder zwei Borsten an der distalen Seite der Mandibelschneidekante begründet. *Ectinosoma* kann vom Schwestertaxon *Halectinosoma* durch die charakteristischen U-förmigen Tuberkel in der

Kutikula des Körpers und der P1-P5 unterschieden werden. *H. porosum* Wells, 1967 weist diese charakteristischen Tuberkel und einen P5 auf, der dem der *Ectinosoma*-Arten entspricht. *H. porosum* wird deshalb zu *Ectinosoma* gestellt.

Acknowledgements

I am especially grateful to Professor Dr. H. K. Schminke for the material from the Pacific region and the constructive criticism on the manuscript. I wish to express my gratitude to Dr. W. Mielke for his comments, which improved the manuscript. Thanks are also due to E. Willen who introduced me to the techniques of modern copepod taxonomy, to Dr. H. Juhl of the Moneculus library and to A. Sievers for the review of the English language, as well as to Dr. A. Ahnert for providing me with the paratypes of *E. litorale*.

References

BOECK, A. (1865): Oversigt over de ved Norges Kyster iagttagne Copepoder henhoerende til Calanidernes, Cyclopidernes og Harpactidernes Familier. Forhandl. Vidensk. - Selsk. Christiana Aar 1864, 226–282.

GEORGE, K. H. (1996): Revisión de los harpacticoídeos marinos (Crustacea: Copepoda) de Chile. Rev. chil. Hist. Nat. **69**, 77–88.

HUYS, R. & G. A. BOXSHALL (1991): Copepod Evolution. The Ray Society London, 1–468.

LANG, K. (1948): Monographie der Harpacticiden. Håkan Ohlson, Lund, 1–1682.

– (1965): Copepoda Harpacticoidea from the Californian Pacific Coast. Kungl. Svenska vetenskaps. Handl. **10**, 1–566.

MIELKE, W. (1979): Interstitielle Fauna von Galapagos. XXV. Longipediidae, Canuellidae, Ectinosomatidae (Harpactiocoida). Mikrofauna Meeresboden **77**, 1–106.

MOORE, C. G. (1976): The form and significance of the hyaline frill in harpacticoid copepod taxonomy. J. nat. Hist. **10**, 451–456.

NOODT, W. (1958): Die Copepoda Harpacticoidea des Brandungsstrandes von Teneriffa (Kanarische Inseln). Akad. Wiss. Lit. Abh. math.-nat. Kl. **2**, 51–116.

Sybille Seifried
Fachbereich 7 (Biologie)
AG Zoomorphologie
Universität Oldenburg
D-26111 Oldenburg

Ultrastructural observations on the "eye spot" of *Halammovortex nigrifrons* (Plathelminthes, Rhabdocoela, "Dalyellioida")

Beate Sopott-Ehlers

Abstract

Halammovortex nigrifrons is characterized by the existence of a virtually x-shaped eye spot consisting of two pairs of eyes. Each eye is composed of one single sensory cell, one cup or mantle cell devoid of pigment granules and an additional multicellular pigment shield. This construction of the eyes is comparable to a type of photoreceptor known for two species of the Proseriata Monocelididae, but does not correspond to the submicroscopical anatomy of light-percepting organs in other representatives of the taxon Rhabdocoela. From these data it is hypothesized that the type of photoreceptor described (1) has been convergently evolved more than once within the Plathelminthes, and (2) is an autapomorphic feature of *H. nigrifrons* (or of the taxon *Halammovortex*).

A. Introduction

Pigmented photoreceptors, so-called "pigment cup ocelli", are a common characteristic of many Plathelminthes. Viewed under the light microscope, all these light-sensing organs appear to have the same structure. However, experiences of the last decades have shown that the dimensions of most of these eyes are such that the structural information the light microscope can provide is limited. The use of the electron microscope has revealed a great diversity in the construction of pigment cup ocelli. These variations concern the number of cells forming the eye spots in general, the nature of the light-sensing organelles, as there are ciliary, rhabdomeric or epigenous organelles, the existence or the lack of dioptrics and the nature of shading devices, as well (for ref. see RIEGER et al. 1991; SOPOTT-EHLERS 1991).

In course of intense research on the submicroscopic anatomy of the free-living "Typhloplanoida" and "Dalyellioida" the eyes of *Halammovortex nigrifrons* (Karling, 1934) were studied to see whether the light-sensing organs correspond to the basic pattern of pigment cup ocelli or whether special variations exist suitable to elucidate the relationships of the non-monophyletic "Typhloplanoida" and "Dalyellioida" within the Rhabdocoela.

B. Materials and Methods

The individuals of *H. nigrifrons* derive from sand samples taken on beaches of the island of Sylt (North Sea). The specimens were extracted from the sediment using the sea water ice method (see PFANNKUCHE and THIEL 1988).

Fixation, dehydration and embedding followed conventional steps (see SOPOTT-EHLERS 1993a). Series of transverse and sagittal sections were stained with uranyl acetate followed by lead citrate and examined using a Zeiss EM 900.

C. Results

Halammovortex nigrifrons (Karling, 1935) is easily to recognize by the existence of an almost virtually x-shaped eye spot. Although the contour of the shading pigment appears not well bordered, but quite diffuse in living animals, the shape of this eye spot gives the impresssion, that four eyes exist. This fact has already been confirmed by meticulous studies of histological sections performed by KARLING (1943). Despite of these insights, however, the light microscope does not provide enough informations on the precise construction of these light-sensing organs. Electronmicroscopical studies have shown, that each of the four eyes consists of three elements: a single sensory cell, an extremely delicate cup or mantle cell free of pigment granules and a special pigment shield made up by several cells.

Fig.1: A. Stained section with four photoreceptors (ph1 - ph4) and electron-lucent pigment. B. Unstained section with three photoreceptors (ph1, ph2, ph4) and electron-dense pigment. Scale in A and B = 10 µm. The asterisks mark rhabdomeric complexes.

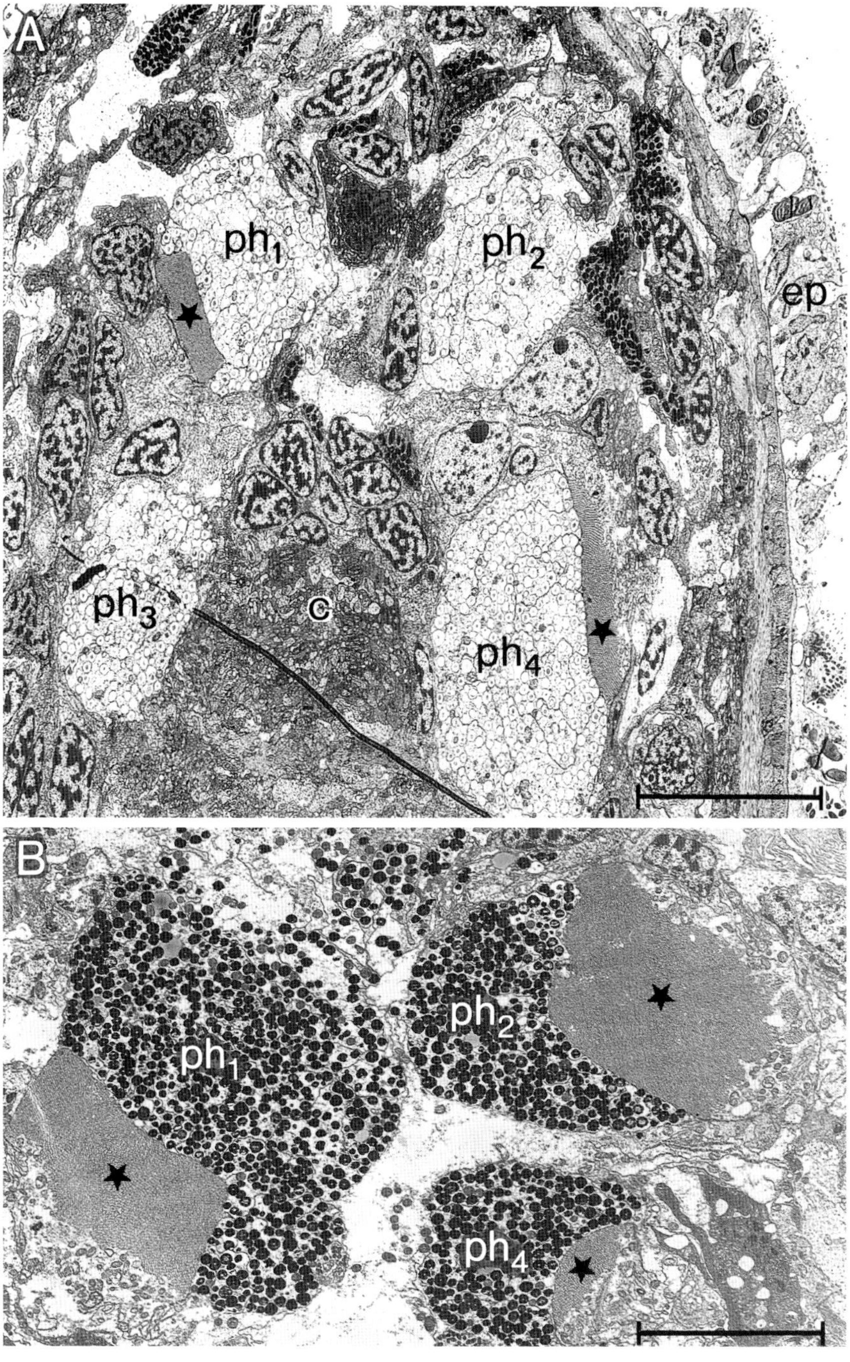
A
ph1
ph2
ep
ph3
c
ph4
B
ph2
ph1
ph4

The whole x-shaped complex lies on the dorsal top of the foremost end of the brain. Each of the sensory cells enters its eye cup through an opening which faces the body surface. Within the eye cup the sensory cells widen to form a flat disk-shaped plate about 6 – 7 µm in diameter. This socket gives rise to the rhabdomeric complex consisting of extensions of the surface membrane. These microvilli are digitiform, very slender and with an extension of about 2.5 – 3 µm comparatively short. In sections the rhabdomeric complexes have a rectangular to triangular contour (Figs. 1 – 3).

The cytoplasm of that part of each retinular cell lying within the eye cavity contains abundant mitochondria, glycogen islets, clear and dense-cored vesicles (Fig. 3). Some evidence of neurotubules was also found. The external segment containing the nucleus exhibits the same cell organelles as the one within the eye cup. The position of the nucleus, however, could not be definitely clarified.

At this side which faces the interior of the animal, each sensory cell is flanked by a cell appearing over-crowded by glycogen deposits (Fig. 4 A). The retinular cells are linked to their adjacent mantle cells by cell junctions (Fig. 4 B).

The mantle cells forming the eye cups sensu stricto have an extremely small diameter (Figs. 2 B,C, 3, 4 A,B,). These components of the light-sensing system in *Halammovortex nigrifrons* are totally inconspicuous on the light-microscopical level and appear, when viewed with the electronmicroscope at low magnifications, like intercellular spaces. Higher magnifications, however, reveal the cellular nature of these "spaces". The cytoplasm of the cup cells enveloping the rhabdomeres is electron-lucent appearing almost free of cell organelles, some dispersed glycogen particles and vesicles excepted. In a few instances neurotubules were observed (Fig. 2 B). There are also some uneven tip-shaped dilatations containing mitochondria (Figs. 2 B, 3, 4 B). The position of the nuclei could not be identified. However, it is supposed that the nuclei and further cell organelles are situated in cytoplasmic pouches outside of the eye region. The cup cells are attached to the sensory cells by septate junctions and zonulae adhaerentes (Fig. 4 B).

The shielding devices of each of the eyes consist of numerous cells including pigment granules. These pigment screens are very large in their extensions and cover the eye bowles like caps. In some instances the pigment cells seem to

---

Fig. 2: A. Photoreceptor with processes of the pigment shield tucking into the brain, sensory cell ▶ and rhabdomeres in longitudinal section. Scale = 2 µm. B. Rhabdomeres, mantle cell with neurotubule. The white asterisk marks intercellular substance between pigment cells. Scale = 0.5 µm. C. Rhabdomeric complex enveloped by the mantle cell (small arrows). Scale = 1 µm.

The "eye spot" of *H. nigrifrons*   63

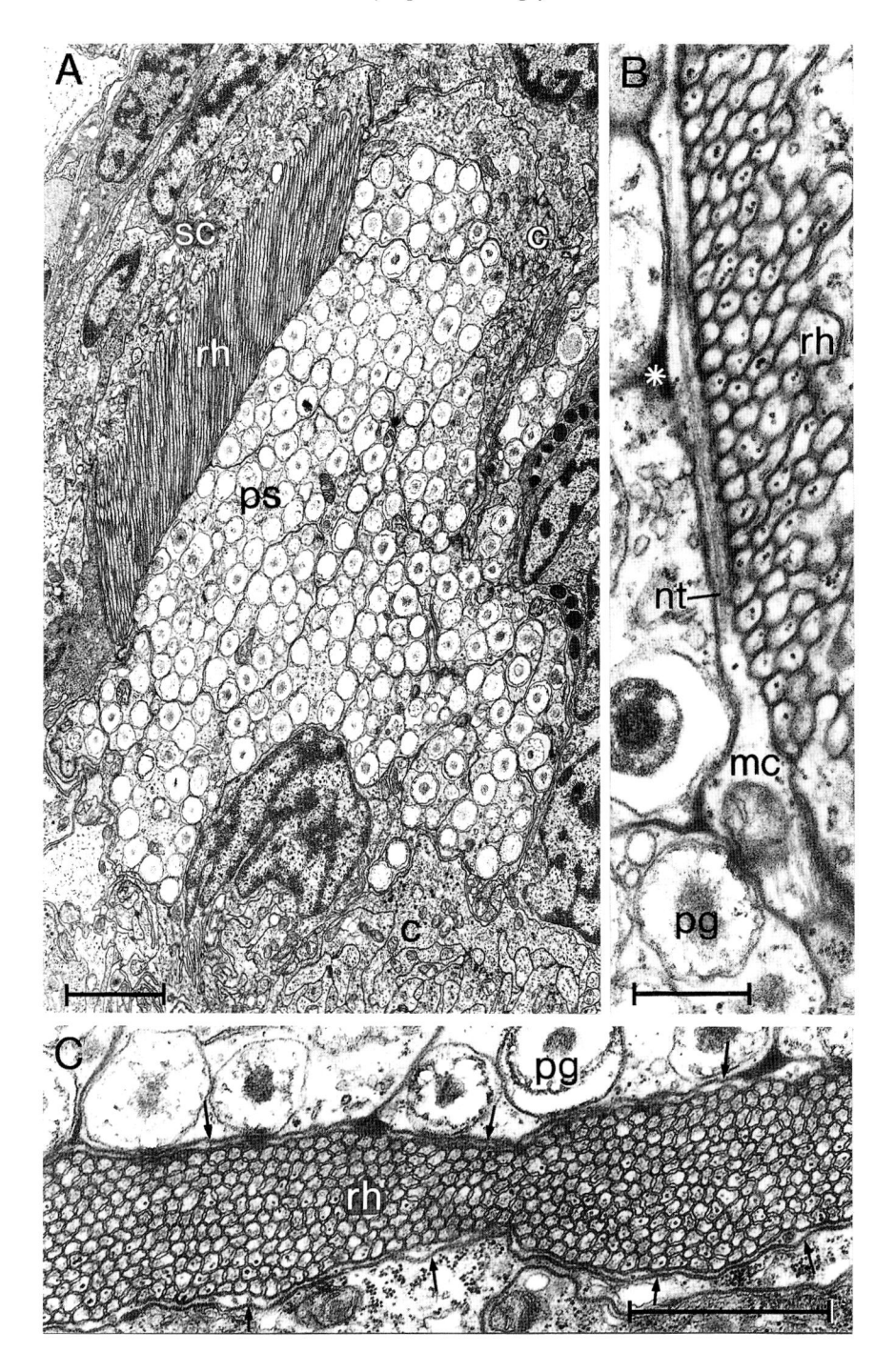

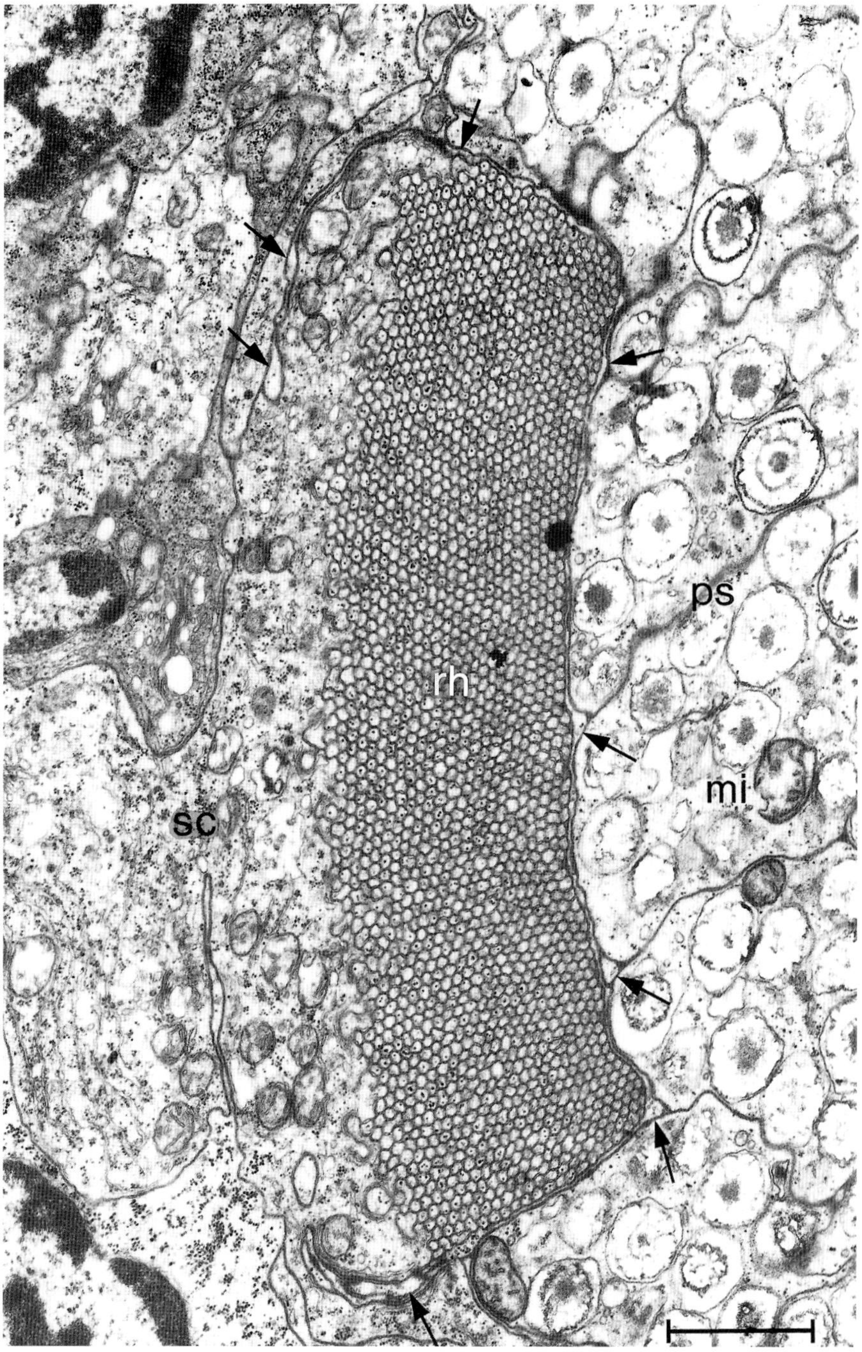
ps
rh
mi
sc

tuck into the brain (Fig. 2 B). In the very center of the x-shaped eye spot the pigment shields of each eye of the photoreceptive system in *H. nigrifrons* touch each other (Fig 1 B). In more peripheral regions processes of cells characterized by electron-light cytoplasm penetrate between the pigment screens.

Most of the lumen of the pigment cells is occupied by relatively small (0.5 – 0.6 µm in diameter) pigment granules limited by a bordering membrane. Strangely enough, the pigment appears lucent with a more or less dense core and a fine dark network in s t a i n e d sections (Fig.1 A), but electron-dark to black in u n s t a i n e d sections (Fig. 1 B) (see also KARLING 1943, p.5). Apart from pigment granules the cytoplasm contains but a few distributed glycogen particles, mitochondria, lipid droplets and short profiles of granular endoplasmic reticulum. The nuclei of the pigment cells are supposed to be situated outside of the screens.

Spaces between neighbouring pigment cells and between pigment cells and cup cell, as well, are filled up by intercellular matrix (Fig. 2 B). In regular distances the outer surface membrane of the pigment cells and the corresponding opposite outer surface membrane of the cup cell thicken (Figs. 2 C, 4 B) to form cell junctions, thus indicating a functional unit.

D. Discussion

The existence of four eyes is rather frequently found in free-living Plathelminthes and in free-living stages of parasitc taxa, as well. From this fact it can be concluded that four eyes have evolved more than once within the Plathelminthes.

So far studied at the EM-level, these eyes have proven to belong to the type of pigment cup ocelli, photoreceptors built up by one or several sensory cells and one or several pigmented cup cells (see i.al. SOPOTT-EHLERS 1988). This type of eyes is considered as a basic feature of the Rhabditophora (EHLERS 1985; see also SOPOTT-EHLERS 1997).

Due to the composition of a single sensory cell, a delicate mantle cell devoid of pigment granules and a special pigment shield the four eyes of *H. nigrifrons* do not correspond to this basic type of a pigment cup ocellus.

To date, light-percepting organs made up by those three components have only been described for two species of the Proseriata Monocelididae, namely

◀ Fig. 3: Sensory cell, mantle cell (small arrows) and segment of the pigment shield. Scale = 2 µm.

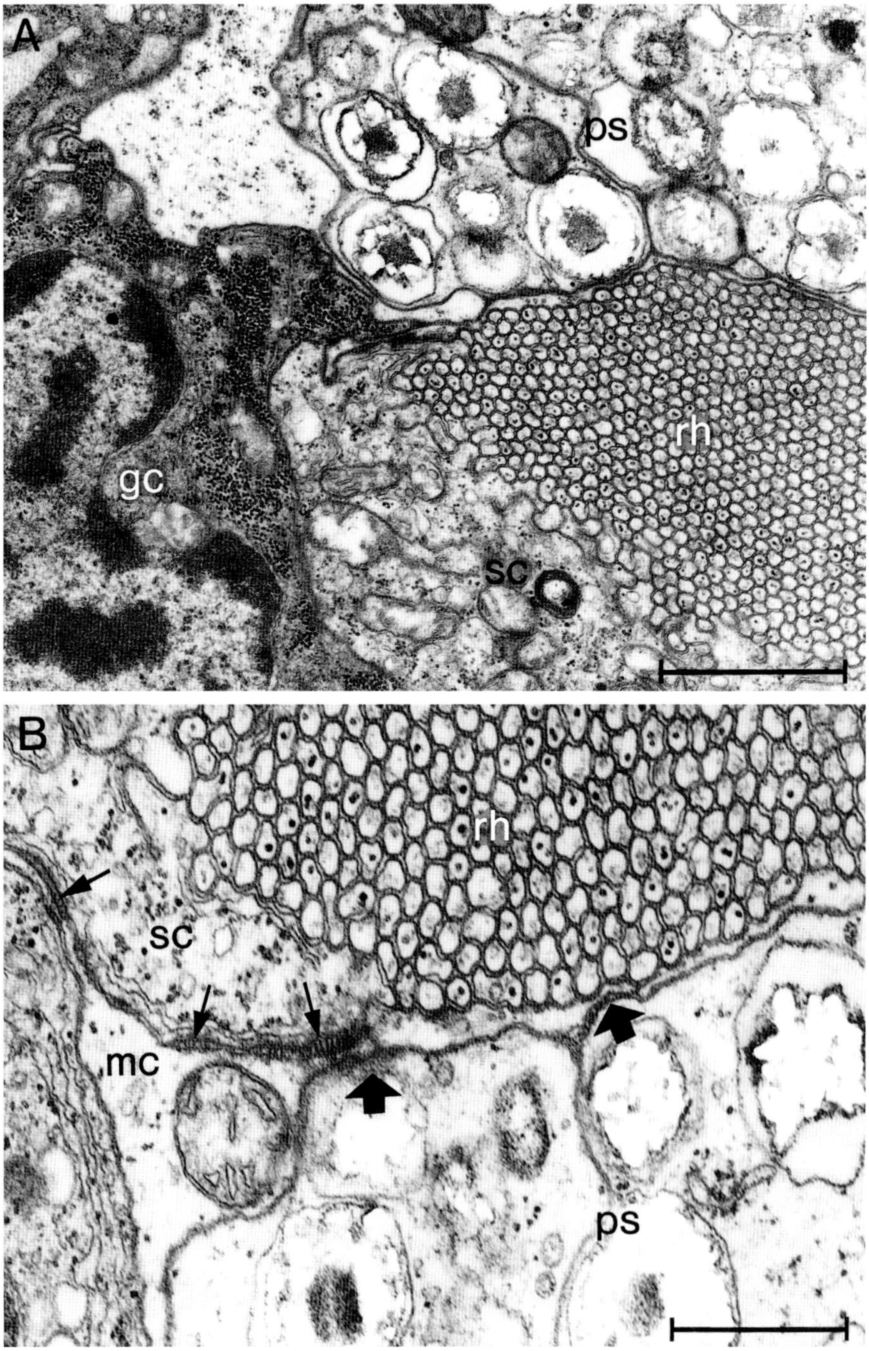
A
ps
rh
gc
sc
B
rh
sc
mc
ps

Monocelis fusca and *Pseudomonocelis agilis* (see Sopott-Ehlers 1984, 1993a). This phenomenon surely is a convergency to the situation existing in *Halammovortex nigrifrons*.

As far as photoreceptors in other representatives of the Rhabdocoela have been studied on the electronmicroscopical level, the light-sensing organs are typical rhabdomeric pigment cup ocelli (see Bedini & Lanfranchi 1990; Lanfranchi & Bedini 1982, 1988; Sopott-Ehlers 1995a), belong to the epigenous type (see Sopott-Ehlers 1995b), are of the rhabdomeric type, but lack a pigmented cup cell (see Sopott-Ehlers 1997) or are equipped with special lenticular differentiations formed by mitochondria (see i.al. Bedini et al. 1973; Sopott-Ehlers 1992, 1993, 1995c, 1996).

As long as further data on eyes of a structure described here for *H. nigrifrons* are laking for other representatives of the Rhabdocoela, this special construction is hypothesized to be an autapomorphic feature for the species considered or for the taxon *Halammovortex*.

Acknowledgements

Financial support was provided by the Akademie der Wissenschaften und der Literatur Mainz. Mrs. S. Gubert is thanked for technical assistance and Mr. B. Baumgart for help with the figures.

Zusammenfassung

H. nigrifrons ist durch einen etwa x-förmigen Augenfleck gekennzeichnet. Dieser besteht aus zwei Augenpaaren. Jedes dieser Augen setzt sich aus einer Sinneszelle, einer pigmentfreien Becher- oder Mantelzelle und einem Pigmentschirm bestehend aus einer Vielzahl von Zellen zusammen. Diese Konstruktion der Augen entspricht zwar dem Aufbau der Photoreceptoren zweier Arten der Proseriata Monocelididae, nicht aber der submikroskopischen Anatomie der Lichtsinnesorgane anderer Vertreter der Rhabdocoela. Aus diesen Befunden wird geschlossen, daß 1. der beschriebene Augentyp innerhalb der Plathelminthes mehr als einmal konvergent entstanden ist und 2. dieser Au-

◀ Fig. 4: A. Photoreceptor with adjacent glycogen cell. Scale = 2 μm. B. Cell junctions between sensory cell and mantle cell and between mantle cell and pigment cells (arrow heads). Scale = 0.5 μm.

gentyp ein autapomorphes Merkmal von *H. nigrifrons* oder dem Taxon *Halammovortex* darstellt.

Abbreviations

c	brain	pg	pigment granule
ep	epidermis	ph	photoreceptor
gc	glycogen cell	ps	pigment shield
mc	mantle cell	rh	rhabdomeric complex
mi	mitochondrion	sc	sensory cell
nt	neurotubule		

References

BEDINI, C., E.FERRERO & A.LANFRANCHI (1973): Fine structure of the eyes in two species of Dalyelliidae (Turbellaria Rhabdocoela). Monit. Zool. Ital. **7**, 51-70.

BEDINI, C. & A. LANFRANCHI (1990): The eyes of *Mesostoma ehrenbergi* (Focke, 1936) (Platyhelminthes, Rhabdocoela). Fine structure and photoreceptor membrane turnover. Acta Zool. (Stockh.). **71**, 125-133.

EHLERS, U. (1985): Das phylogenetische System der Plathelminthes. Fischer, Stuttgart, New York, pp. 1-317.

KARLING, T.G. (1943): Studien an *Halammovortex nigrifrons* (Karling) (Turbellaria Neorhabdocoela). Acta Zool. Fenn. **37**, 3-23.

LANFRANCHI, A. & C. BEDINI (1982): The ultrastructure of the sense organs of some Turbellaria Rhabdocoela. I. The eyes of *Polycystis naegelii* Kölliker (Eukalyptorhynchia Polycystididae). Zoomorphology **101**, 95-102.

– (1986): Electron microscopic study of larval eye development in Turbellaria Polycladida. Hydrobiologia **132**, 121-126.

– (1988): The ultrastructure of the eyes of *Rhynchomesostoma rostratum* (Müller, 1774) (Turbellaria, Rhabdocoela). Fortschr. Zool./Prog. Zool. **36**, 235-241.

PFANNKUCHE, O. & H. THIEL (1988): Sample processing. In: HIGGINS, R.P. & H. THIEL (eds). Introduction to the study of meiofauna. Smithonian Institution Press, Washington, London, pp. 134-145.

RIEGER, R.M., S. TYLER, J.P.S. SMITH III & G.E. RIEGER (1991): Platyhelminthes: Turbellaria. In: Microscopic anatomy of invertebrates, Vol. 3. F.W. HARRISON & B.J. BOGITSH (eds). Wiley Liss, New York, pp. 7-140.

SOPOTT-EHLERS, B. (1984): Feinstruktur pigmentierter und unpigmentierter Photoreceptoren bei Proseriata (Plathelminthes). Zool. Scr. **13**, 9-17.

– (1988): Fine structure of two species of the Prolecithophora. Fortschr. Zool./ Prog. Zool. **36**, 221-227.

– (1991): Comparative morphology of photoreceptors in free-living plathelminths – a survey. Hydrobiologia **227**, 231-239.

– (1992): Photoreceptors with mitochondrial lenses in *Pogaina suecica* (Plathelminthes, Rhabdocoela). Zoomorphology **112**, 11-15.

– (1993a): Ultrastructural features of the pigmented eye spot in *Pseudomonocelis agilis* (Plathelminthes, Proseriata). Microfauna Marina **8**, 77-88.

– (1993b): Mitochondrial lenses in the eyes of the graffillid species *Pseudograffilla arenicola* (Plathelminthes, "Dalyellioida"). Microfauna Marina **8**, 89-98.

– (1995a): Electronmicroscopical studies on the photoreceptors of *Provortex tubiferus* (Plathelminthes, Rhabdocoela). Microfauna Marina **10**, 79-88.

- (1995b): A new type of photoreceptor in *Anthopharynx sacculipenis* (Plathelminthes, Solenopharyngidae). Hydrobiologia **305**, 177-182.
- (1995c): Ultrastructural features of *Bresslauilla relicta* (Plathelminthes, Rhabdocoela). The eyes. Microfauna Marina **10**, 31-40.
- (1996): First evidence of mitochondrial lensing in two species of the "Typhloplanoida" (Plathelminthes, Rhabdocoela): phylogenetic implications. Zoomorphology **116**, 95-101.
- (1997): First report on the fine structure of unpigmented rhabdomeric photoreceptors in a free-living species of the "Dalyellioida" (Plathelminthes, Rhabdocoela). Microfauna Marina **11**, 59-69.

Dr. Beate Sopott-Ehlers
II. Zoologisches Institut und Museum der Universität Göttingen
Berliner Straße 28, D-37073 Göttingen

Mielkiella spinulosa gen.n. sp.n., a new taxon of the Laophontidae (Copepoda, Harpacticoida) from Porvenir (Tierra del Fuego, Chile).

Kai Horst George

Abstract

Male and female of *Mielkiella spinulosa* gen.n. sp.n. are described. Specimens were collected in a sandy beach in Porvenir (Tierra del Fuego, Chile). The new species belongs to the family Laophontidae. It exhibits a combination of apomorphies in the A1 (male), A2, the peraeopods and the genital field (female), so that a new genus is erected. The phylogenetic relationship between *Mielkiella spinulosa* gen.n. sp.n. and *Stygolaophonte arenophila* Lang, 1965 is discussed.

Keywords: Taxonomy, Harpacticoida, *Mielkiella spinulosa*, Chile

A. Introduction

In material collected along a sandy beach in Porvenir (Tierra del Fuego, Chile), a few specimens of a peculiar laophontid copepod were found. Subsequent search in the literature revealed, that it closely resembles a male of a new species that MIELKE (1987) had collected in Punta Arenas (Chile). He found only one male specimen, so he did not give a complete description of the species and named his specimen "Laophontidae spec. 1". In spite of some slight differences, the material from Porvenir doubtlessly belongs to the same species of which a detailed description is presented here.

B. Material and methods

Eight males, 25 females and 1 copepodid have been sorted out from material of a sandy beach at Porvenir, Tierra del Fuego, Chile, on 13. 09. 91.

Locality: 53° 17' 50"S, 70° 22' 45"W. Samples were taken at low-tide directly at the waterline.

Three males and 8 females have been dissected, and 6 specimens were utilized for Scanning Electron Microscopy. The holotype has been preserved in 5% buffered formaldehyde and was later dissected and transferred into lactophenol embedding medium. Drawings were made with the aid of a camera lucida on a Leitz-Dialux EB 22 microscope equipped with an interference contrast 100times objective. All specimens are provisionally in the collection of the AG Zoomorphologie of Oldenburg University.

The terminology is adopted from HUYS & BOXSHALL (1991), as well as the numeration of the setae of caudal rami. The terminology related to systematics is used according to AX (1984). Abbreviations used in the text: cphth: cephalothorax, A1: antennule, A2: antenna, md: mandible, mxl: maxillule, mx: maxilla, mxp: maxilliped, enp: endopodite, exp: exopodite, exp1: first segment of exp, CR: caudal ramus, P1-P6: swimming legs 1–6, benp: baseoendopodite.

C. Results

Description of female
Body length (including CR): 380 μm
Cphth length: 110 μm
Cphth width: 160 μm

Rostrum (Figs. 2A, 8A) triangular, as long as wide, fused with cphth, but nevertheless presenting a well defined chitinous reap at ist base, with 2 small setules anteriorly.

Body (Figs. 1A, B) dorsoventrally flattened. Cphth wider than long. Thoracic somites bearing P2-P4 as wide as cphth, the following somites tapering posteriorly. Telson as long as preceding 2 somites. Whole cphth covered with small sensillae. Posterior border of cphth and free somites except the last 2 ones bearing dorsally a row of small sensillae of which the number decreases posteriorly. Posterior border of all thoracic and abdominal somites with a line of small spinules. Ventrally and laterally, abdominal somites with several rows of long spinules (Figs. 1B, 2B). Cphth and thoracic somites laterally covered with fine cuticular "hairs" (Fig. 1B), in thoracic somites accompanied by several short spinules. Anal somite with spinulose anal operculum flanked by 2 setules arising from small knobs.

CR (Fig. 2B) short, slightly longer than broad. Sizes and positions of setae: I very small, inserting at halflength of outer margin of ramus. II located

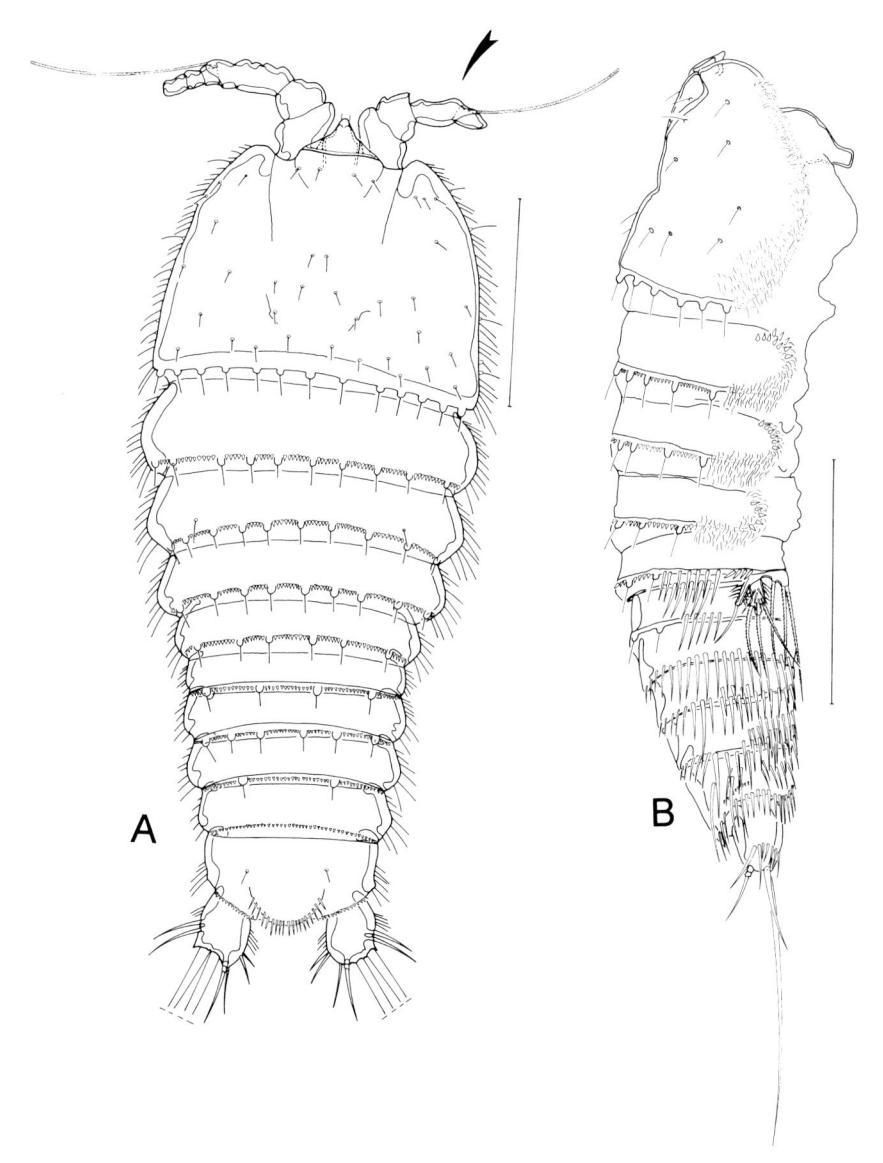

Fig. 1: *Mielkiella spinulosa* gen.n. sp.n., female, A) dorsal view (paratype 1), B) lateral view (Paratype 2). Black arrow indicates deformed A1. Scales: 100 µm.

directly posteriorly of I, nearly five times longer than the first. III directly posteriorly of II, of approximately the same size. IV and V terminal, V approximately 3 times longer than IV and reaching the same length as first 3 thoracic somites. VI short and strong, subterminally on the inner side of CR. VII biarticulated at base, situated in midline of dorsal surface near the posterior end of CR. IV, V and VI accompanied by strong short spinules. At bases of I, II and III with a group of short slender spinules, a further one subterminally at the inner side of CR.

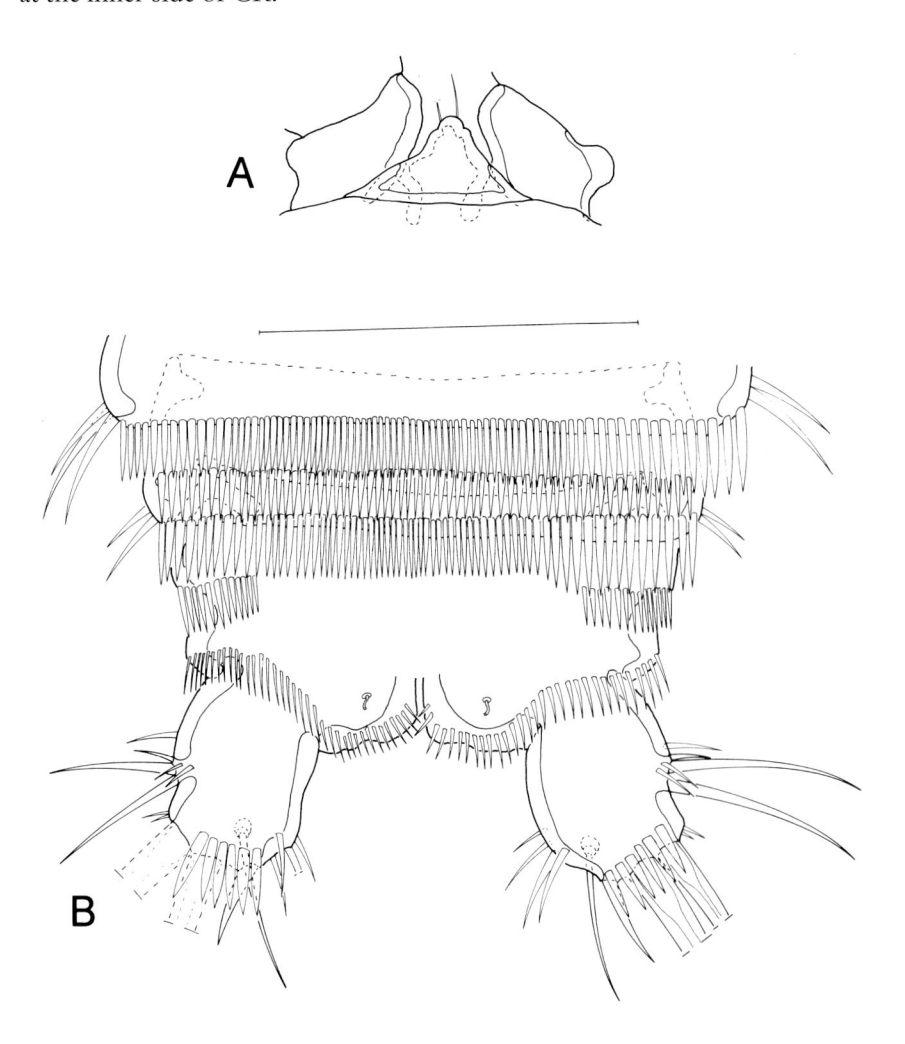

Fig. 2: *Mielkiella spinulosa* gen.n. sp.n., female, A) rostrum (paratype 1), B) furca, ventral view (holotype). Scale: 50 µm.

Genital field (Figs. 3A, 9A, B): last thoracic and first abdominal somites not completely fused to genital double somite. The dorsal view (Fig. 1A) shows clearly the separation between genital and first abdominal somite. Nevertheless, a lateral view reveals that the 2 somites are fused ventrally (Fig. 1B). Copulatory porus located ventrally in middle of somite, showing 2 seminal ducts to seminal receptacles. P6 small, with 2 setae, forming genital operculum which covers the genital apertures. Posteriorly with 2 strong sclerotized tubes,

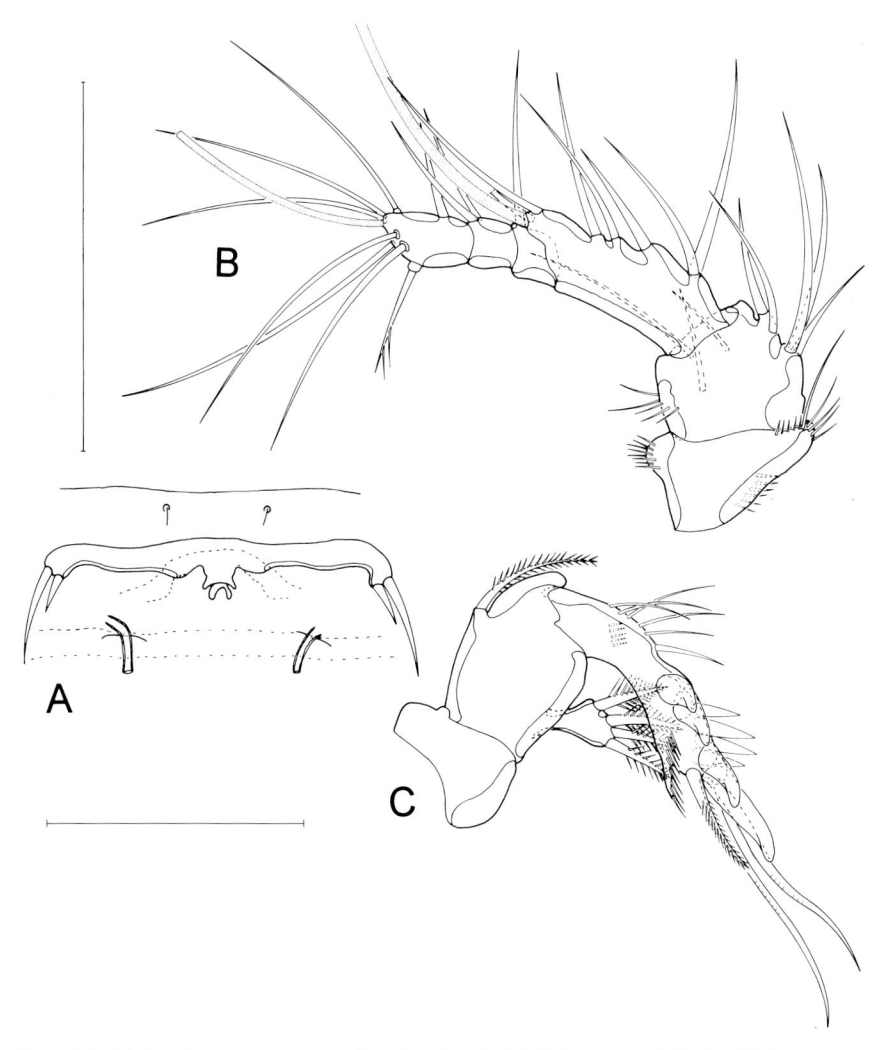

Fig. 3: *Mielkiella spinulosa* gen.n. sp.n., female, A) genital field (paratype 3), B) A1, C) A2 (paratype 4). Scales: 50 µm.

crossing a chitinous reap. Observations by Scanning Electron Microscopy revealed that each tube inserts at the upper part of a hole (Figs. 9A, B). The chitinous reap is passing at the inner side of the somite, being originally the limit of the genital and first abdominal somite.

A1 (Fig. 3B) 6-segmented, short, outwardly directed. First segment with 1 small seta and a row of long spinules on inner side near base of segment, and 1 row on the outer, protruding corner. Second segment with 5 setae on inner side and 2 short setae and 1 long slender seta on ventral surface. Outer border with a group of long spinules. Third segment longest, along inner margin with 5 setae, terminally with 2 setae and 1 aesthetasc (aes.). Fourth segment small, bearing 1 seta at inner corner. Fifth segment as long as fourth, with 2 setae on inner corner. Sixth segment approximately 1,5 times longer than fifth, with 1 seta at inner base and 4 long setae subterminally, all arising from small knobs.

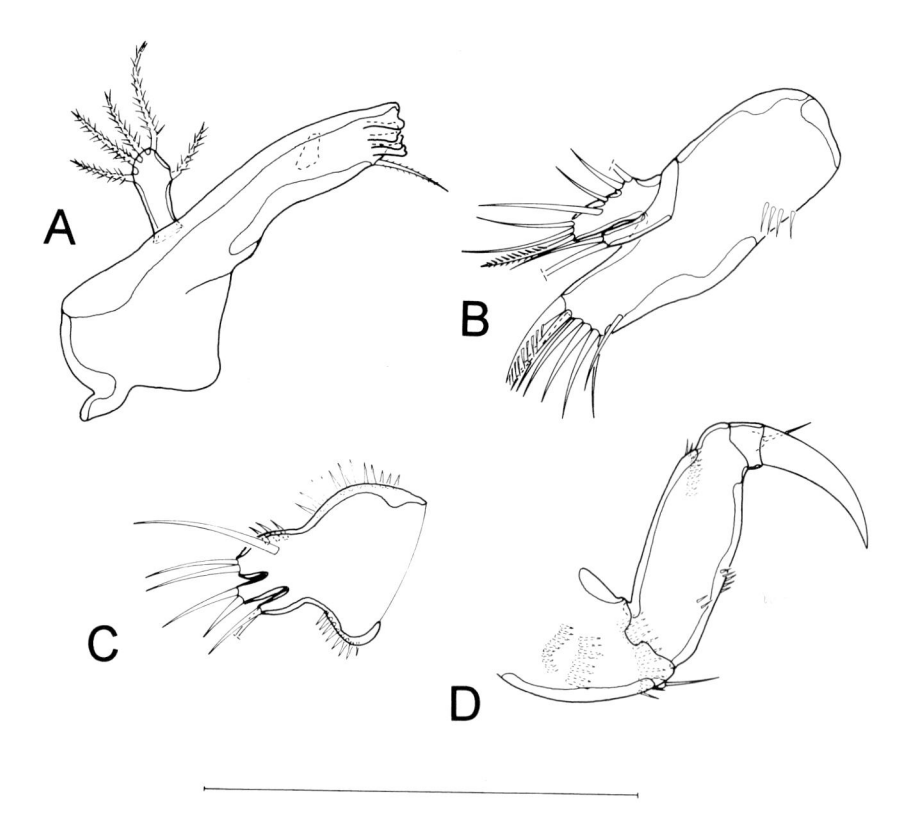

Fig. 4: *Mielkiella spinulosa* gen.n. sp.n., female, A) md (paratype 4); B) mxl, C) mx (paratype 5), D) mxp (paratype 4). Scale: 50 µm.

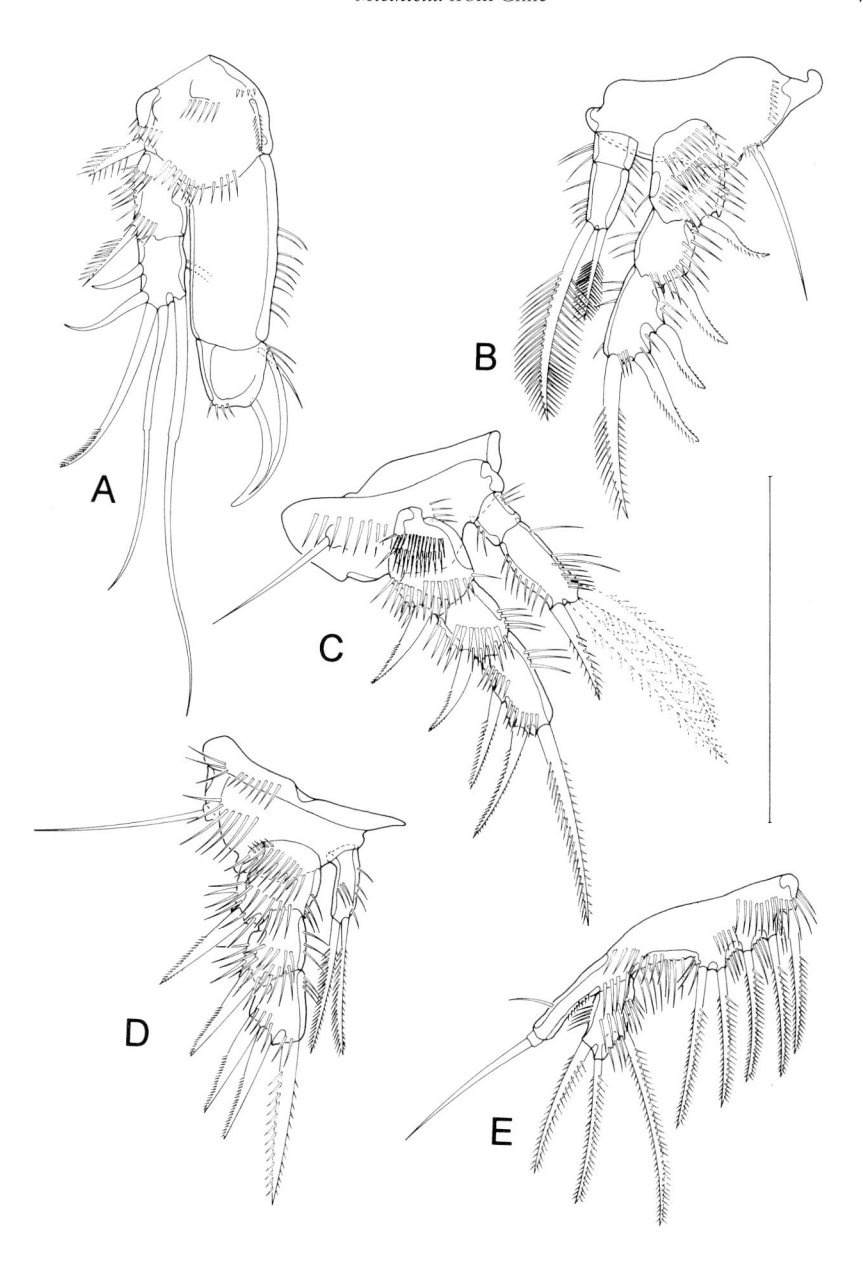

Fig. 5: *Mielkiella spinulosa* gen.n. sp.n., female, A) P1, B) P2, C) P3 (holotype), D) P4 (paratype 6), E) P5 (paratype 4). Scale: 50 µm.

Terminally with 2 long setae and 1 small slender aes. Outside with 1 additional seta arising from a knob.

Setal formula: I-1; II-8; III-7 + aes; IV-1; V-2; VI-8 + aes.

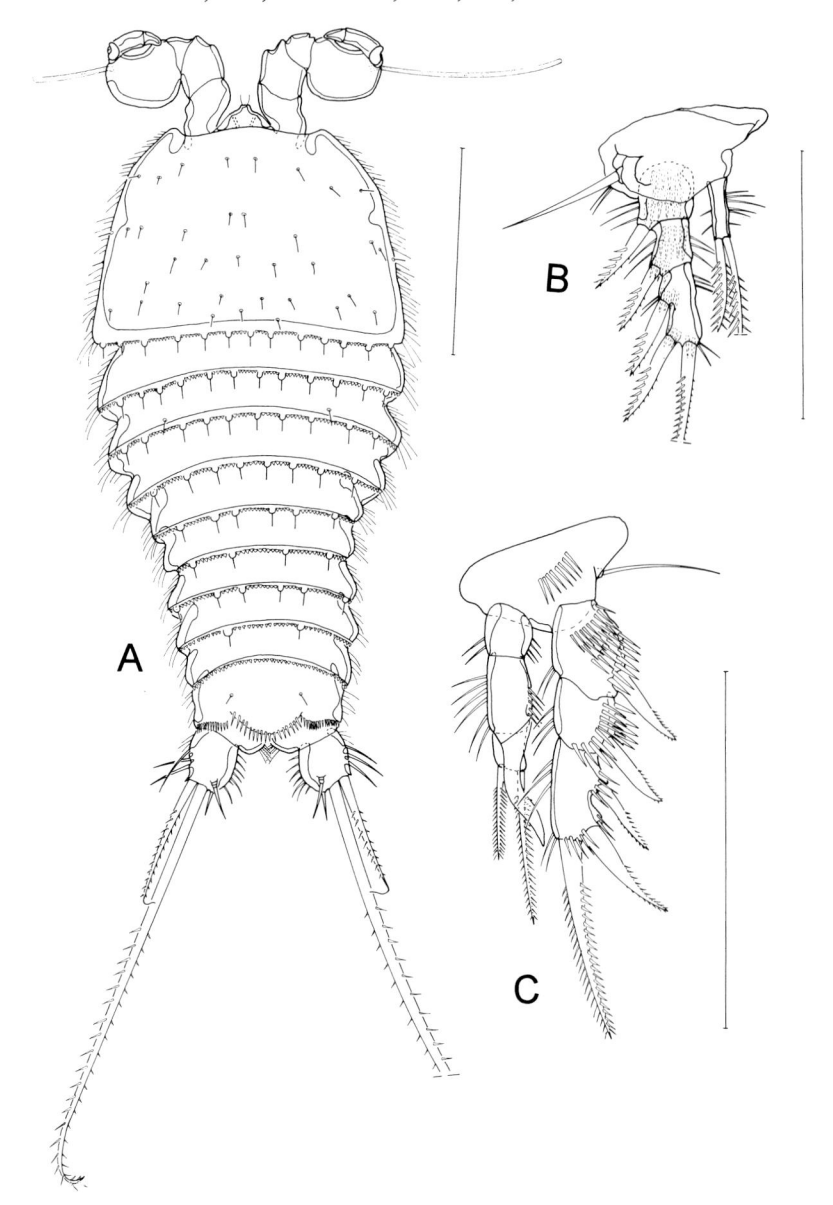

Fig. 6: *Mielkiella spinulosa* gen.n. sp.n., male, A) dorsal view (paratype 11), Scale: 100 µm, B) P2, C) P3 (allotype). Scales: 50 µm.

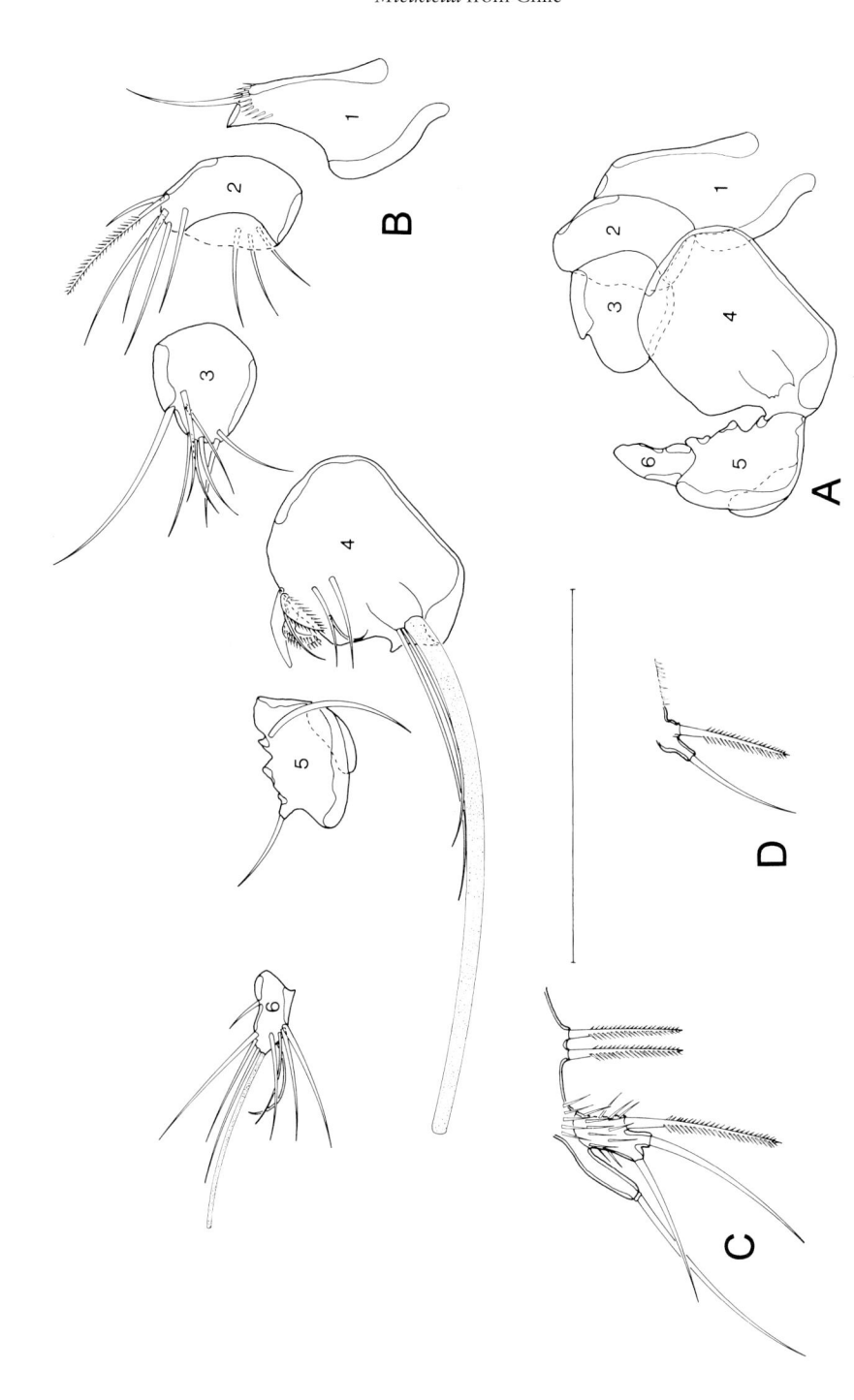

Fig. 7: *Mielkiella spinulosa* gen.n. sp.n., male (paratype 12), A) complete A1, B) segments 1 to 6 with corresponding setation, c) P5, d) P6 (paratype 13). Scale: 50 μm.

A2 (Fig. 3C) with unarmed coxa. Allobasis compressed, with 2 plumose seta. Enp. on inner side with 6 long spinules, ventrally with a row of short spinules. Laterally with 3 claw-like setae, apically with 1 claw-like seta, 2 geniculated setae and 1 plumose seta. Subapically there is a cuticular frill bearing spinules. Exp. 1-segmented, bearing 4 plumose setae.

Md (Fig. 4A) with unarmed coxa, except of 1 pinnate seta at base of cutting edge consisting in 4 teeth. Basis, enp and exp fused to 1-segmented palp carrying 5 plumose setae.

Mxl (Fig. 4B): Arthrite of praecoxa apically with 8 setae, one of them plumose, on inner side with 4 small spinules. Coxal endite bearing 2 setae, basis without exp and enp, terminally with 2 bare setae and 1 plumose seta, along outer margin with 4 setae.

Mx (Fig. 4C) derived. Syncoxa with several spinules, bearing 2 endites and fused with basis. Proximal endite terminally with 2 bare setae, second endite bearing 1 smaller and 1 longer seta. Basis sturdy, terminally without claw-like projection and bearing 2 setae, subterminally with 1 long seta representing the enp.

Mxp (Fig. 4D) prehensile, syncoxa proximally with 2 rows of spinules, subterminally with 1 inner bare seta and an additional row of long spinules. Basis with 2 rows of spinules and a few single ones. Enp bearing 1 small seta, arising from a process, and a big, claw-like seta.

P1 (Fig. 5A) prehensile, segments short and sturdy. Basis with 1 outer plumose seta, accompanied by a row of large spinules at ist base; subterminally with a row of strong spinules, and another row in the middle of the segment. Inner basal seta accompanied by a row of small spinules. Exp 2-segmented, inserting near outer basal seta, first segment with 1 plumose seta and with 2 rows of long, slender spinules on outer margin. Second segment as long as first, not reaching end of enpl. with 5 setae: 3 lateral ones, the distal one plumose, and terminally 2 long setae, of which the proximal half is broader than the distal half. Enp 2-segmented, enp1 longer than exp, with a row of long spinules on inner margin. Enp2 small, medially with 1 slender and 1 massive claw-like seta, terminally with 4 small spinules.

P2-P4 (Figs. 5B-D): bases with 1 long bare outer seta, exp 3-segmented, segments with 1 up to 4 rows of long spinules. Enp P2 and P3 2-segmented, enp P4 1-segmented. Basis P2 with 2 rows of short spinules, bases P3 and P4 with several long spinules. Exp1 and exp2 with 1 strong seta on outer margin, exp3 with 2 plumose outer setae, and terminally with 1 plumose seta. Enp1 P2 and P3 without setae, enp2 P2 with 1 long and 1 small plumose seta. Enp2 P3 with 2 long plumose setae and 1 smaller plumose seta. Enp P4 with 2 slender plumose setae.

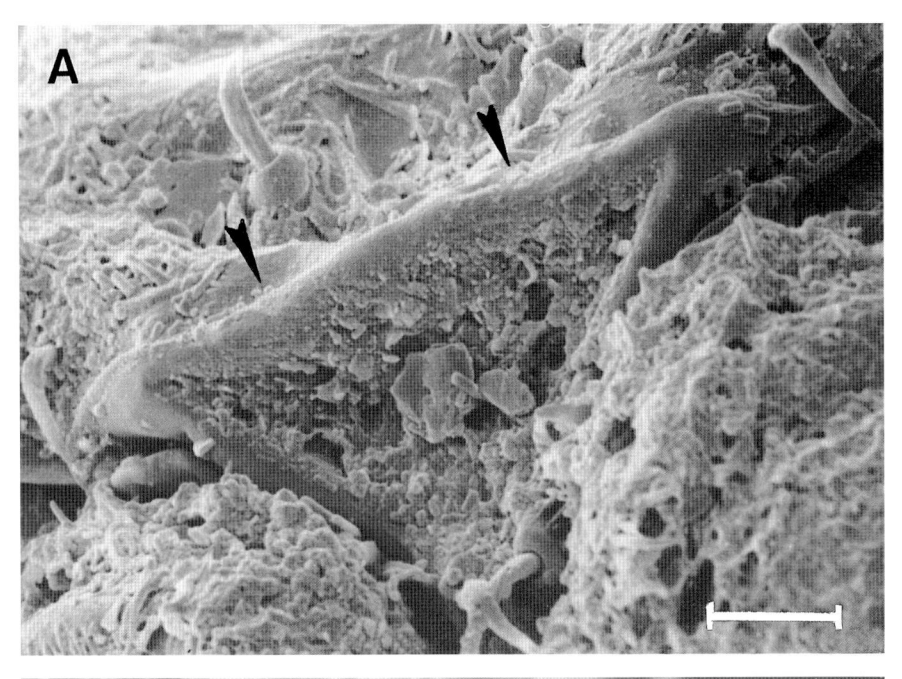

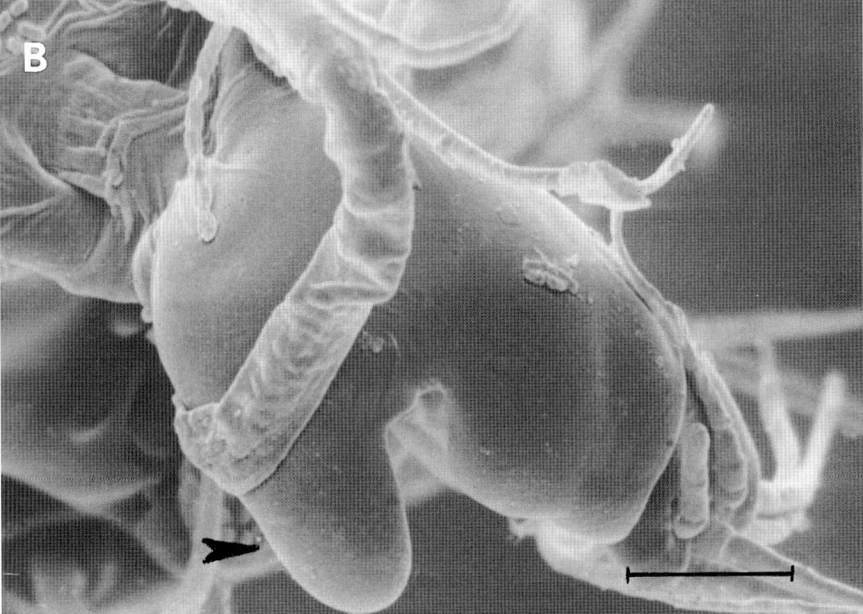

Fig. 8: *Mielkiella spinulosa* gen.n. sp.n., A) female (paratype 14), rostrum. Arrows indicate basal chitinous reap. Scale: 5 µm, B) male (paratype 16), fifth segment of A1 showing cuticular process (arrow). Scale: 5 µm.

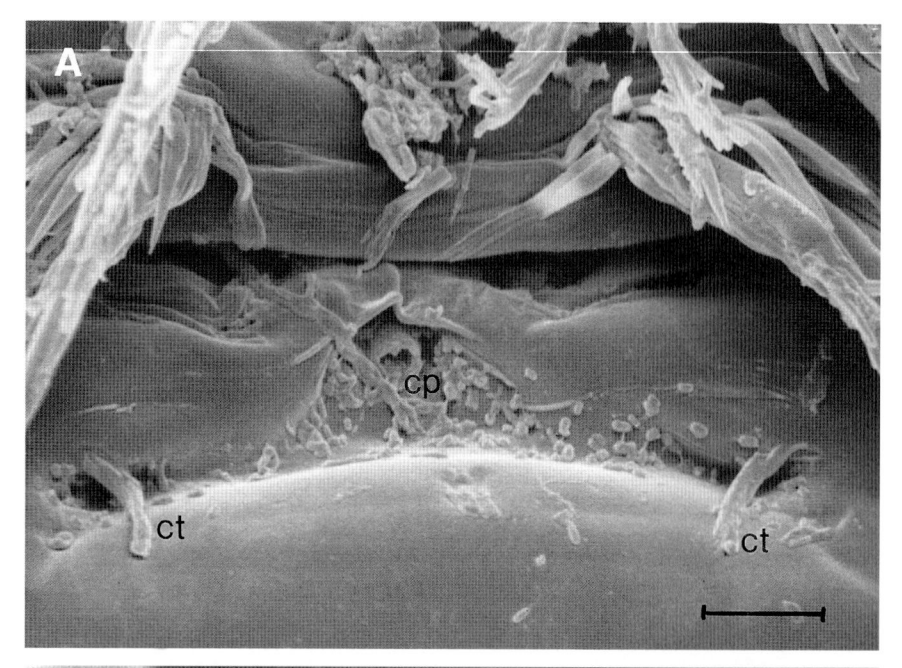

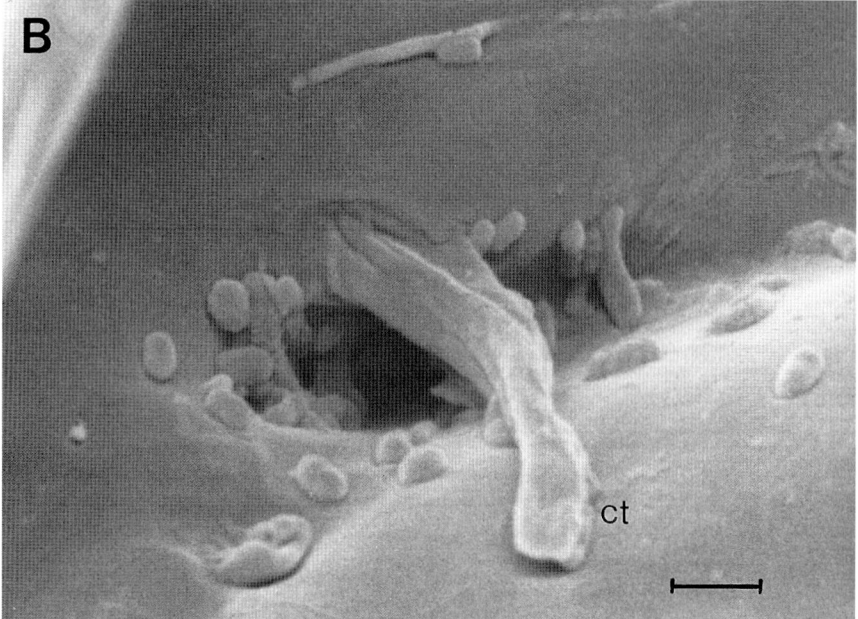

Fig. 9: *Mielkiella spinulosa* gen.n. sp.n., female (paratype 15), ventral view showing A) genital field, scale: 5 µm, B) cuticular tube, scale: 1 µm. Abbreviations: cp: copulatory pore, ct: cuticular tube.

Setal formula:

	Exp1	Exp2	Exp3	Enp1	Enp2
P2	0	0	0; 1; 2	0	0; 2; 0
P3	0	0	0; 1; 2	0	0; 2; 1
P4	0	0	0; 1; 2	0; 2; 0	–

P5 (Fig. 5E): Benp wider than long, covered with several rows of long spinules. Outer margin with 1 bare seta, arising from a process. Distally with 5 long plumose setae. Exp small and sturdy, with 2 rows of long spinules and 3 long plumose setae.

Description of male

Body length:	310 µm
Cphth length:	100 µm
Cphth width:	150 µm

Body (Fig. 6A) smaller than female, presenting the following differences: posterior border of Cphth with a line of small spinules, as well as in the thoracic and abdominal somites; transition from thoracic to abdominal somites is more pronounced than in female; 2 rows of spinules running outwards from anal operculum; seta I of CR as long as setae II and III; V reaching the same length as all free thoracic and abdominal somites.

A1 (Figs. 7A, B, 8B) 6-segmented, sturdy, subchirocer Geniculation between fourth and fifth segment. Fourth segment swollen, with 3 modified flat setae at inner side. Fifth segment dorsally with a strong cuticular process, and with indentated inner margin. Aes on furth and sixth segment.

Setal formula: I-1; II-9; III-8; IV-9 + aes; V-2; VI-9 + aes.

P2 (Fig. 6B) with 1-segmented enp, bearing 2 plumose setae.

P3 (Fig. 6C) with 3-segmented enp, second segment with apophysis, reaching end of exp. Enp3 with 2 plumose setae.

P5 (Fig. 7C) smaller than in female. Benp. with 1 row of long spinules, 1 bare outer seta, distally with 2 plumose setae. Exp with 2 rows of spinules, 2 long bare setae and 1 plumose seta.

P6 (Fig. 7D) very small, with 1 plumose seta at inner side and 1 bare seta at outer side.

D. Discussion

In his fragmentary description of a male of a new species ("Laophontidae spec. 1") found in Punta Arenas (Chile) MIELKE (1987) noticed that this species did not fit into any of the so far known Laophontidae genera. He made illustrations of the A2, of P1-P6 and of one caudal ramus. When comparing these drawings with the specimens found in Porvenir (Tierra del Fuego, Chile) there is no doubt that both refer to the same species. *Mielkiella spinulosa* gen.n. sp.n. belongs to the family Laophontidae as recognized by MIELKE (1987) before. It shows all diagnostic characters that had been interpreted by HUYS (1990) as apomorphies of the family: Rostrum fused with cphth (nevertheless it shows a well defined chitinous reap at ist base, not comparable with the less sclerotized hyaline band of other Laophontidae mentioned by HUYS (1990)), A1 setae and spines without ornamentation, 1-segmented mandibular palp (basis and exp each represented by 1, enp. represented by 3 setae), P1 with displacement of inner basal seta into centre of segment, basis produced into pedestal for insertion of enp, loss of inner seta enp1, posterior seta of enp2 reduced to setule, modification of front seta enp2 into a strong claw. Nevertheless it is not possible to ally the new species to one of the known laophontid genera, even though it shares some characters with some of them:

a) A1 6-segmented, without thorn (*Arenolaophonte* Lang, 1965, *Hemilaophonte* Jakubisiak, 1932, *Klieonychocamptus* Noodt, 1958 (part.), *Phycolaophonte* Pallares, 1975, *Xanthilaophonte trispinosa* (Sewell, 1940));

b) Allobasis A2 strongly tossed (*Stygolaophonte* Lang, 1965);

c) Enp. A2 with 7 apical setae (*Afrolaophonte* Chappuis, 1960, *Coullia* Hamond, 1973, *Harrietella* T. Scott, 1906, *Hemilaophonte*, *Lobitella* Monard, 1934, *Loureirophonte* Jakobi, 1953, *Platychelipus* Brady, 1880, *Raptolaophonte* Cottarelli & Forniz, 1989, *Stygolaophonte*);

d) Exp P1 2-segmented (exp2 with 4 or 5 setae) (widerspread within Laophontidae, e.g. *Archilaophonte* Willen, 1995, *Arenolaophonte*, *Esola hirsuta* Thompson & A. Scott, 1903, *Folioquinpes* Fiers & Rutledge, 1990, *Heterolaophonte furcata* Noodt, 1958, *H. serratula* Mielke, 1981, *Hoplolaophonte* Hamond, 1973, *Klieonychocamptus* (part.), *Laophonte* Philippi, 1840 (part.), *Loureirophonte*, *Microlaophonte* Vervoort, 1964, *Myctyricola* Nicholls, 1957, *Phycolaophonte*, *Platychelipus*, *Psammolaophonte* Wells, 1967, *Quinquelaophonte* Wells, Hicks & Coull, 1982 (part.), *Raptolaophonte*, *Stygolaophonte*, *Xanthilaophonte carcinicola* Fiers, 1991);

e) P2-P4 exp3 with 3 setae (*Robustunguis* Fiers, 1992, *Stygolaophonte*);

f) P2-P4 short and sturdy (*Stygolaophonte*);

g) Inner setae exp P2-P4 totally reduced (*Stygolaophonte*, except P2 male);

h) P5 small, benp broader than long, with cuticular spines, exp small (*Stygolaophonte*);

i) Ventral side of abdomen and peraeopods with long cuticular spines (*Arenolaophonte*, *Heterolaophonte furcata*, *H. serratula*, *Laophonte* (part.), *Myctyricola*, *Stygolaophonte*).

On the other hand there is a number of autapomorphic characters which seem to justify the establishment of a new genus:
- 5th segment of A1 (male) with strong cuticular process;
- Basis of Mx obtuse and shortened;
- Mxp short and compact;
- P1 short and sturdy;
- Terminal setae exp P1 with abrupt transition between broad proximal and slender distal half;
- Female with 2 cuticular "tubes" at the genital field.

Looking at the characters a-i it becomes evident that *Mielkiella spinulosa* gen.n. sp.n. shares the gratest number of characters with *Arenolaophonte* Lang, 1965 (characters a, d, i) and *Stygolaophonte* Lang, 1965 (characters b, c, d, e, f, g, h, i). While some of them (a, c, d, i) may be convergent there are two which are interpreted here as synapomorphic for *Mielkiella spinulosa* gen.n. sp.n. and *Stygolaophonte arenophila* Lang, 1965:
- The similarity of P5 female as well as of male (see character h). Within Laophontidae only *Stygolaophonte* and *M. spinulosa* gen.n. sp.n. present such a particular P5. It is not likely that this particular form is the result of convergent evolution due to an adaptation to the same environment.
- The strongly compressed allobasis of A2, which is unique for all Laophontidae.

Despite the synapomorphies there are important differences between both species so that they cannot be united in one genus. *S. arenophila* shows the following autapomorphies in respect to *M. spinulosa* gen.n. sp.n.:
  1. Mandibular palp with 3 setae;
  2. A2 exp with 2 setae;
  3. P1 exp2 with 4 setae;
  4. P3 enp3 of male with 1 seta;
  5. P4 exp 2-segmented;
  6. P5 benp of female with 4 setae.

M. spinulosa gen.n. sp.n. on the other hand shows the following autapomorphies in respect to *S. arenophila*:
  1. Body short and sturdy;
  2. A1 6-segmented;
  3. A1 of male with cuticular pedestal base on fifth segment;
  4. Basis of Mx fused with syncoxa, obtuse and shortened;
  5. Mxp short and compact;
  6. P1 short and sturdy;
  7. Terminal setae of P1 exp2 with abrupt transition between broad proximal and slender distal half;

8. P2 exp male without inner setae;
9. P2 male with 1-segmented enp;
10. Female with 2 cuticular "tubes" at the genital field.

As a result of this discussion it can be said that *Mielkiella* gen.n. is distinct from *Stygolaophonte* but that both genera appear to be closely related.

Zusammenfassung

Weibchen und Männchen von *Mielkiella spinulosa* gen.n. sp.n. werden beschrieben. Die Art wurde in Sandproben aus Porvenir (Tierra del Fuego, Chile) gefunden. Sie gehört in die Familie Laophontidae, weist allerdings eine Reihe von Apomorphien an der A1 (Männchen), der A2, den Peraeopoden und dem Genitalfeld (Weibchen) auf, so daß eine neue Gattung errichtet wurde. Die phylogenetische Beziehung zwischen *Mielkiella spinulosa* gen.n. sp.n. und *Stygolaophonte arenophila* Lang, 1965 wird diskutiert.

Acknowledgements

I would like to thank Dr. W. Mielke (Göttingen) and Prof. Dr. H. K. Schminke (Oldenburg) for helpful criticism on the manuscript. Thanks are also due to Dipl.-Biol. P. Martinez Arbizu (Oldenburg) for stimulating and constructive discussions, and to Mrs. D. George-Henning for revising and improving the manuscript.

Reference

Ax, P. (1984): Das Phylogenetische System. Gustav Fischer Verlag, Stuttgart, 349 pp.

Fiers, F. (1990): *Abscondicola humesi* n.gen. n. sp. from the gill chambers of land crabs and the definition of the Cancrincolidae n. fam. (Copepoda, harpacticoida). Bull. Inst. r. Sci. nat. Belg. (Biologie) **60**, 69–103.

Huys, R. (1990): A new family of harpacticoid copepods and an analysis of the phylogenetic relationships within the Laophontoidea T. Scott. Bijdragen tot de Dierkunde 60, 79–120.

Huys, R. & G. A. Boxshall (1991): Copepod Evolution. the Ray Society, London, 468 pp.

Lang, K. (1965): Copepoda Harpacticoidea from the Californian Pacific coast. K. svenska vetensk. Akad. Handl. **10** (2), 1–566.

Mielke, W. (1987): Interstitielle Copepoda von Nord- und Süd-Chile. Microfauna Marina **3**, 309–361.

Kai Horst George
Fachbereich (7) Biologie, Arbeitsgruppe Zoomorphologie, Carl von Ossietzky-Universität, D-26111 Oldenburg, Germany, e-mail: george@biologie.uni-oldenburg.de

Ultrastruktur und Morphogenese der Hakenborsten bei *Psammodrilus balanoglossoides* – Bedeutung für die Stellung der Psammodrilida (Annelida)

Rudolf Meyer & Thomas Bartolomaeus

Abstract

Until today, only three species of the Psammodrilida are known. All individuals of this taxon bear hooked setae aligned in a pair of transversal rows in each abdominal segment. Such setae are also known from other annelids, like Maldanidae, Arenicolidae and Oweniida. Hooked setae are composed of a strong, curved rostrum, which is surmounted by several rows of smaller spines, the so-called capitium. Both merge to form the setal shaft (manubrium). Rostrum and spines of the capitium are bent towards the manubrium. Each hooked seta of *Psammodrilus balanoglossoides* arises from an ectodermal follicle which is composed of three cells, i.e. the basally situated chaetoblast and two follicle cells which subsequently surround the seta like rings. The setae are generated in a formative site at the medio-lateral edge of each row. Formation starts from a group of three epidermal cells lying underneath the epidermal surface. The basalmost cell of this anlage bears apical microvilli that extend into a central compartment leading to the exterior. A group of slender microvilli preformes the rostrum, while each spine of the capitium is determined by a single stout microvillus. Setal material surrounds the microvilli and, thus, forms the seta. While the chaetoblast generates additional microvilli on its surface, all microvilli are continuously reoriented. This reorientation leads to the curvation characteristic for the apical structures of hooked setae. During this process the microvilli are continuouly withdrawn from the older parts of the setal anlage and the canals left by the microvilli are refilled by electron-dense material. Initial during the formation of the manubrium the chaetoblast constantly forms additional microvilli on the rostral side of the setal anlage, while microvilli continuously merge at its adrostral side. This leads to the formation of a wide subrostral process, characteristic to the hooked setae of *Psammodrilus balanoglossoides*. Later during chaetogenesis the microvilli merge to form the slender basal section of the manubrium. At the end of setal formation the remains of the microvilli are left inside the setae and the chaetoblast forms intermediate filaments that attach the seta to the follicle. During formation the seta has been twisted almost 180°. The special mode of differentiating each part of the hooked seta, its twisting during chaetogenesis and the position of the formative site have been observed almost identically during formation of the uncini in Sabellida and Terebellida, the hooded hooks of the Capitellidae and the hooked setae in Oweniida, Arenicolidae and Maldanidae; although the formative site lies ventrally in all members of the latter two taxa. This correspondence substantiates the hypothesis of a homology of uncini, hooded hooks and hooked setae. This homology hypothesis allows the assumption that Psammodrilida are the sister group of the Maldanomorpha (Arenicolidae + Maldanidae), sharing the reduction of a pelagic larval stage as a synapomorphous characterstic.

A. Einleitung

Die Psammodrilida sind derzeit mit den drei interstitiellen Arten *Psammodrilus balanoglossoides* Swedmark, 1952, *Psammodrilus aedificator* Kristensen und Nørrevang, 1982, und *Psammodriloides fauveli* Swedmark, 1958 bekannt. Psammodriliden sind in vier Tagmata gegliedert, dem Kopf aus Prostomium und erstem Segment, dem Kragensegment, dem Thorax aus sechs Segmenten und dem aus zahlreichen Segmenten bestehenden Abdomen. Die paarigen Cirren der Thoraxsegmente sind durch innere Borsten, Aciculae, gestützt. Äußerlich sichtbare Borsten sind indes nur im Abdomen zu finden. Hier weisen alle drei Psammodriliden-Arten transversale Reihen von Hakenborsten auf, die, paarig angeordnet, sich auf den ventralen Abschnitt eines jeden Segments beschränken (BARTOLOMAEUS 1995a; KRISTENSEN & NØRREVANG 1982; SWEDMARK 1952, 1958). Solche transversalen Reihen von Hakenborsten sind ebenfalls von den Sabellida, Pogonophora, Terebellida, Oweniida, Arenicolidae und Maldanidae bekannt und teilweise auf ihre Morphogenese hin untersucht worden (s. BARTOLOMAEUS 1995b; KNIGHT-JONES 1981; K. MEYER & BARTOLOMAEUS 1996). Ihre lichtmikroskopisch erkennbare Struktur läßt die Hypothese einer Homologie dieser Hakenborsten mit denen der genannten sessilen und hemisessilen polychaeten Anneliden, insbesondere denen der Maldaniden, zu (SWEDMARK 1958). Die vorliegende Arbeit stellt die Ultrastruktur und Morphogenese der Hakenborsten von *Psammodrilus balanoglossoides* vor und erhärtet die Hypothese einer Homologie der Hakenborsten innerhalb der Anneliden. Die Homologie-Hypothese be-deutet gleichzeitig, daß die Hakenborsten von Psammodriliden in einer ihnen und den Sabellida, Pogonophora, Terebellida, Oweniida, Arenciolidae und Maldanidae gemeinsamen Stammlinie evolviert sein müssen. Auf dieser Basis lassen sich unter Berücksichtigung weiterer Merkmalskomplexe Hypothesen zur Phylogenie der Psammodrilida formulieren.

B. Material und Methoden

Die bearbeiteten Exemplare von *Psammodrilus balanoglossoides* stammen aus dem Sandwatt von List/Sylt, wo sie etwa 2 m vom Knick in den oberen 2-3 cm des Sedimentes siedeln (WESTHEIDE 1966). Die Tiere wurden im Juni 1995 gesammelt und vor Ort für die Elektronenmikroskopie in 0,1 M Natriumcacodylat gepufferter 2,5% Glutaraldehydlösung (pH 7,2; 4°C) für 60 min fixiert. Das Fixativ enthielt Rutheniumrot. Anschließend wurde das Fixativ mit 0,1 M Natriumcacodylatpuffer mehrmals ausgewaschen, und die Tiere wurden für 60 min bei 4°C in einer 0,1 M Natriumcacodylat gepufferten 1% Osmiumtetroxidlösung nachfixiert. Die Entwässerung erfolgt in einer aufsteigenden Acetonreihe, aus der die Tiere über Propylenoxid in Araldit eingebettet wurden. Im Ultramikrotom REICHERT Ultracut 2 wurden mit einem Diamantmesser Serien von etwa 70 nm dicken, silbern reflektierenden Längs- und Querschnitten angefertigt, im LKB Ultrostainer mit wäßriger Uranylacetat- und Bleiacitratlösung kontrastiert und im ZEISS EM 900 Transmissionselektronenmikroskop bei 50 kV untersucht.

Die Struktur der Hakenborsten und ihre Entwicklungsstadien wurde anhand von Fotoserien von Längs- und Querschnitten zweier Tiere rekonstruiert. Die Rekonstruktionen wurden als Tuschezeichnungen ausgeführt, mit einem MUSTEK GS 6000 in einen Personalcomputer digitalisiert und mit dem Grafikprogramm MIKROGRAFX Picture Publisher bearbeitet.

Für Untersuchungen am Raster-Elektronenmikroskop (REM) wurden die Tiere nach den in BARTOLOMAEUS (1995a) angegebenen Methoden behandelt und am Zeiss Novascan 30 Rasterelektronenmikroskop ausgewertet.

C. Ergebnisse

Terminologie

Die **Borstentasche** ist eine epidermale Einsenkung, in der sich die Follikel ausgebildeter und in Bildung stehender Borsten befinden. Ein **Borstenfollikel** ist eine funktionelle Einheit, die aus dem Chaetoblasten, den Follikelzellen und einer angrenzenden Epidermiszelle besteht. Der **Chaetoblast** liegt an der Basis des Follikels, seine Mikrovilli ragen in die Borste und sind für die Ausbildung des inneren Röhrensystems verantwortlich. **Follikelzellen** schließen sich apikal an den Chaetoblasten an und umgeben die Borste. Ihre Ordnungszahl richtet sich nach der Reihenfolge, in der sie sich distad an den Chaetoblasten anschließen. Der **Borstenkanal** ist ein mit der Umgebung in Verbindung stehender Hohlraum, in den die Borste hineinwächst. Die Borste füllt ihn im Bereich des Chaetoblasten und der Follikelzellen aus. Die Bezeichnung der einzelnen Teile der Hakenborste geht auf HOLTHE (1986) zurück. Apikal ist ein Hauptzahn, das **Rostrum**, ausgebildet, dessen Struktur von mehreren Mikrovilli determiniert wird. Dem Rostrum sind adrostral mehrere Reihen von Nebenzähnen aufgelagert, deren Gesamtheit das **Capitium** bildet. Jeder Nebenzahn wird von einem Mikrovillus determiniert. Rostrum und Capitium stehen gewinkelt zum Borstenschaft, dem **Manubrium**, das sich nach basal verjüngt. Rostral ist das Manubrium zu einem **subrostralen Fortsatz** verbreitert.

Struktur der Hakenborste und des Follikels

Hakenborsten sind ausschließlich im Abdomen zu finden. Das untersuchte Exemplar wies 18 Abdominalsegmente auf, die sich nach caudal verjüngen. Auf ihnen befinden sich medio-ventral cilienlose Wülste (Tori), in denen sich eine Transversalreihe von Hakenborsten befindet (Abb. 1A, B). Die Tori der ersten beiden Abdominalsegmente sind kleiner und tragen keine sichtbaren Borsten. Die Spitzen von Capitium und Rostrum weisen frontad; eine Bewegung der Borsten im Follikel ist möglich. Die Borsten können bis auf Höhe der Körperoberfläche eingezogen werden (Abb. 1C, D). Die Anzahl der Hakenborsten in einem Torus nimmt vom ersten bis zum letzten borstentragenden Abdominalsegment von 10 bis auf 1 bis 2 graduell ab (Abb. 1A). In den vorderen Segmenten des Abdomens liegen die Borstenreihen auf dem frontalen Teil, in den hinteren zentral bis caudal auf der ventralen Seite des Torus.

Jede Hakenborste ist deutlich sigmoid (Abb. 2). Der Winkel zwischen Capitium und dem subrostralen Fortsatz beträgt 40°–55°, der zwischen dem subrostralen Fortsatz und dem Schaft des Manubriums variiert von 115° bis 130°. Die gesamte Länge der Borste beträgt von der Basis bis zum Capitium im Mittel 26,7 µm ± 2,1 µm (n=5). Der Abstand von der Basis bis zur Biegung des subrostralen Fortsatzes beträgt 19,6 µm ± 2,9 µm. Der subrostrale Fortsatz hat eine Breite von 6,6 µm ± 1,6 µm, das Rostrum hat eine Länge von 3,4 ± 1,3 µm (n=4). Darüber staffeln sich halbkreisförmig etwa 4 Reihen mit insgesamt 14–18 Einzelzähnen (Abb. 1D). Zur adrostralen Seite steigt die Anzahl der

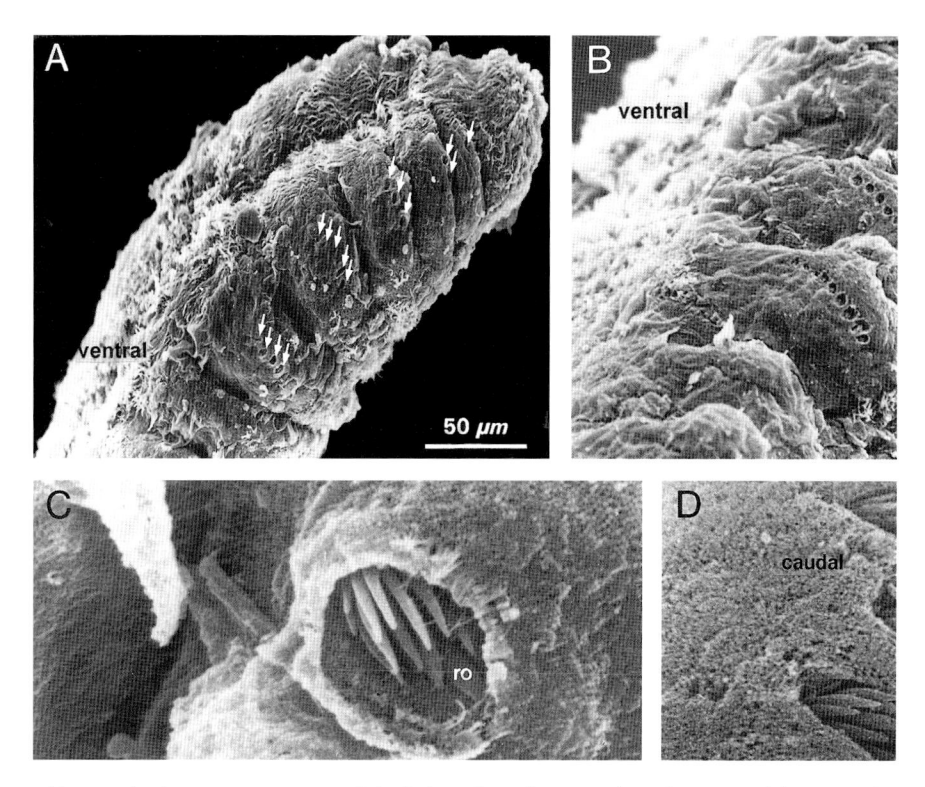

Abb. 1: Hakenborsten von *Psammodrilus balanoglossoides*. Ventralansicht. A. Caudale Region des Abdomens; die Zahl der Borsten (Pfeile) nimmt nach caudal ab. B. Abdominalsegmente 1–6. Den ersten beiden fehlen Hakenborsten, die in den folgenden Segmenten tief in den Borstenkanal zurückgezogen sind und nur durch Poren markiert sind. C. Detail des vorletzen Abdominalsegmentes. Rostrum (*ro*) und Capitium der Hakenborsten befinden sich im Borstenkanal. D. Abdominalsegment 5. Die Hakenborsten ragen kaum über das Niveau der Epidermis hinaus.

Zähne in jeder Reihe von ca. 3 auf 8 an. Die Länge und der Durchmesser der Zähne nehmen in gleicher Richtung ab. Die rostralen Einzelzähne sind im Mittel 1,9 µm, die Zähne der äußersten adrostralen Reihe sind dagegen nur 0,5 µm lang. Bei allen untersuchten Exemplaren reichen die Zähne nur wenig über die Körperoberfläche hinaus (Abb. 1C, D, 3A). Die Variabilität von Größe und Form der Borsten ist zwischen frontalen und caudalen Segmenten des Abdomens nicht größer als die zwischen den Borsten eines Segmentes.

Die Borste wird von einer unterschiedlichen Zahl von Kanälen durchzogen. Im basalen Teil des Manubriums befinden sich 14 ± 4 Kanäle, ihr Durchmesser beträgt 0,8 µm und ist über den Querschnitt konstant. Direkt an der Basis der Borste sind die Kanäle mit 1,3 µm erweitert. Im basalen Teil des Manubri-

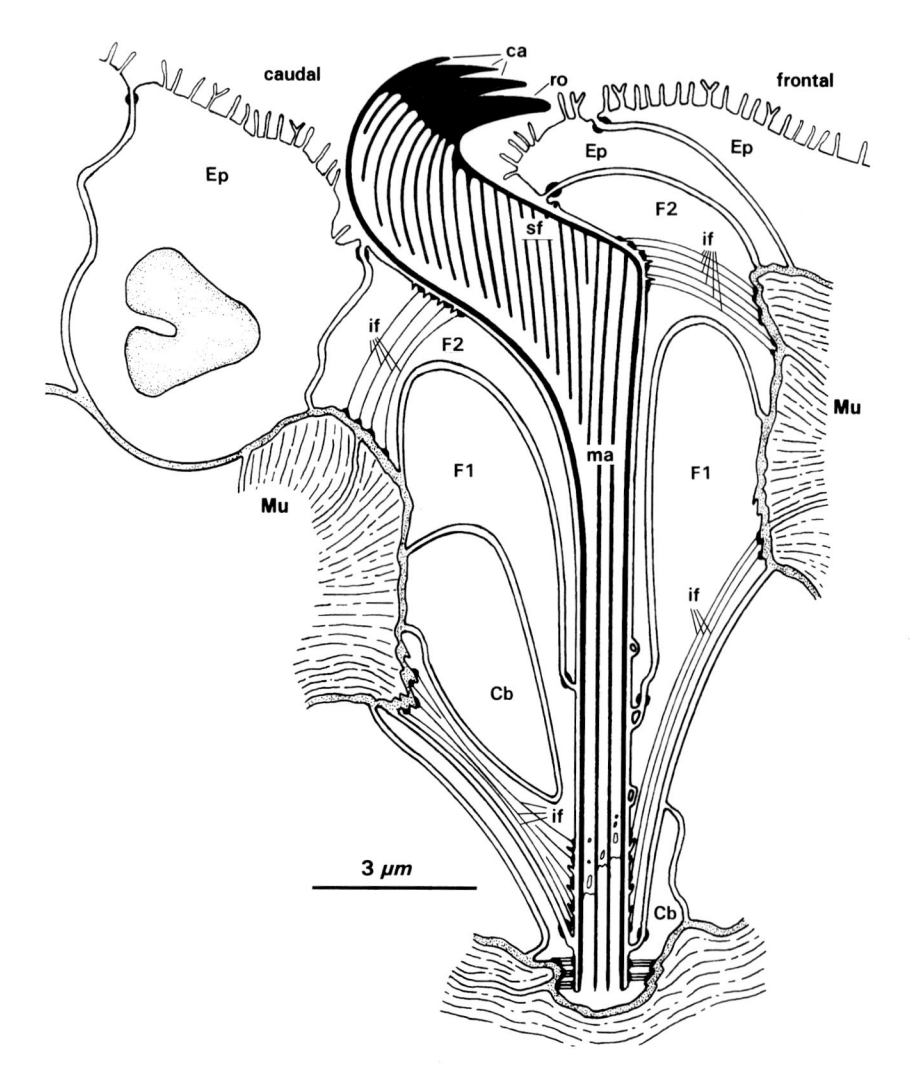

Abb. 2: Ausdifferenzierte Hakenborste von *Psammodrilus balanoglossoides*. Rekonstruierter Längsschnitt durch die Hakenborste und den Follikel. *ca* Capitium; *Cb* Chaetoblast; *Ep* Epidermiszelle; *F1*, *F2* Follikelzelle 1 und 2; *if* intermediäre Filamente; *ma* Manubrium; *Mu* Längsmuskulatur; *ro* Rostrum; *sf* subrostraler Fortsatz.

ums ist der Aufbau des Kanalwandmaterials aus parallel zu den Kanälen verlaufenden Fibrillen erkennbar. Vom basalen Abschnitt des Manubriums laufen einige adrostrale Kanäle bis in die Zähne des Capitiums (Abb. 3A). Rostrad erhöht sich die Zahl der Kanäle. Sie ziehen parallel in einem Winkel von

70° nach apikal und enden im subrostralen Fortsatz. Die Anzahl der Kanäle beträgt hier maximal 137, deren Durchmesser vom Manubrium zum subrostralen Fortsatz hin zunimmt. Im oberen Teil des subrostralen Fortsatzes und im Übergang vom oberen Manubrium zum Capitium und Rostrum sind die Kanäle nach distal zunehmend von elektronendichten Material ausgefüllt. Dennoch läßt sich erkennen, daß die einzelnen Zähne des Capitium von jeweils einem Kanal durchzogen sind. Die Anzahl der Kanäle im Rostrum beträgt an dessen Basis maximal 18 und nimmt zur Spitze auf etwa 4 Kanäle ab.

Die Zellen eines Borstenfollikels liegen in einer Einsenkung der Längsmuskulatur des Tieres. Alle Zellen haben auf der frontalen und caudalen Seite der Borstentasche großflächigen Kontakt mit der begrenzenden Basalmembran (Abb. 3A). An der dorsalen und ventralen Seite liegen die Follikel der benachbarten Borsten, ohne daß ECM zwischen ihnen liegt. Ein Follikel besteht aus einem basalen Chaetoblasten, zwei Follikelzellen und einer assoziierten Epidermiszelle. Die Zellen erscheinen elektronenhell und leer, nur im basalen Teil des Chaetoblasten und der Follikelzellen treten häufiger Mitochondrien und Vesikel auf. Im distalen Teil des Follikels sind die Follikelzellen untereinander stark verzahnt und im Bereich der Borste über Adhaerenszonen verbunden. Die Basen der Borsten einer Transversalreihe liegen meist dicht nebeneinander und ungeordnet in der Mitte der Borstentasche direkt an der Basalmembran. Dort sind auch einzelne Stränge von Ringmuskulatur zu erkennen, die mit den Borstenbasen durch eine ECM verbunden sind (Abb. 3C). Der überwiegende Teil des Chaetoblasten und der Follikelzellen umgibt die Borste in Höhe des subrostralen Fortsatzes. Im Längsschnitt durch den Borstenfollikel ist zu erkennen, daß die Zellen regelmäßig dachziegelartig angeordnet sind und ihre Zellgrenzen gerundet verlaufen (Abb. 2).

Der Chaetoblast umgibt das basale Drittel der Borste und senkt sich direkt an der Borstenbasis etwa 2 µm tief in die Basalmembran ein. In diesem stark komprimierten, tassenförmigen Bereich treten von allen Seiten Cytoplasmafortsätze an die Borste heran, die von Intermediärfilamenten durchzogen sind. Diese sind mit Hemidesmosomen mit der Borste und der basal den Chaetoblasten begrenzenden Basalmembran verbunden (Abb. 3C). Der Chaetoblast liegt der Borste direkt an. In einen Teil der Kanäle der Borste reichen bis zu 2,3 µm lange und elektronenleer erscheinende Zellausläufer des Chaetoblasten, denen das für Mikrovilli charakteristische Filamentgeflecht fehlt (Abb. 3C). Im basalen Teil der Borstenkanäle befinden sich mitunter dunkle Vesikel mit einem Durchmesser von 30-100 nm. Der Zellkern des Chaetoblasten liegt distal der Borstenbasis und kann die Borste hufeisenförmig umgeben.

Die sich distal anschließenden Follikelzellen umgeben die Borste manschettenförmig. Den entstehende Kontaktbreich schließen parallel zur Borsten-

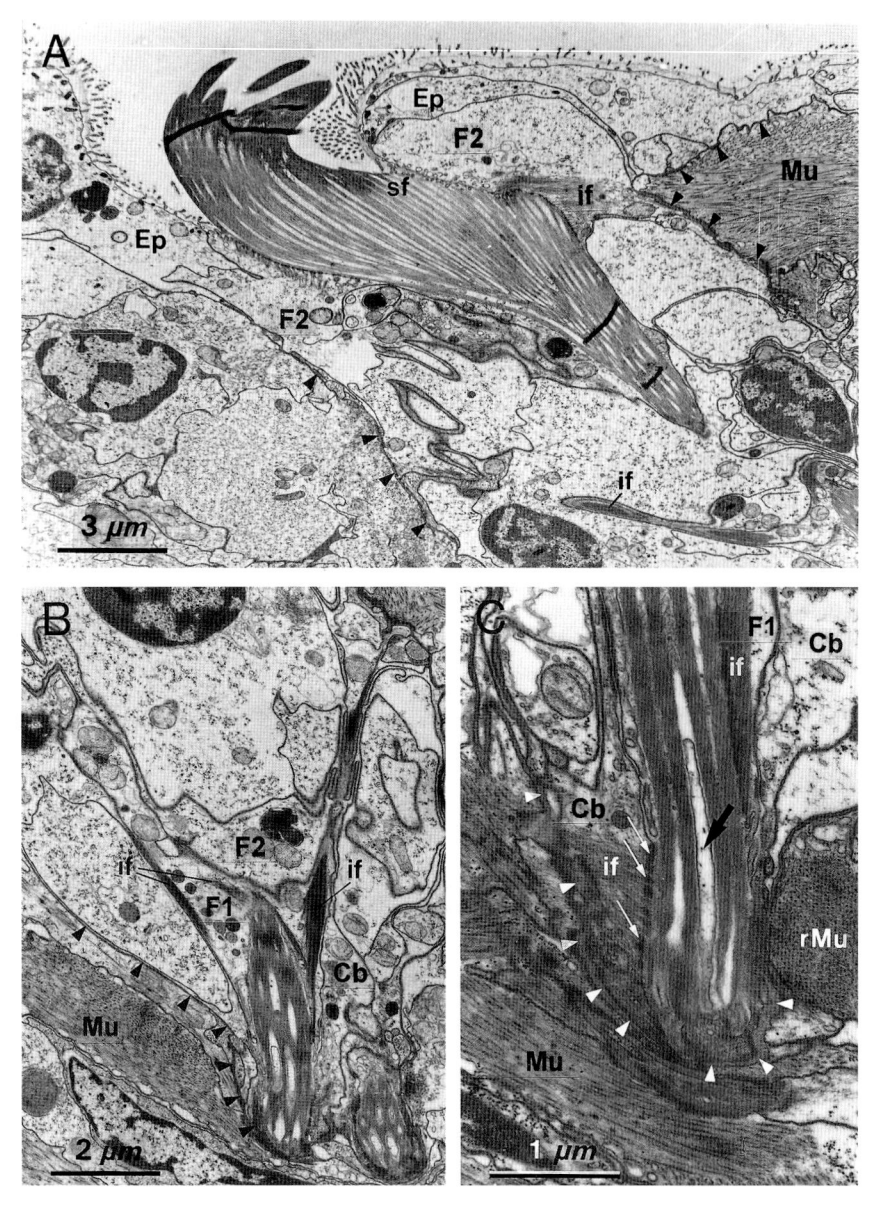

Abb. 3: Ausdifferenzierte Hakenborste von *Psammodrilus balanoglossoides*. Pfeilspitzen markieren die extracelluläre Matrix. A. Apikaler Bereich mit Rostrum (*ro*), Capitium (*ca*) und subrostralem Fortsatz (*sf*). Bündel intermediärer Filamente (*if*) in Follikelzelle 2 (*F2*) verbinden die Borste mit der Längsmuskulatur (*Mu*). B. Borstenbasis. Folikelzelle 1 (*F1*) enthält Bündel intermediärer Filamente, die zur Längsmuskulatur ziehen. C. Cytoplasmafortsatz (Pfeil) des Chaetoblasten ragt in die Borste. Hemidesmosomen (weiße Pfeile) verbinden die intermediären Filamente des Chaetoblasten (*Cb*) mit der Borste. *Ep* Epidermiszelle; *rMu* Ringmuskulatur.

achse verlaufende Adhaerenszonen. In beiden Zellen wird die Borste in ähnlicher Weise wie im Chaetoblasten über Intermediärfilamente und Hemidesmosomen mit der Basalmembran verbunden (Abb. 3A, B). Die Filamente sind jedoch erheblich länger und verlaufen von der rostralen und adrostralen Seite der Borste zu den die Borstentasche begrenzenden Ausläufern der Längsmuskulatur. Dort, wo Hemidesmososmen fehlen, befindet sich ein 40 bis 75 nm breiter Spalt zwischen Follikelzellen und Borste.

Follikelzelle 1 umgibt das untere Manubrium auf einer Länge von etwa 5–8 µm. Die Intermediärfilamente sind in dieser Zelle bis zu 10 µm lang und in den direkt auf den Chaetoblasten folgenden 3 µm des Manubriums über Hemidesmosomen mit der Borste über deren gesamten Umfang verbunden. Sie verlaufen durch enge Fortsätze der Follikelzelle und verdichten sich zu etwa 1 µm starken Bündeln, die nach distal zu den seitlichen Ausläufern der Längsmuskulatur ziehen (Abb. 3B). Das Perikaryon von Follikelzelle 1 liegt mit dem Hauptteil des Cytoplasmas distal und umschließt die Follikelzelle 2, die etwa von der Mitte des unteren Manubriums an die Borste zunächst nur mit einer cytoplasmitschen Manschette umgibt. Die Filamente in Inneren dieser Zelle verlaufen hier vom subrostralen Fortsatzes und vom adrostralen oberen Manubrium zu den frontalen bzw. caudalen Ausläufern der Längsmuskulatur (Abb. 2, 3A).

An die Follikelzellen schließt sich apikal eine ringförmig um die Borste liegende Epidermiszelle an; sie bildet den erweiterten Borstenkanal. Auf dieser Epidermiszelle fehlt, ebenso wie auf den Epidermiszellen des Torus und des gesamten Abdomens, eine mit Kollagenfasern durchzogene Kutikula. Die Epidermis ist hier aciliär und dicht mit bis zu 1µm langen, verzweigten Mikrovilli besetzt. In der Peripherie des Borstenkanals ist der Mikrovillisaum dichter und geringfügig höher. Im proximalen Teil der Epidermiszelle sind membranständige oder dicht unter der Membran liegende, elektronendichte Vesikel zu erkennen, deren Durchmesser 100 bis 250 nm beträgt. Besonders häufig treten diese Vesikel am Borstenkanal auf (Abb. 3A).

Entwicklung der Hakenborsten

Die kurz vor der praeanalen Sprossungszone liegenden jüngeren Segmente des Abdomens weisen eine geringere Anzahl von Borsten auf, die im Vergleich zu den vorderen Segmenten weiter ventral liegen. An der dorsalen Seite der Borstenreihe sind hier durchbrechende Borsten zu erkennen. Im Transmissionselektronenmikroskop zeigt sich, daß die Borstenbildung im Unterschied zu BARTOLOMAEUS (1995b: 175) ausschließlich in einer dorsal gelegenen Zone der Borstentasche stattfindet. Da sich in 4 von insgesamt 24 untersuchten Rei-

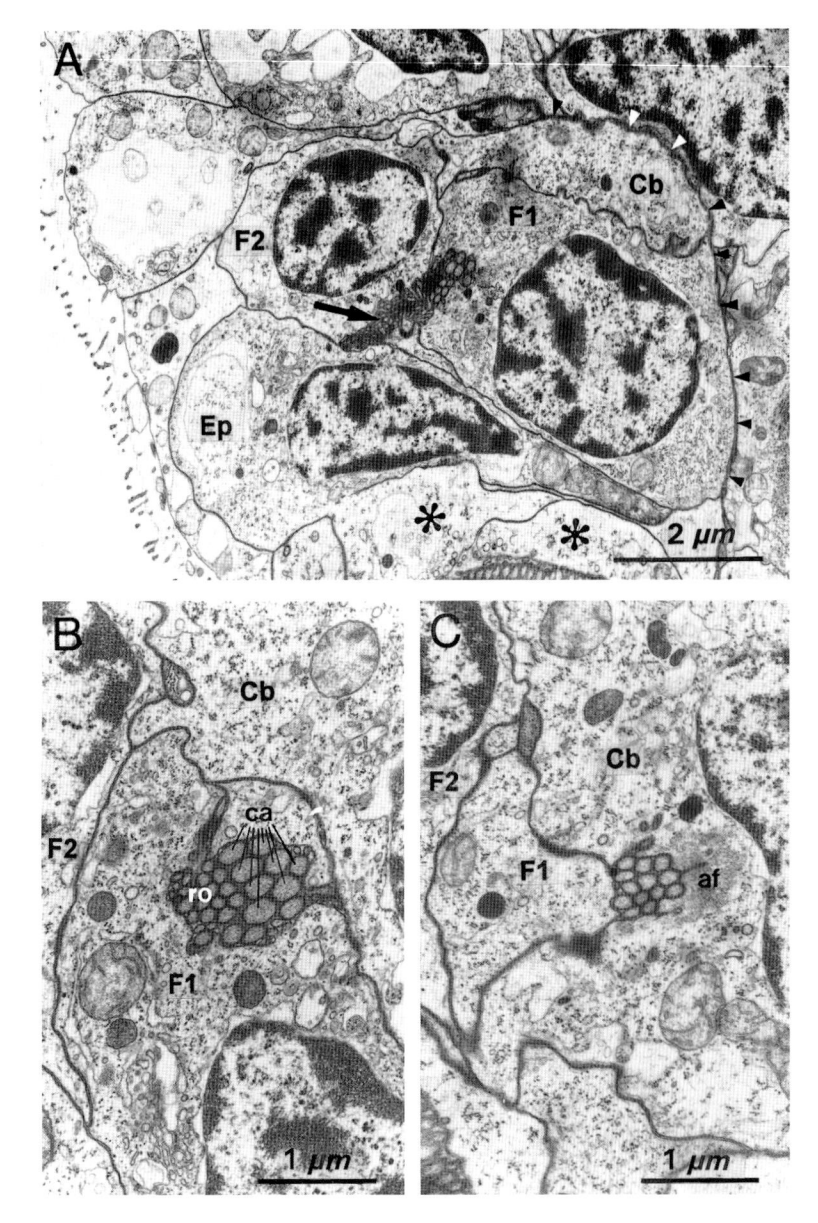

Abb. 4: Chaetogenese bei *Psammodrilus balanoglossoides*. Bildung von Rostrum und Capitium. A. Der Borstenkanal (Pfeil), in den sich die differenzierende Borste schiebt, wird von den Follikelzellen 1 (*F1*) und 2 (*F2*) und einer Epidermiszelle (*Ep*) begrenzt. (Sterne markieren Zellen des angrenzenden Follikels, Pfeilköpfe die extracelluläre Matrix.) B. Rostrum (*ro*) und Zähne des Capitum (*ca*) werden von Mikrovilli unterschiedlichen Durchmessers praeformiert. C. Aktinfilamente (*af*) im Inneren der Mikrovilli ragen tief in den Chaetoblasten (*Cb*).

hen jeweils ein Entwicklungsstadium befindet, muß man die Bildung der Borsten bei *Psammodrilus balanoglossoides* als einen diskontinuierlichen Prozeß ansehen.

Die Borstenbildung beginnt mit der Ausbildung einer von 4 Zellen begrenzten epidermalen Einstülpung, dem prospektiven Borstenkanal (Abb. 4A). In diesen entsendet der basale Chaetoblast auf seiner fronto-apikalen Seite ein axiales Bündel von 30 Mikrovilli (Abb. 4B, 6A). Zwischen den Mikrovilli liegt Hüllmaterial der späteren Borste in Form einer amorphen Matrix. Die Zellen des Borstenfollikels liegen auf Höhe des oberen Manubriums der ventrad benachbarten Borste. Die Zellinhalte von Follikeln ausdifferenzierter Borsten erscheinen elektronenheller und die Zellgrenzen verlaufen weniger geradlinig als die der Anlage (Abb. 4A).

Die Mikrovilli lassen sich anhand ihres Durchmessers in zwei Gruppen unterteilen, aus denen sich in der weiteren Entwicklung durch die Abscheidung von Borstenmaterial das Rostrum und die Einzelzähne des Capitiums bilden (Abb. 4B). Die Anlage des Rostrums wird von 15 an der proximalen Seite liegenden, 80 nm dicken Mikrovilli gebildet. Die Anzahl nimmt zur Spitze hin ab, sie verlaufen über ihre gesamte Länge sehr dicht gebündelt. Sie bilden im Querschnitt ein hexagonales Muster; die Mikrovilli liegen dicht gepackt. Die prospektiven Einzelzähne des Capitium liegen basad vom Rostrum und weisen mit 0,16 µm einen erheblich größeren Durchmesser auf. Der Abstand zwischen ihnen und dem Rostrum nimmt von der Basis zur Spitze hin zu (Abb. 6A). Die an der adrostralen Seite liegenden Mikrovilli sind kürzer, die rostralen ebenso lang wie diejenigen, die das Rostrum praeformieren. Die Borstenanlage liegt, verglichen mit den adulten Borsten der gleichen Reihe, um 180° nach caudal gedreht.

Die Basis der prospektiven Borste ist an der rostalen Seite 0,6 µm und der adrostralen Seite 0,3 µm vom Chaetoblasten umgeben. Der Zellkern liegt an der Basis der Anlage (Abb. 4C). Wie in späteren Differenzierungsstadien sind an der Borstenbasis bereits die aus den Mikrovilli in das Zellinnere ziehenden Aktinfilamente sowie mit elektronendichtem Material gefüllte Vesikel zu erkennen. Das Cytoplasma enthält darüber hinaus Golgi-Stapel, rauhes Endoplasmatisches Reticulum und mehrere Mitochondrien. Die Follikelzelle 1 umgibt den größten Teil der prospektiven Borste manschettenförmig. Ihre Zellkompartimente entsprechen denen des Chaetoblasten. Kurz vor der Spitze der Borste erweitert sich der Borstenkanal rostrad. Er wird nun von Follikelzelle 2 gebildet, die ihn mit einem schmalen Zellausläufer umschließt. In den Borstenkanal entsenden beide Follikelzellen Mikrovilli (Abb. 4A). Die Zellen des Borstenfollikel sind in der Umgebung der prospektiven Borste und des Borstenkanals über Adhaerenszonen miteinander verbunden.

In der Laufe der weiteren Entwicklung lagert sich Borstenmaterial um die Mikrovilli und bildet das Rostrum und die Zähne des Capitiums. Aufgrund des geringen Abstandes verschmelzen die Basen der Zähne des Capitiums bald untereinander und mit der Basis des Rostrums. Gleichzeitig erfolgt die Differenzierung weiterer Mikrovilli an der Chaetoblastenoberfläche, so daß sich die Zahl der Zähne des Capitiums erhöht und das Rostrum an Umfang zunimmt (Abb. 6B). Dabei kommt es zu einer verstärkten Ablagerung von Borstenmaterial auf der adrostralen Seite. Diese Vorgänge werden von einer Umorientierung der Mikrovilli begleitet, so daß die Borstenanlage ein Biegung von 115°–130° zum Rostrum erfährt. Die Orientierung der Chaetoblastenoberfläche ändert sich dabei nicht, und der fertiggestellte Haken erfährt eine Drehung um den gleichen Winkel nach frontad (Abb. 5B).

Die Borste liegt nun um etwa 30° nach caudal gekippt zu den ausdifferenzierten Borsten der Borstenreihe. Zwischen den Basen der Mikrovilli erfolgt

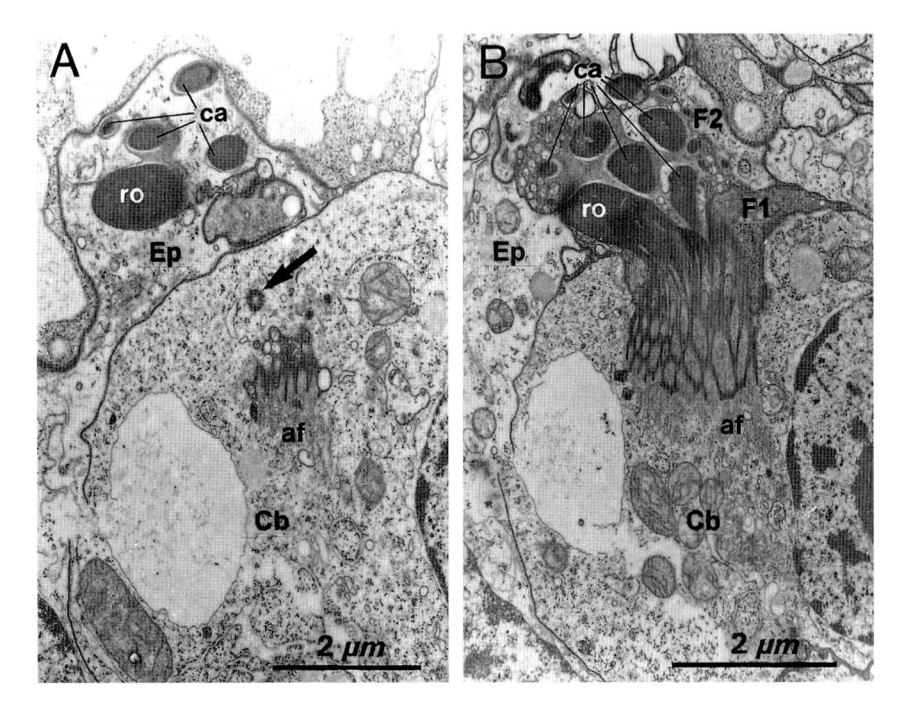

Abb. 5: Chaetogenese bei *Psammodrilus balanoglossoides*. Bildung des subrostralen Fortsatzes. A. Bildung des rostralen Abschnittes des subrostralen Forstsatzes. (Pfeil deutet auf Centriol des Chaetoblasten-Diplosoms). B. Mikrovilli des adrostalen Abschnittes des oberen Manubriums stehen um 90° gedreht zu den Zähnen des Capitiums (ca). Die adrostralen Mikrovilli, die in Verlängerung des Capitiums liegen, haben einen größeren Durchmesser als die rostral gelegenen. *af* Aktinfilamente; *Cb* Chaetoblast; *Ep* Epidermiszelle; *F1*, *F2* Follikelzellen 1 und 2; *ro* Rostrum.

Borstenstruktur und Chaetogenese bei den Psammodrilida 99

Abb. 6: Chaetogenese bei *Psammodrilus balanoglossoides*. Rekonstruierte Längsschnitte dreier Differenzierungsstadien der Hakenborsten A. Praeformation von Capitium und Rostrum durch Mikrovilli unterschiedlichen Durchmessers. B. Beginn der Differenzierung des subrostralen Fortsatzes. C. Bildung der Basis des Manubriums (Follikelzellen und Epidermiszellen sind unterscheidlich gerastert; das elektronendichte Material, das in die Borstenspitze eingelagert wird, ist dunkel gerastert). *Cb* Chaetoblast, *Ep* Epidermiszelle.

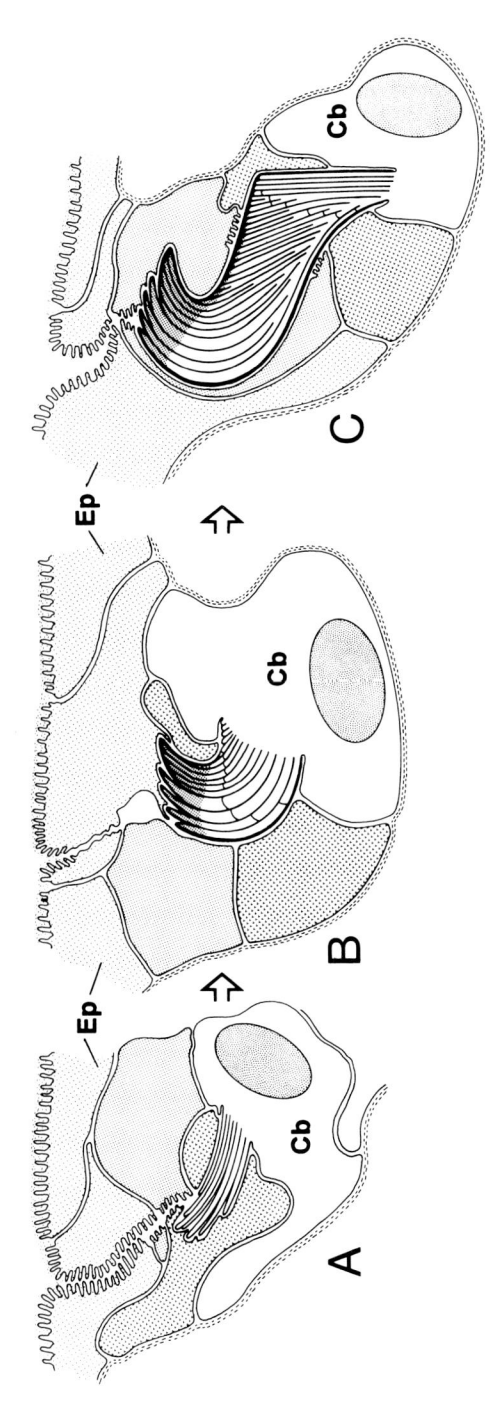

die Polymerisation weiteren Borstenmaterials, wobei die Dicke der Kanalwände nach apikal rasch zunimmt. Im Laufe der Borstendifferenzierung ziehen sich die Mikrovilli aus den fertigen Bereichen der Borste zurück. Aktinfilament-Bündel ragen aus den Mikrovilli weit in das Cytoplasma hinein (Abb. 5A). Die verbleibenden Kanäle werden mit elektronendichtem Material verfüllt (Abb. 5B). Jeweils ein Kanal zieht in jeden Einzelzahn und mehrere in das Rostrum. Die adrostral liegenden Mikrovilli des oberen Manubriums sind 1,5 µm bis 2,2 µm dick und ziehen bis zu 4 µm in den gebogenen Haken der Borste. Rostrad liegen die Mikrovilli des subrostralen Fortsatzes. Sie weisen einen Durchmesser von 0,1 µm auf und haben sich noch nicht aus ihren Kanälen zurückgezogen (Abb. 5A, 6B). Rostral schreitet die Differenzierung des subrostralen Fortsatzes voran. Hier stülpt der Chaetoblast eine Vielzahl zusätzlicher Mikrovilli aus, so daß die Borstenbasis erweitert wird. Dabei erfolgt neben der bisher beobachteten Wachstumsrichtung der Borste von apikal nach basal eine laterale Verschiebung der vom Chaetoblasten ausgestülpten Mikrovilli (Abb. 6B, 7B). Diese Verschiebung wird durch die kontinuierliche Ausbildung von Mikrovilli verursacht. Da die Borste gleichzeitig in die Länge wächst, resultiert hieraus eine Neigung des subrostralen Fortsatzes. Gleichzeitig fusionieren die zentral gelegenen, ältesten Mikrovilli der Anlage des subrostralen Fortsatzes miteinander.

Im dem dritten untersuchten Bildungsstadium ist die juvenile Borste bis auf den schlanken basalen Abschnitt des Manubriums fertiggestellt (Abb. 6C). Das Bildungsstadium liegt um etwa 15° nach caudal gekippt zu den benachbarten Borsten. Die Borste erreicht etwa die Hälfte der Länge ausdifferenzierter Borsten. Die Spitzen vom Capitium liegen 2,6 µm unter der Körperoberfläche (Abb. 7A, C).

Der Durchmesser an der Borstenbasis beträgt 1,4 µm, sie liegt unterhalb des subrostralen Fortsatzes. Die Mikrovilli an der rostralen Seite sind bis zu 0,2 µm dick und reichen 4,1 µm in die Schulter des subrostralen Fortsatzes. Sie liegen dort vollständig in ihren Kanälen. An der adrostralen Seite haben sich nur wenig apikal der Borstenbasis die Mikrovilli bereits zurückgezogen und den distalen Abschnitt des Manubriums fertiggestellt.

Follikelzelle 1 und 2 sind stark miteinander verzahnt und umschließen die Borste manschettenförmig. Follikelzelle 1 umgibt die Borste mit vielen kleinen Cytoplasmafortsätzen bis über die Schulter des subrostralen Fortsatzes. Die Verankerung der Borste an der ECM beginnt mit der Ausbildung von Intermediärfilamenten in Follikelzelle 2. Noch bevor die Zelle die Anheftungspunkte an der Schulter auf der rostralen und dem oberen Manubrium auf der adrostralen Seite der Borste umgibt, bilden sich intrazellulär Intermediärfilamente von der Basalmembran der frontalen und caudalen Längsmuskulatur

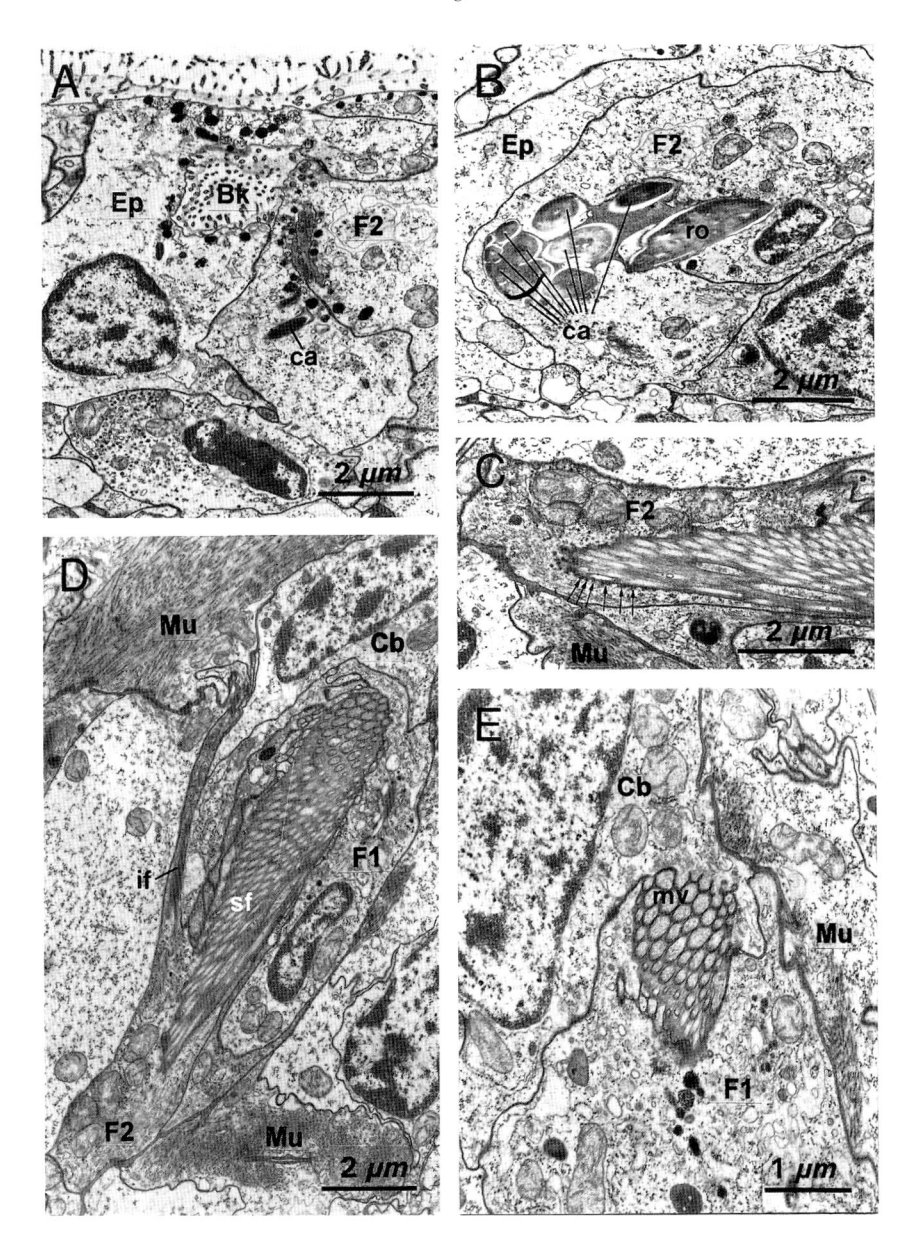

Abb. 7: Chaetogenese bei *Psammodrilus balanoglossoides*. Bildung der Basis des Manubriums. A. Epidermaler Borstenkanal (*Bk*) mit peripheren elektronendichten Vesikeln. B. Follikelzelle 2 (*F2*) umgibt Capitium (*ca*) und Rostrum (*ro*) manschettenförmig. C. Oberes Manubrium mit Hemidesmosomen (Pfeile), von denen intermediäre Filamente caudad ziehen. D. Rostraler Abschnitt des subrostralen Fortsatzes (*sf*) mit praeformierenden Mikrovilli. Intermediäre Filamente (*if*) ziehen in einem Ausläufer der Follikelzelle 2 frontad. E. Bildung des proximalen Abschnittes des Manubriums. *Cb* Chaetoblast; F1 Follikelzelle 1; *Mu* Längsmuskulatur; *mv* Mikrovilli.

zur Borste hin aus (Abb. 4B). Von Follikelzelle 2 verläuft ein 5 µm langer und 0,35 µm schmaler Zellausläufer zur frontalen Wand der Borstentasche, an die sich die Längsmuskulatur anschließt. In dem Ausläufer ziehen Intermediärfilamente von der Basalmembran in den Bereich des oberen subrostralen Fortsatzes der Borste (Abb. 7B). Eine Befestigung der Filamente an der Borste über Hemidesmosomen ist nicht zu erkennen. Bei ausdifferenzierten Borsten sind die nach frontal ziehenden Filamente in Follikelzelle 2 an der Schulter des subrostralen Fortsatzes befestigt. Der dieser Befestigungszone entsprechende Bereich wird im dritten untersuchten Bildungsstadium von Follikelzelle 1 umgeben. Die distale Verbindung zwischen Manubrium und den caudalen Ausläufern der Längsmuskulatur ist weiter differenziert. An der Borste sind Hemidesmosomen zu erkennen, die im Vergleich zu ausdifferenzierten Borsten kleiner sind. Die Filamente ziehen relativ ungeordnet zu der 0,5 µm entfernten Basalmembran und sind mit dieser ebenfalls über Hemidesmosomen verbunden. Diese frühe Verankerung könnte der Grund für die Ausdehnung des Manubriums ausdifferenzierter Borsten bis tief in die Basalmembran und die darunterliegende Längsmuskulatur sein. Der Chaetoblast und die Follikelzellen folgen der Verlängerung des unteren Manubriums mit schmalen, manschettenförmigen Zellausläufern, während ihre Kerne die apikale Lage in der Borstentasche beibehalten.

Mit dem Ende der Differenzierung neuer Mikrovilli und deren Fusion wächst die Borste wieder entlang ihrer zukünftigen baso-apikalen Achse; es beginnt die Bildung des geraden, basalen Abschnittes des Manubriums (Abb. 7E). Die Borste hat sich nun um weitere 45° gedreht und die gleiche Position wie die ausdifferenzierten Borsten in derselben Reihe eingenommen. Damit hat die Borste ihre endgültige Lage erreicht und insgesamt eine Drehung um ca. 180° erfahren. Mit Fertigstellung der Borste erfolgt auch ihre Verankerung im Chaetoblasten. Kurze Intermediärfilamente verlaufen seitlich zu der die Borstenbasis eng umschließenden Basalmembran, mit der sie über Hemidesmosomen verbunden sind. Die Kanäle der nun ausdifferenzierten Borste sind entweder leer oder enthalten Cytoplasmafortsätze des Chaetoblasten, die jedoch nicht der Verankerung dienen und als Membranreste zurückgezogener Mikrovilli angesehen werden können.

D. Diskussion

Die Monophylie der Psammodrilida läßt sich durch die spezielle Körpergliederung in Kopf, Kragensegment, Thorax und Abdomen begründen. Das Kragensegment caudal des Kopfes birgt einen mit Sphinktern ausgestatteten Saugpumpenmechanismus. Das Fehlen der Saugpumpe bei *Psammodriloides fauveli* wird als Reduktion bewertet, die möglicherweise mit der progenetischen Evolution dieser Art zusammenhängt (GOULD 1977). An den 6 Segmenten des Thorax sind die neuropodialen Borsten reduziert; jedes Segment trägt ein Paar von Cirren, die jeweils mit einem Aciculum verstärkt sind. Am Abdomen sind die notopodialen Borsten reduziert. Der subrostrale Fortsatz

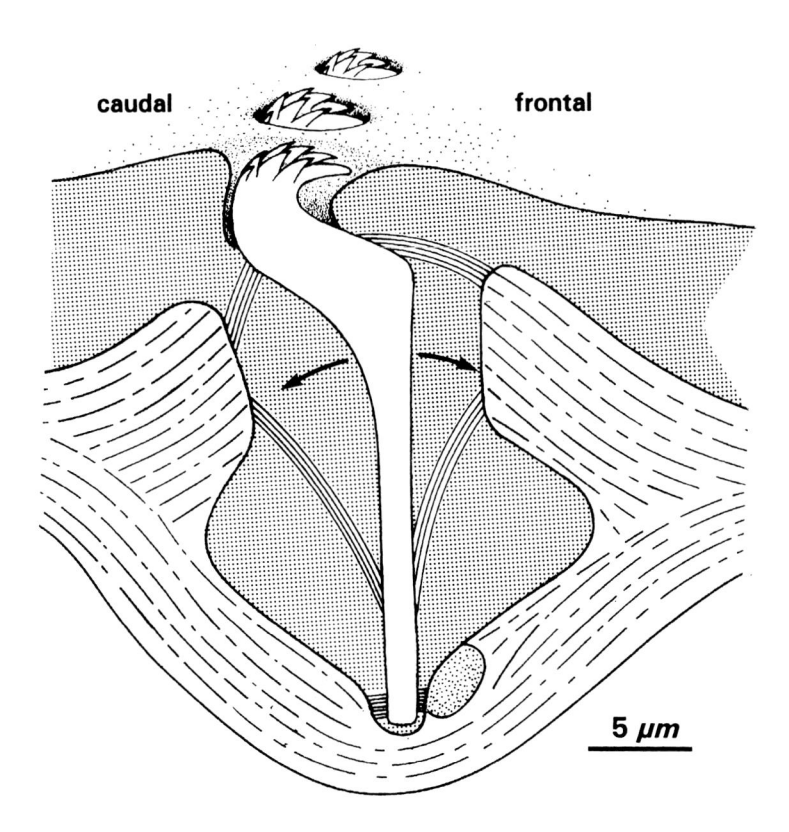

Abb. 8: Schematischer Längsschnitt durch die Borstentasche zur Verdeutlichung der Borstenbewegung. Die Borstentasche ist eine epidermale Einsenkung (grau gerastert) in die subepidermale Längsmuskulatur (gestrichelt). Die Borste kann über frontale und caudale Ausläufer dieser Muskulatur nach caudal und frontal gekippt werden. Bündel intermediärer Filament schaffen eine mechanische Verbindung zwischen Hakenborste und Muskulatur.

der Hakenborste ist stark verbreitet. Die nahezu vollständige Bewimperung des Körpers und die thoracalen Aciculae werden kontrovers diskutiert, da die Bewertung dieses Merkmals von den vorgeordneten Verwandtschaftshypothesen abhängt (BARTOLOMAEUS 1995a).

Borsten der Psammodrilida

Bei den Psammodrilida sind Hakenborsten ausschließlich auf das Abdomen beschränkt, wo sie in transversalen, ventrolateral gelegenen Reihen dicht nebeneinander stehen. Sie zeigen bei *Psammodriloides fauveli* und *Psammodrilus aedificator* in Größe, Anordnung und sigmoider Form mit stark verbreitertem subrostralen Fortsatz Übereinstimmungen mit den hier untersuchten Hakenborsten von *Psammodrilus balanoglossoides* (Abb. 9).

Psammodriloides fauveli besitzt bei einer Körpergröße von 1 mm Hakenborsten mit einer Länge von bis zu 33 µm. Jeweils nur ein Paar von Hakenborsten befinden sich ventral an jedem Abdominalsegment. Wülste oder Tori sind nicht vorhanden, die Borsten liegen an der Grenze zum folgenden Segment. Das Rostrum der Hakenborste ist im Vergleich zu *P. balanoglossoides* verlängert. Die Zähne des Capitiums sind auf drei reduziert und umgeben das Rostrum. An der Schulter des subrostralen Fortsatzes, in Verlängerung des unteren Manubriums, befinden sich 5-6 ca. 4 µm lange gerade Stäbe, die weit aus dem Tier herausragen und als Bart bezeichnet werden (SWEDMARK 1958).

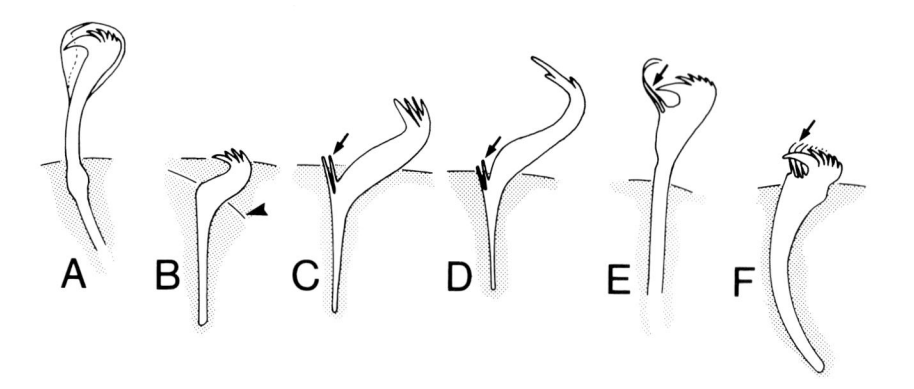

Abb. 9: Hakenborsten von Psammodriliden, Maldaniden, Arenicoliden und Capitelliden im Vergleich. A. *Capitella capitata* (Fabricius, 1780), (Capitellidae); B. *Psammodrilus balanglossoides* (Psammodrilida), (Pfeilkopf markiert die Position der Borste relativ zur Epidermis bei juvenilen Tieren); C: *Psammodrilus aedificator* (Psammodrilida); D. *Psammodriloides fauveli* (Psammodrilida); E. *Clymenura clypeata* (Staint-Joseph, 1894), (Maldanida); F. *Arenicola marina* (Linné, 1758), (Arenicolida), Postlarve. Die Pfeile in C.- F. deuten auf die Haare des Bartes, die als homolog hypothetisiert werden.

Bei den 40 μm langen Hakenborsten von *Psammodrilus aedificator* variiert das Vorkommen des Bartes. Dieser besteht, wenn vorhanden, aus 1-2 Stäben und reicht in den rostrad erweiterten Borstenkanal. Das Rostrum steht wie bei *Psammodriloides fauveli* ungefähr im rechten Winkel zum oberen Manubrium. Die ersten 2-3 Segmente des Abdomens tragen keine Borsten. Die Anzahl der Borsten in den folgenden caudalen Segmenten liegt bei maximal 5 je Seite und nimmt caudad bis auf eine ab (KRISTENSEN & NØRREVANG 1982).

Für *P. balanoglossoides* ist die Veränderung der Borstenverteilung im Verlauf der Ontogenese bekannt. Die ersten Hakenborsten treten in den 350 μm langen Juvenilstadien auf und gleichen denen der Adulti. In der weiteren Entwicklung wandeln sich die drei frontalen Abdominalsegmente in Thoraxsegmente um. Dabei erscheinen die dorsalen Rudimente der Cirren mit den Aciculae erst nach dem vollständigen Verlust der Hakenborsten (SWEDMARK 1955). In Unterschied dazu sind die Aciculae der erranten Polychaeten die allerersten Borstenstrukturen, die im Laufe der segmentalen Differenzierung angelegt werden (Bartolomaeus, unveröff.). Dieser Unterschied wird als Hinweis auf eine konvergente Evolution der Aciculae bei den erranten Polychaeten und den Psammodriliden gewertet.

Die Aciculae als Verstärkungselement der dorsalen, als Cirren bezeichneten, fadenförmigen Körperanhänge des Thorax werden komplett von der Epidermis umgeben. Nach KRISTENSEN & NØRREVANG (1982) haben die Cirren von bei *Psammodrilus balanoglossoides* die Funktion eines Propellers bei der Fortbewegung im offenen Wasser. Nach SWEDMARK (1955) erzeugen sie einen kontinuierlichen Wasserstrom um das Tier und dienen der Respiration. Der innere Aufbau der Aciculae von *Psammodrilus aedificator* weicht von dem der Hakenborsten und von dem bei anderen Polychaeten-Borsten beschriebenen Aufbau (BOULINGAND 1967; ORRHAGE 1971) in einigen Aspekten ab. Die Kanäle im Inneren der Aciculae von *Psammodrilus aedificator* sind spiralig gewunden und das Kanalwandmaterial fusioniert nicht mit dem benachbarter Kanäle. Die entstehende gedrehte, seilartige Struktur erhöht die Flexibilität des Aciculums (KRISTENSEN & NØRREVANG 1982). In den Kanälen der bis zu 300 μm langen Aciculae verbleiben nach der Chaetogenese mit bis zu 60 μm ungewöhnlich lange Mikrovilli. Die Aciculae von *Psammodrilus balanoglossoides* und *Psammodriloides fauveli* sind ebenfalls spiralig gewunden (BARTOLOMAEUS 1995a; SWEDMARK 1958). Ein vergleichbarer innerer Aufbau der Aciculae ist daher auch für diese beiden Arten anzunehmen.

Die Funktion der ausdifferenzierten Borste und des Follikels

Die Hakenborsten von *Psammodrilus balanoglossoides* dienen in Verbindung mit der Struktur des Follikels dem Tier zur Lokomotion und Verankerung in einer auf beiden Seiten offenen Schleimröhre (Abb. 2, 8). Der Aufbau von Borste und Follikel spiegelt diese mechanische Beanspruchung wider. Der Haken der Borste dient der Verankerung im Substrat (WOODDIN & MERZ 1987). Die gezahnte Struktur von Capitium und Rostrum erhöht dabei die Friktion und verhindert das Abrutschen am Substrat (SPECHT 1988). Die auf diese im Vergleich zur gesamten Borste kleinen Strukturen einwirkenden Kräfte erfordern eine erhöhte Steifigkeit, die durch Einlagerung von Borstenmaterial in die Kanäle des Hakens erreicht wird. KRYVI und SØRVIG (1990) haben eine ähnliche Verstärkung an den 'limbate setae' von *Sabella penicillus* (Linné, 1767) nachgewiesen: Bei gleichem Durchmesser besitzen kompakt aufgebaute Bereiche der Borste eine weit höhere Biegefestigkeit, als von Kanälen oder Microstäben durchzogene Teile der Borste. Letztere Struktur ist unter dem Aspekt der Materialersparnis vorteilhaft, da sie bei gleicher Materialmenge und einem vergrößerten Durchmesser eine höhere Steifigkeit und größere Elastizität aufweist. Eine weitere Verstärkung der sonst komplett von Kanälen durchzogenen Borste befindet sich am ebenfalls mechanisch stark beanspruchten Übergang vom oberen, distalen zum schlanken, basalen Teil des Manubriums. Die Lumen der Kanäle sind dort enger und das Borstenmaterial entsprechend stärker ausgebildet.

Die Zellen des Borstenfollikels verändern im Laufe der Genese ihre Funktion und dienen der adulten Borste der Verankerung und der Verbindung mit der Muskulatur. Nach SCHERF (1970) ist die Ausbildung von Tonofibrillen (intermediäre Filamante) in den Borstentaschen von *Pectinaria koreni* (Malmgren, 1866) abhängig von den Muskelansatzstellen und Ausdruck des herrschenden Kräftespiels. Überträgt man diese Vorstellung auf die Befunde bei *Psammodrilus balanoglossoides*, so ergibt sich folgendes Bild.

In den Follikelzellen verbinden Intermediärfilamente die Borste mit den frontalen und caudalen Ausläufern der Längsmuskulatur. Die Filamente verlaufen in Follikelzelle 2 zu den proximalen Ansatzstellen an der Borste. In Follikelzelle 1 verlaufen die Filamente gebündelt zur distalen Borstenbasis. Die Borste ist damit über zwei Dreieckskonstruktionen aus Intermediärfilamenten indirekt mit einem Protraktor und einem Retraktor verbunden (Abb. 8). Dabei dient die apikale Befestigung der Übertragung der Drehbewegung auf die Borste und die basale Befestigung der Ableitung der entstehenden Kräfte auf die fixierte Borstenbasis. Der Vorgang der Austülpung der Borste läßt sich damit folgendermaßen erklären. Die basale Befestigung dient als

Drehpunkt der Borste. Durch Kontraktion des frontalen Ausläufers der Längsmuskulatur und der basalen Längsmuskulatur dreht sich die Borste nach frontal und damit aus ihrem Kanal. Die caudalen Ausläufer stellen den Retraktor dar und ziehen die Borste in die Borstentasche ein. Dabei dreht sich die Borste und dringt in den nach caudal ansteigenden Torus ein. Die Follikelzellen mit ihrer im apikalen Teil der Borste kissenförmigen Anordnung haben bei dieser Drehbewegung eine Pufferfunktion (SPECHT 1988). Ihr elektronenhelles Erscheinungsbild kann so mit einem, bis auf die oben genannten Filamentbündel, schwach entwickelten Cytoskelett erklärt werden. Eine Beteiligung des als hydrostatisches Skelett fungierenden Körpercoeloms an der Bewegung der Borsten oder des gesamten Torus erscheint unwahrscheinlich, da dieses bei der Lokomotion von *Psammodrilus balanoglossoides* nicht mitwirkt (BARTOLOMAEUS 1995a).

Homologie von Hakenborsten, Kapuzenhakenborsten und Uncini

Neben den Psammodrilida besitzen Oweniida, Maldanida und Arenicolida in neuropodialen Transversalreihen angeordnete Hakenborsten (ASHWORTH 1912, BARTOLOMAEUS & K. MEYER 1997, HARTMANN-SCHRÖDER 1971). Die Kapuzenhakenborsten der Capitelliden sind nach Schweigkofler et al. (1998) den Hakenborsten homolog. Sie zeichnen sich durch die zusätzliche zum Borstenkern gebildete Struktur der Kapuze aus. Diese bedeckt das obere Manubrium und läuft an den Spitzen von Capitium und Rostrum zu einer schlitzförmigen Öffnung aus (Abb. 9A). Die Uncini von Terebellida, Sabellida und Pogonophora haben mit den Hakenborsten die Anordnung in Transversalreihen und den Aufbau der Borste gemeinsam (BARTOLOMAEUS 1995b; GEORGE & SOUTHWARD 1973; NØRREVANG 1970; KNIGHT-JONES 1981; HOLTHE 1986): Das Rostrum wird von mehreren Mikrovilli gebildet, die einzelnen Zähne des Capitiums nur von einem Mikrovillus. Capitium und Rostrum liegen in einem Winkel von 70° bis 110° zum Manubrium. In der Genese entsteht dieser Winkel durch die Verschiebung der Mikrovillibasen nach rostral. Durch die weitere Anlage von Mikrovilli entsteht der subrostrale Fortsatz. Uncini erreichen jedoch ihre endgültige Position im ausdifferenzierten Zustand durch Zellbewegungen und nicht wie Hakenborsten durch basales Wachstum. Aufgrund der weitgehenden Übereinstimmungen im Aufbau und der Bildungsweise nimmt BARTOLOMAEUS (1995b) die Homologie von Uncini und Hakenborsten an. Uncini sind durch die Verkürzung des Manubriums und der Abwesenheit von Mikrovilli in den Basen der ausgebildeten Borsten als abgeleitete Hakenborsten anzusehen. Diese Hypothesen zur Homologie haben zur Folge, daß es eine gemeinsame Stammart der obengenannten Taxa gegeben haben muß, in dessen Stammlinie diese Borstenstrukturen evolviert wurden.

Ein Unterschied zu den bisher untersuchten Bildungsmodi (BARTOLOMAEUS 1995b, BARTOLOMAEUS & K. MEYER 1997, K. MEYER & BARTOLOMAEUS 1997, SCHWEIGKOFLER et al.1997) besteht lediglich in der zeitgleichen Anlage von Capitium und Rostrum. Dieses Merkmal könnte eine Folge des geringen Größen- und Längenunterschiedes von Capitium und Rostrum der Hakenborsten von *Psammodrilus balanoglossoides* sein. Es steht einer Homologisierung jedoch nicht im Wege.

Die neuropodialen Hakenborsten werden bei Psammodriliden von einer medio-lateral gelegenen Zone aus gebildet (nachgewiesen nur für *Psammodrilus balanoglossoides*), so daß die ältesten Hakenborsten ventral anzutreffen sind. Eine entsprechende Lage der Borstenbildungzone wurde für Oweniidae, Terebellida, Capitellidae und Spionida (K. MEYER & BARTOLOMAEUS 1996; SCHWEIGKOFLER et al. 1997; HAUSEN & BARTOLOMAEUS 1997) nachgewiesen. Diese Position erweist sich als plesiomorphes Merkmal im Vergleich zu anderen Annelidentaxa (K. MEYER & BARTOLOMAEUS 1996). Nur Maldaniden und Arenicoliden haben eine abweichende ventrale Lage der Borstenbildungszone in den Querreihen der neuropodialen Borsten, die als abgeleitetes Merkmal des beide Taxa umfassenden Taxons Maldanomorpha angenommen (BARTOLOMAEUS & K. MEYER 1997).

In den Erstbeschreibungen von *Psammodriloides fauveli* (SWEDMARK 1958) und *Psammodrilus aedifictor* (KRISTENSEN & NØRREVANG 1982) weisen die Autoren auf die Ähnlichkeit der Hakenborsten mit denen der Maldaniden hin (Abb. 9B-E). Sie vermuten eine engere Verwandtschaft, da Psammodriliden und Maldaniden den Besitz eines aus Anhängen des Subrostrums bestehenden Bartes gemeinsam haben. Demzufolge müßte der Bart ein Grundmustermerkmal der Psammodriliden sein und bei *Psammodrilus balanoglossoides* eine sekundäre Reduktion vorliegen. Dies wird durch die Beobachtung unterstützt, daß bei Hakenborsten von Maldaniden und barttragenden Psammodriliden ebenso wie bei juvenilen *P. balanoglossoides* (SWEDMARK 1958) zumindest der subrostrale Fortsatz außerhalb der Körperoberfläche liegt. Bei adulten Exemplaren von *P. balanoglossoides* reicht nur der apikale Teil des Hakens über die Epidermis und die Epidermis bedeckt das Subrostrum (Abb. 9B). Eine Reduktion des Bartes kann man im Zusammenhang mit der inneren Lage annehmen und somit als abgeleitetes Merkmal von *Psammodrilus balanoglossoides* bezeichnen. Arenicoliden besitzen nach den Untersuchungen von BARTOLOMAEUS und K. MEYER (1997) an den Juvenilborsten ebenfalls einen sehr ähnlich ausgebildeten Bart (Abb. 9F). Aufgrund der Gemeinsamkeiten in der Lage und der Struktur aus relativ feinen elastischen

Stäben wird im Stand der Untersuchungen eine Homologie des Bartes bei Maldanomorpha und Psammodrilida angenommen.

Hypothesen zur Stellung der Psammodrilida

Aufgrund der Hypothese einer Homolgie von Hakenborsten, Uncini und Capitelliden-Kapuzenhakenborsten muß eine einmalige Evolution einer Borste aus Rostrum, Capitium und dazu gewinkelt liegendem Manubrium in der Stammlinie eines Taxons erfolgt sein, das alle Anneliden mit solchen Borsten umfaßt. Da die Psammodriliden neben Hakenborsten auch Aciculae besitzen, beeinflussen diese Strukturen die Hypothesen zur Stellung der Psammodriliden nachhaltig, wenn Aciculae innerhalb der Anneliden homolog sein sollten (BARTOLOMAEUS 1995a). Aciculae müßten dann entweder bereits im Grundmuster der Anneliden vorhanden oder erst innerhalb der Anneliden entstanden sein. In der Stammlinie eines alle übrigen hakenborstentragenden Anneliden umfassenden Taxons wären Aciculae sekundär reduziert, da diesen Anneliden Aciculae fehlen. In dem Fall wären die Psammodriliden die ursprünglichsten Vertreter der hakenborstentragenden Anneliden.

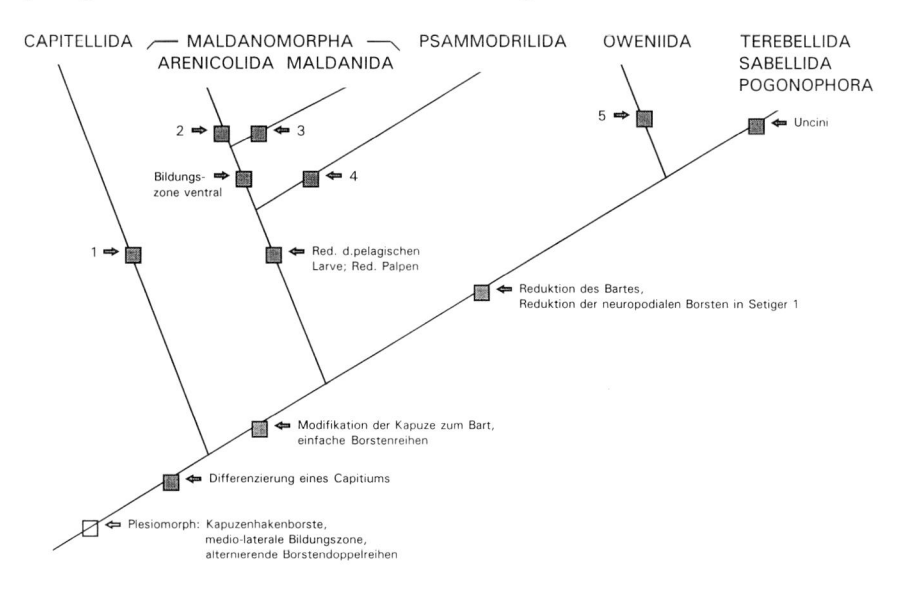

Abb. 10: Hypothese zur Stellung der Psammodrilida innerhalb der hakenborstentragenden Annelida. 1: Neben den Metanephridien ein Paar Gonodukte pro Segment. 2: Segmentale Anordnung unverzweigter Kiemen. 3: Verlängerte anteriore Segmente, Eingrabung in das Sediment mit dem Prostomium. 4: Kragensegment mit Saugpumpenmechanismus, Thorax ohne neuropodiale Borsten und mit Cirren mit jeweils einem Aciculum, Abdomen ohne notopodiale Borsten, fast vollständige Bewimperung. 5: Neuropodiale Borstenfelder, Mitrarialarve.

Die oben genannten Besonderheiten der Aciculae bei Psammodriliden scheinen jedoch gegen eine generelle Homologie dieser inneren Borsten bei Anneliden sprechen. Damit wird der Weg zu einer Hypothese frei, die nicht nur sich auf Borsten beziehende Merkmale berücksichtigt und die thoracalen Aciculae als Autapomorphie der Psammodrilida ausweist. Im Stand der Untersuchungen wird diese Hypothese favorisiert. Damit erweisen sich die Psammodrilida als die Schwestergruppe der Maldanomorpha aufgrund der synapomorphen Reduktion der Palpen, die innerhalb der Anneliden ein ursprüngliches Merkmals darstellen (WESTHEIDE 1997), sowie des Fehlens eines pelagischen Larvenstadiums (NEWELL 1948, 1951; BOOKHOUT & HORN 1949; ROUSE 1992), (s. Abb. 10).

Die Kapuzenhakenborsten der Capitelliden sind Hakenborsten, die im apikalen Bereich eine Hülle ausbilden, die das obere Manubrium, das Rostrum und das Capitium hufeisenförmig umgibt, so daß ein rostraler Spalt übrig bleibt, der den Blick auf die letztgenannten Strukturen freigibt (Abb. 9A). Solche Kapuzen sind ebenfalls für Spioniden beschrieben und tatsächlich weisen Struktur und Entwicklung der Kapuze bei Capitelleiden und Spioniden Übereinstimmungen auf, die für einen Homologie dieser umhüllenden Struktur sprechen (HAUSEN & BARTOLOMAEUS 1998, SCHWEIGKOFLER et al. 1998). Den Kapuzenborsten der Spioniden fehlen jedoch die für Hakenborsten charakteristischen Strukturen, und es gibt derzeit keinen Hinweis darauf, diesen Mangel als sekundär aufzufassen. Aufgrund der spezifischer Übereinstimmungen in der Organisation der Kapuze bei Spioniden und Capitelliden müssen für das Grundmuster der hakenborstentragenden Anneliden Kapuzenhakenborsten angenommen werden.

Im Rahmen dieser Hypothese kann der Bart bei Psammodrilida und Maldanomorpha entweder als ursprüngliches oder abgeleitetes Merkmal gedeutet werden. Falls der Bart eine Autapomorphie eines aus Psammodriliden und Maldanomorpha bestehenden Taxons darstellt, wurde die Kapuze erst in der gemeinsamen Stammlinie von Oweniida, Terebellida, Sabellida und Pogonophora reduziert. Es ist jedoch wahrscheinlicher, eine Modifikation der Kapuze zum Bart in der gemeinsamen Stammlinie aller genannten Taxa anzunehmen. Eine Reduktion des Bartes wäre dann erst in der gemeinsamen Stammlinie der Oweniida, Tebellida, Sabellida und Pogonophora erfolgt. Diese Auffassung wird hier favorisiert, so daß innerhalb der hakenborstentragenden Taxa eine sukzessive Reduktion der Kapuze erfolgt sein muß (Abb. 10).

Danksagung

Herrn Prof. Dr. P. Ax gilt unser Dank für die kritische Durchsicht des Manuskriptes. Die vorliegende Arbeit wurde durch Sachmittel von der Akademie der Wissenschaften und der Literatur, Mainz, unterstützt.

Zusammenfassung

Die Psammodrilida sind derzeit mit drei interstitiellen Arten bekannt. Alle drei weisen im caudalen Körperabschnitt, dem Abdomen, segmentale Transversalreihen von Hakenborsten auf, die auch bei anderen Teilgruppen der Anneliden vorkommen. Die Hakenborsten zeichnen sich durch einen kräftigen Hauptzahn (Rostrum) aus, dem adrostral zahlreiche kleine Zähne aufsitzen, die in ihrer Gesamtheit als Capitium bezeichnet werden. Alle Zähne stehen gewinkelt zum Schaft (Manubrium) der Borste. Bei *Psammodrilus balanoglossoides* befindet sich jede dieser Hakenborsten in einem dreizelligen Follikel, der neben dem basalen Chaetoblasten aus zwei Follikelzellen besteht, die die Borste ringförmig umgeben. Die Borsten werden diskontinuierlich in einer medio-lateral gelegenen Bildungszone generiert. Dabei senkt sich eine Gruppe von drei Zellen aus der Epidermis ein. Die basale Zelle dieser Anlage bildet apikale Mikrovilli aus, die in ein zentrales, apikal offenes Kompartiment ragen. Eine Gruppe schlanker Mikrovilli praeformiert das Rostrum, während kräftigere, adrostral gelegene Mikrovilli jeweils einen Zahn des Capitiums formen. Im Laufe der Addition weiterer Mikrovilli verlagert sich deren Ausrichtung im Verhältnis zur Zelloberfläche, wodurch es zu der charakteristischen Krümmung der apikalen Strukturen der Hakenborste kommt. Anschließend differenziert der Chaetoblast zusätzliche Mikrovilli auf der rostralen Seite, während auf der adrostralen Seite ständig Mikrovilli fusionieren. Als Folge dieser lateralen Verschiebung der Borstenbasis wird ein sehr breiter subrostraler Fortsatz ausgebildet, der für *Psammodrilus balanoglossoides* charakteristisch ist. Später fusionieren diese Mikrovilli, wodurch sich der Durchmesser der Borste verringert. Am Ende der Chaetogenese bildet der Chaetoblast intermediäre Filamente aus, die die Borste im Follikel verankert. Während der Borstendifferenzierung dreht sich die Borste um 180°. Die spezielle Differenzierung der einzelnen Elemente der Hakenborste, ihre Drehung im Laufe der Chaetogenese und ihre Genese in einer medio-lateral gelegenen Bildungszone stimmen nahezu vollständig mit Vorgängen während der Differenzierung der Uncini bei Sabelliden und Terebelliden, der Kapuzenhakenborsten bei Capitelliden sowie der Hakenborsten bei Oweniiden und, mit Ausnahme der Position der Bildungszone, bei Arenicoliden und Maldaniden überein. Diese Übereinstimmungen erlauben die Hypothese einer Homologie von Uncini, Kapuzenhakenborsten und Hakenborsten. Auf dieser Basis wird die Hypothese formuliert, daß die Psammodrilida die Schwestergruppe der Maldanomorpha (Arenicolidae + Maldanidae) sind, deren gemeinsames abgeleitetes Merkmal die Reduktion der pelagischen Larve darstellt.

Literaturverzeichnis

ASHWORTH, J.H. (1912): A catalogue of the Chaetopoda of the British Museum. Part 1 Arenicolidae, Bd. **7**. London.

BARTOLOMAEUS, T (1995a): Zur Ultrastruktur von *Psammodrilus balanoglossoides*: Hypothesen zur Stellung der Psammodrilida innerhalb der Annelida. Microfauna mar. **10**, 295-303.

BARTOLOMAEUS, T. (1995b): Structure and formation of the uncini in larval *Pectinaria koreni*, *Pectinaria auricoma* (Terebellida) and *Spirorbis spirorbis* (Sabellida): Implications for annelid phylogeny and the position of the Pogonophora. Zoomorphology **115**, 161-177.

BARTOLOMAEUS, T. & K. MEYER (1997): Morphogenesis and phylogenetic significance of hooked setae in Arenicolidae (Polychaeta, Annelida). Inv. Biol. **116**, 227-242.

BOOKHOUT C. G. & E. C. HORN (1949): The development of *Axiothella mucosa* (Andrews). J. Morph. **84**, 145-181.

BOULIGAND, Y. (1967): Les soies et les cellules associées chez deux Annélides Polychètes. Étude en microscopie photonique à contraste de phase et en microscopie électronique. Z. Zellforsch mikrosk. Anat. 79, 332-363.

GEORGE, D. & E. SOUTHWARD (1973): A comparative study of the setae of Pogonophora and polychaetous Annelida. J. mar. biol. Ass. UK **53**, 403-424.

GOULD, S.J. (1977): Ontogeny and Phylogeny. The Belknap Press of the Harvard University. Cambridge. Massachusetts.

HARTMANN-SCHROEDER, G. (1971): Annelida, Borstenwürmer, Polychaeta. In: DAHL, M. & F. PEUS (Hrsg.) Die Tierwelt Deutschlands, Bd. **58**. G. Fischer, Jena.

HAUSEN, H. & T. BARTOLOMAEUS (1998): Setal structure and chaetogenesis in *Scolelepis squamata* and *Malacoceros fuliginosus* (Spionidae, Annelida). Acta Zool. (im Druck).

HOLTHE, T. (1986): Evolution, systematics, and distribution of the Polychaeta Terebellomorpha, with a catalogue of the taxa and a bibliography. Gunneria **55**, 1-236.

KNIGHT-JONES, P (1981): Behaviour, setal inversion and phylogeny of Sabellida (Polychaeta). Zool. Scr. **10**, 183-202.

KRISTENSEN, R.M. & A. NØRREVANG (1985): Description of *Psammodrilus aedificator* sp.n. (Polychaeta), with notes on the arctic interstitial fauna of Disko Island, West Greenland. Zool. Scr. **11**, 265-279.

KRYVI, H., SØRVIG, T. (1990): Internal organization of limbate polychaete setae (*Sabella penicillus*), with notes on bending stiffness. Acta. Zool. **71**, 25-31.

MEYER, K & T. BARTOLOMAEUS (1996): Ultrastructure and formation of the hooked setae in *Owenia fusiformis* delle Chiaje, 1842: Implications for annelid phylogeny. Can. J. Zool. **74**, 2143-2153.

NEWELL, G. E. (1949): A contribution to our knowledge of the life-history of *Arenicola marina* L. J. Mar. Biol. Ass. U. K. **27**, 554-580.

NEWELL, G. E. (1951): The life history of *Clymnella torquata*, Leidy. (Polychaeta). Proc. Zool. Soc. London. **121**, 561-586.

NØRREVANG, A. (1970): The position of Pogonophora in the phylogenetic system. Z. Zool. Evolutionsforsch. **8**, 161-172.

ORRHAGE, L. (1971): Light and electron microscope structure of some annelid setae. Acta Zool. 52, 157-169.
ROUSE, G. W. (1992): Oogenesis and larval development in *Micromaldane* spp. (Polychaeta: Capitellida: Maldanidae). Invertebrate Reproduction and Development **21**, 215-230.
SCHERF, H. (1970): Struktur und Insertion der Paleen von *Pectinaria auricoma*, Malmgren. Zool. Jb. Anat. **87**, 386-401.
SCHWEIGKOFLER, M., T. BARTOLOMAEUS & L.v. SALVINI-PLAWEN (1998): Ultrastructure and formation of hooded hooks in *Capitella capitata* (Fabricius, 1780) (Capitellida, Annelida). Zoomorphology (im Druck).
SPECHT, A. (1988): Chaetae. In: WESTHEIDE W. & C.O. HERMANS (Hrsg.). The ultrastructure of the Polychaeta, Microfauna Marina **4**, 45-59.
SWEDMARK, B. (1952): Note préliminaire sur un polychete sedentaire aberant, *Psammodrilus balanoglossoides* n. gen., n. sp. Ark. Zool. **4**, 159-161.
SWEDMARK, B. (1955): Recherches sur la morphologie, de développment et la biologie de *Psammodrilus balanoglossoides* Polychaete sédentaire de la microfaune des sables. Arch. Zool. Exp. Gen. **92**, 141-220.
SWEDMARK, B. (1958): *Psammodriloides fauveli* n. gen., n. sp. et la famille des Psammodrilidae (Polychaeta, Sedentaria). Ark. Zool. **12**, 55-64.
WESTHEIDE, W. (1966): Zur Polychaetenfauna des Eulitorals der Nordseeinsel Sylt. Helgol. wiss. Meeresunters. **13**, 203-209.
WESTHEIDE, W. (1997): The direction of evolution within the Polychaeta. J. Nat. Hist. **31**, 1-15.
WOODIN, S. A. & R. A. MERZ (1987): Holding by their hooks: anchors for worms. Evolution **41**, 427-432.

Dipl.-Biol. Rudolf Meyer und Priv.-Doz. Dr. Thomas Bartolomaeus
II. Zoologisches Institut und Museum der Universität Göttingen
Berliner Straße 28, D-37073 Göttingen

Interstitial Fauna of Galapagos. XXXIX. Copepoda, part 7

Wolfgang Mielke

Abstract

Five species of benthic copepods are recorded from different sandy beaches of the Galápagos Archipelago. Four of them belong to the Ameiridae, one species to the Diosaccidae.

Nitocra bisetosa Mielke, 1993 was first found in Costa Rica. The animals from various sites of Galápagos exhibit some variability and are united to one species and ascribed to the Costa Rican species only with great reservation.

Another broadly distributed *Nitocra*-species proved to be new to science because of a unique combination of several morphological features: *N. galapagoensis*.

Stenocopia is represented by at least two species. One can be identified as *S. limicola*, which was first reported from Bermuda by WILLEY (1935). The second *Stenocopia*-population is described as a new species: *S. sarsi*.

Neither the two species of *Nitocra* nor the pair of *Stenocopia*-species can be derived from a respective hypothetic stem species, which, after its immigration, would have evolved into the species presented here. All four species must have reached the archipelago independently. Nevertheless, there is distinct evidence concerning obvious evolutionary modifications, especially of *N. bisetosa*, which developed subsequent to the species expansion within the Galápagos Islands.

The only diosaccid representative reported on in this paper proved to be a variable species. Its designation as *Balucopsylla triarticulata* Wells & Rao, 1987 seems to be rather questionable.

Fauna intersticial de Galápagos. XXXIX. Copepoda, parte 7

Resumen

Se describen cinco especies de copépodos bentónicos de diferentes playas arenosas del Archipiélago de Galápagos. Cuatro de ellas pertenecen a los Ameiridae y una especie a los Diosaccidae.

Nitocra bisetosa Mielke, 1993 ha sido ya reportada en Costa Rica. Los animales de diferentes lugares de Galápagos muestran cierta variedad y pueden atribuirse sólo con gran reserva a una sola especie, en este caso a la especie de Costa Rica.

Una especie adicional de *Nitocra* con amplia distribución resultó nueva para la ciencia a causa de una combinación única de algunas características morfológicas: *N. galapagoensis*.

Stenocopia está representada al menos por dos especies. La una ha sido identificada como *S. limicola*, que fue encontrada anteriormente en Bermuda por WILLEY (1935). La segunda *Stenocopia* es descrita como nueva especie: *S. sarsi*.

Ninguna de las especies de *Nitocra* ni de las de *Stenocopia* pueden atribuirse a una especie primitiva común, que, posterior a su inmigración se habría divergido en las especies aquí presentadas. Se postula que las cuatro especies tienen que haber llegado al archipiélago independientemente. Sin embargo, existen diferentes indicios que concerniendo modificaciones evolutivas, especialmente en *N. bisetosa*, hablan en favor de una especiación posterior a su expansión en las Islas Galápagos.

El único representante de los Diosaccidae aquí mencionado, mostró gran variabilidad. Su designación como *Balucopsylla triarticulata* Wells & Rao, 1987 debe considerarse como bastante incierta.

A. Introduction

More than 20 years ago, an extensive, qualitatively (**ql**) and quantitatively (**qn**) collected material containing far more than 30 000 copepods from various islands and beaches of the Galápagos Archipelago (AX & SCHMIDT 1973) was placed at my disposal. Individuals of relevant taxa were laboriously sorted out and subsequently described.

The present study is a continuation of, thus far, six publications dealing with different groups of benthic copepods of Galápagos (MIELKE 1979 etc). Now five species of the still comprehensive material are introduced which, in part, were provisionally treated long ago. The pencil drawings of *Nitocra bisetosa* and *Balucopsylla ? triarticulata* were made in 1976, the ones of *Nitocra galapagoensis* nov. spec. in 1984. Yet, their publication has been delayed by reason of an advanced treatment of other systematic groups. Furthermore, two *Stenocopia*-species could be established for the Galápagos Islands.

The material has been deposited in the collections of the Zoological Museum of the University of Göttingen.

The interpretation of body, mouth parts and thoracopods is adopted from LANG (1948, 1965). With respect to the mouth parts, the interpretation of the components by HUYS & BOXSHALL (1991) is given in parentheses.

As to the spelling of *Nitocra/Nitokra* see my remark in MIELKE 1993, p.265.

B. Results

Ameiridae Monard, 1927, emend. Lang, 1936
Nitocra Boeck, 1865
Nitocra bisetosa Mielke, 1993

(Figs. 1 – 3A)

Localities and material. **qn**: Fernandina: Cabo Douglas (I,2; 27.9.72); 63 ♀♀, 4 ovigerous ♀♀, 46 ♂♂, 17 copepodites. Marchena: South-west beach (IV,1; 28.7.72); 22 ♀♀, 5 ovigerous ♀♀, 18 ♂♂, 3 copepodites. Tower: Bahía Darwin (V,1; 22.1.73); 3 ♀♀, 1 ♂, 7 copepodites. Bartholomé: Northern Bartholomé (VII,1; 30.4.72); 18 ♀♀, 15 ♂♂, 4 copepodites. Santa Cruz: Bahía Academy (IX,6; monthly transects from March 1972 to February 1973); 318 ♀♀, 36 ovigerous ♀♀, 318 ♂♂, 174 copepodites. Barrington: Exposed beach (XI,1; 1.4.72); 5 ♀♀, 3 ovigerous ♀♀, 11 ♂♂, 15 copepodites. Floreana: Punta Cormorant (XII,2; 30.5.72); a few animals. Hood: Bahía Gardner (XIV,2; 30.6.72); 81 ♀♀, 9 ovigerous ♀♀, 71 ♂♂, 116 copepodites.

ql: Fernandina: Cabo Hammond (I,1b; 3.11.72); frequent. Fernandina: Cabo Douglas (I,2b; 4.11.72); 20 ♀♀, 14 ovigerous ♀♀, 3 ♂♂, 4 copepodites. Fernandina: Punta Espinosa (I,3; 25.9.72); 1 ♀. Isabela: Bahía Urbina (II,1; 22.10.72); 2 ♀♀. Isabela: Caleta Black (II,3; 23.9.72); 1 ♀. Pinta (III,1; 28.11.72); 2 ♀♀, 5 copepodites. Jervis (VIII,1; 4.12.72); 2 ♀♀. Santa Cruz: Private beach (IX,5d; 17.8.72); 13 ♀♀, 4 ovigerous ♀♀, 14 ♂♂. Floreana: Black Beach (XII,1; 10.2.73); 1 ♂. Floreana: Punta Cormorant (XII,2; 30.5.72); a few animals. San Cristóbal: Bahía Wreck (XIII,1; 20.11.72); 18 ♀♀, 3 ovigerous ♀♀, 7 ♂♂, 6 copepodites. San Cristóbal: Punta Bassa (XIII,2; 17.11.72); 19 ♀♀, 4 ovigerous ♀♀, 7 ♂♂. San Cristóbal: Bahía Tortuga (XIII,3; 19.12.72); 4 ♀♀, 1 ovigerous ♀, 3 ♂♂. Hood: Punta Suarez (XIV,1; 18.11.72); 20 ♀♀, 1 ♂. Hood (XIV); 1 ♂.

Remark: Although (or because) 59 specimens (40 ♀♀, 19 ♂♂) were dissected, no clear conception could be gained about the status of the different beach populations within the Archipelago (see Fig. 3A). It is almost impossible to decide whether it deals with one variable species or a species complex resulting from processes of speciation. The majority of the animals agrees well with the description of the specimens from Costa Rica (MIELKE 1993, 1994; see Figs. 1 and 2). Occasionally, deformed pereiopodal rami, as well as missing or supernumerary setae/spines, are found. These defects are obvious and at best reveal some genetic inconstancy. However, there are individuals which differ in the length relations of their setae, e.g. enp. P.3 ♂ or enp. P.4 ♂ (sometimes corresponding to enp. P.4 ♀ or corresponding to enp. P.3 ♂). Furthermore, an

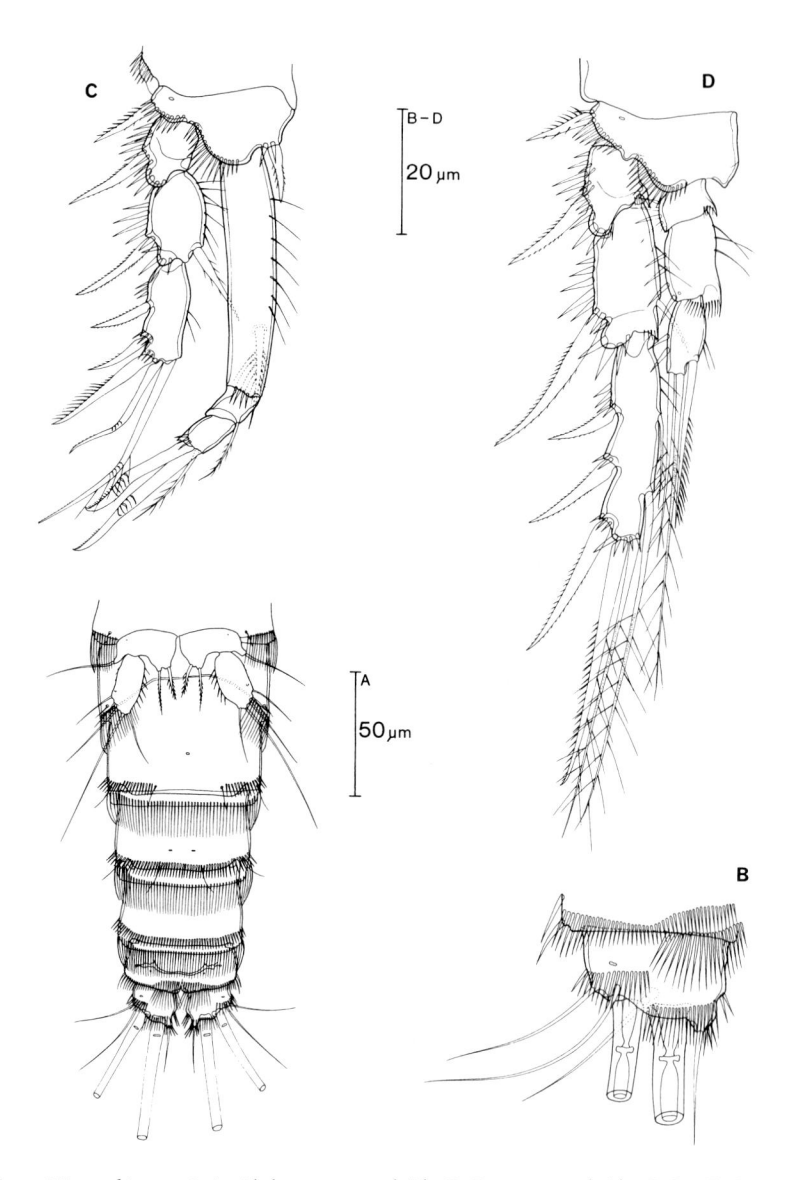

Fig. 1. *Nitocra bisetosa* ♀. A. Abdomen, ventral side. B. Furca, ventral side. C. P.1. D. P.2.

indication of the existence of at least two species is the presence of smaller animals (less than 0.3 mm), which have weaker contours and middle seta/setae on distal segments of enp. P.2 – P.4 that are distinctly shorter (do not reach

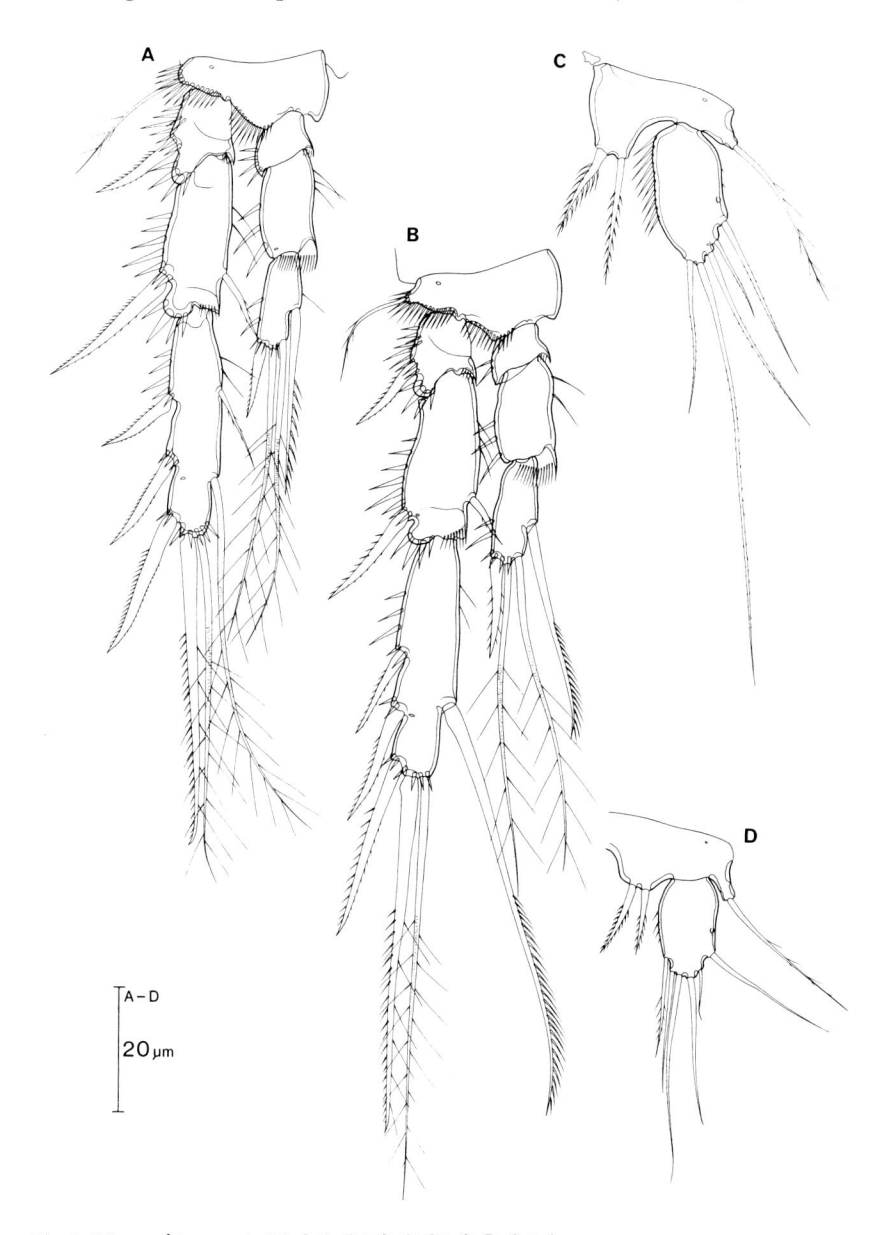

Fig. 2. *Nitocra bisetosa*. A. P.3 ♀. B. P.4 ♀. C. P.5 ♀. D. P.5 ♂.

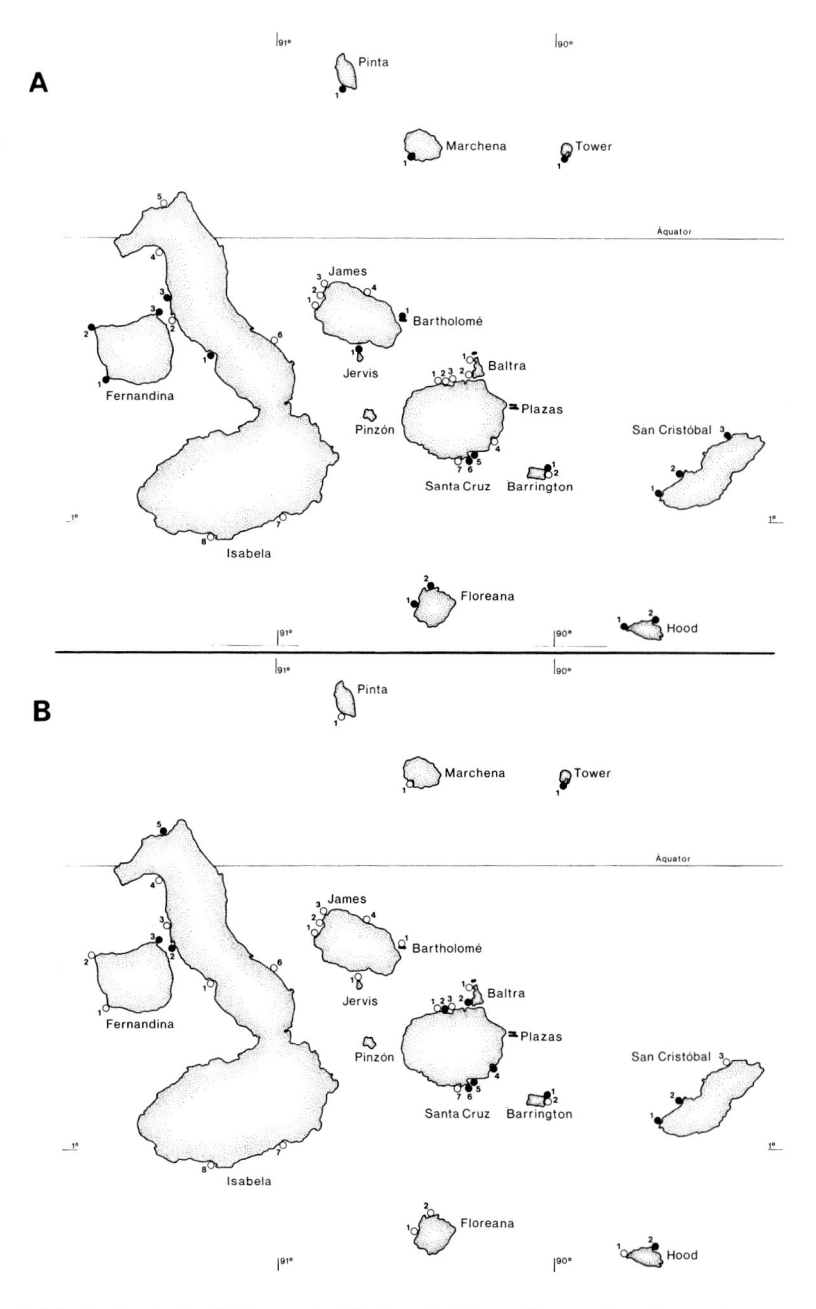

Fig. 3. Distribution in the Galápagos Archipelago. A. *Nitocra bisetosa*. B. *Balucopsylla ? triarticulata*. The circles correspond to the sampling sites in AX & SCHMIDT 1973; black circles mark the localities of the species.

beyond the end of the exopodites) than in the „true" *N. bisetosa*. On the other hand, seta and spine formula, ornamentation of the body, structure of the furca, etc. seemingly correspond in both forms. Occasionally (e.g. Floreana XII,2 and Hood XIV,2), both forms were found in the same sample suggesting the coexistence of two morphologically very similar species. At present, nothing can be said about their microdistribution, life cycles, etc. In any case, this is another example of the difficulties of systematists delimiting species solely on a morphological level. Morphological and taxonomical distance must not correspond. Subpopulations of widely distributed species may be morphologically more different than species which result from recent speciation processes. The superficial controversy between „splitters" and „lumpers" is nonsense. Each case has to be valued and discussed on its own merits.

Nitocra galapagoensis nov. spec.
(Figs. 4 – 8)

Localities and material. **qn**: Santa Cruz: Bahía Academy (IX,6; **Locus typicus**. Monthly transects from March 1972 to February 1973); 174 ♀♀, 142 ovigerous ♀♀, 264 ♂♂, 367 copepodites. Fernandina: Cabo Douglas (I,2; 27.9.72); 5 ♀♀, 5 ♂♂, 6 copepodites. Tower: Bahía Darwin (V,1; 22.1.73); 1 ♀, 3 ♂♂, 3 copepodites. Barrington: Exposed beach (XI,1; 1.4.72); 22 ♀♀, 1 ovigerous ♀, 24 ♂♂, 16 copepodites.

ql: Fernandina: Cabo Douglas (I,2b; 4.11.72); 2 ♀♀, 4 ovigerous ♀♀, 12 ♂♂, 2 copepodites. Fernandina: Punta Espinosa (I,3; 21.2.72); 11 ♀♀, 12 ovigerous ♀♀, 6 ♂♂. Fernandina: Punta Espinosa (I,3; 25.9.72); 1 ♀, 1 ovigerous ♀, 8 ♂♂. Isabela: Bahía Urbina (II,1; 22.10.72); 5 ♀♀, 4 ovigerous ♀♀, 8 ♂♂. Isabela: South Tagus (II,2; 25.10.72); 16 ♀♀, 26 ovigerous ♀♀, 18 ♂♂, 20 copepodites. Isabela: Caleta Black (II,3; 23.9.72); 1 ♀, 1 ovigerous ♀, 2 ♂♂. Isabela: Puerto Bravo (II,4; 25.10.72); 2 ♀♀, 2 ♂♂, 10 copepodites. Pinta (III,1); 3 ♀♀, 9 ovigerous ♀♀, 10 ♂♂. Tower (V,1a; 14.3.72); 1 ♀, 1 ♂. Santa Cruz: North side, beach 1 (IX,1; 28.3.72); 20 ♀♀, 35 ovigerous ♀♀, 35 ♂♂. Santa Cruz: Bahía Academy (IX,6a; 15.2.72); 2 ♀♀, 9 ovigerous ♀♀, 2 ♂♂. Baltra: Northern harbour (X,2b; 30.3.72); 3 ♀♀, 14 ovigerous ♀♀, 7 ♂♂. Floreana: Black Beach (XII,1; 24.6.72); 2 ♂♂. San Cristóbal: Punta Bassa (XIII,2; 17.11.72); 3 ♀♀, 1 ovigerous ♀, 2 ♂♂.

43 specimens (26 ♀♀, 17 ♂♂) were dissected. Holotype female (partitioned upon 11 slides), reg. no. II Gal 73a-k. Paratypes are 4 ♀♀ and 3 ♂♂, reg. no. I Gal 728 – 734. 1st and 2nd antennae, mandible, maxilliped, P.1 - P.5 are drawn from holotype.

Description

Female: Body length from tip of rostrum to end of furcal rami 0.27 – 0.41 mm (holotype 0.33 mm). Rostrum slender, hyaline, extending to end of first segment of 1st antenna. Genital double-somite with a dorsolateral line of subdivision, which is accompanied by a row of spinules. The hyaline frill of genital double-somite and the two following abdominal somites either striated or divided into single, very slender lobes. Proximal to the hyaline frill, the abdominal somites are encircled by a row of spinules. These rows are interrupted ventrolaterally in the genital double-somite and in the preanal somite. Ventrally the spinules are shorter than laterally and dorsally. Dorsally, the length of the spinules is non-uniform. The anal somite, which is likewise furnished with an encircling row of spinules, covers the proximal part of the furcal rami

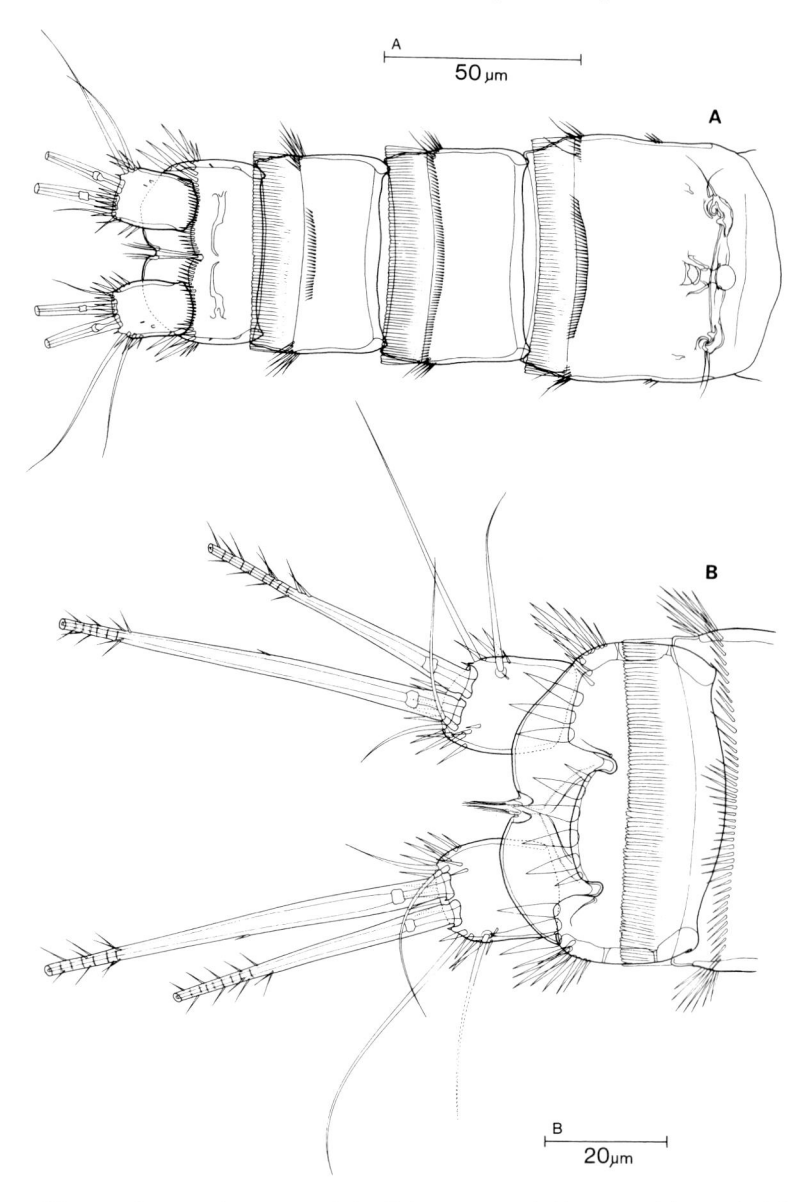

Fig. 4. *Nitocra galapagoensis* nov. spec. ♀. A. Abdomen, ventral side. B. Caudal part of Abdomen, dorsal side.

dorsally. Anal operculum with several coarse spinules and a fine setule near its lateral notch. Furcal rami longer than broad. Laterally with 2 slender setae, apically 2 plumose setae and an inner short one. On dorsal inner part a seta in-

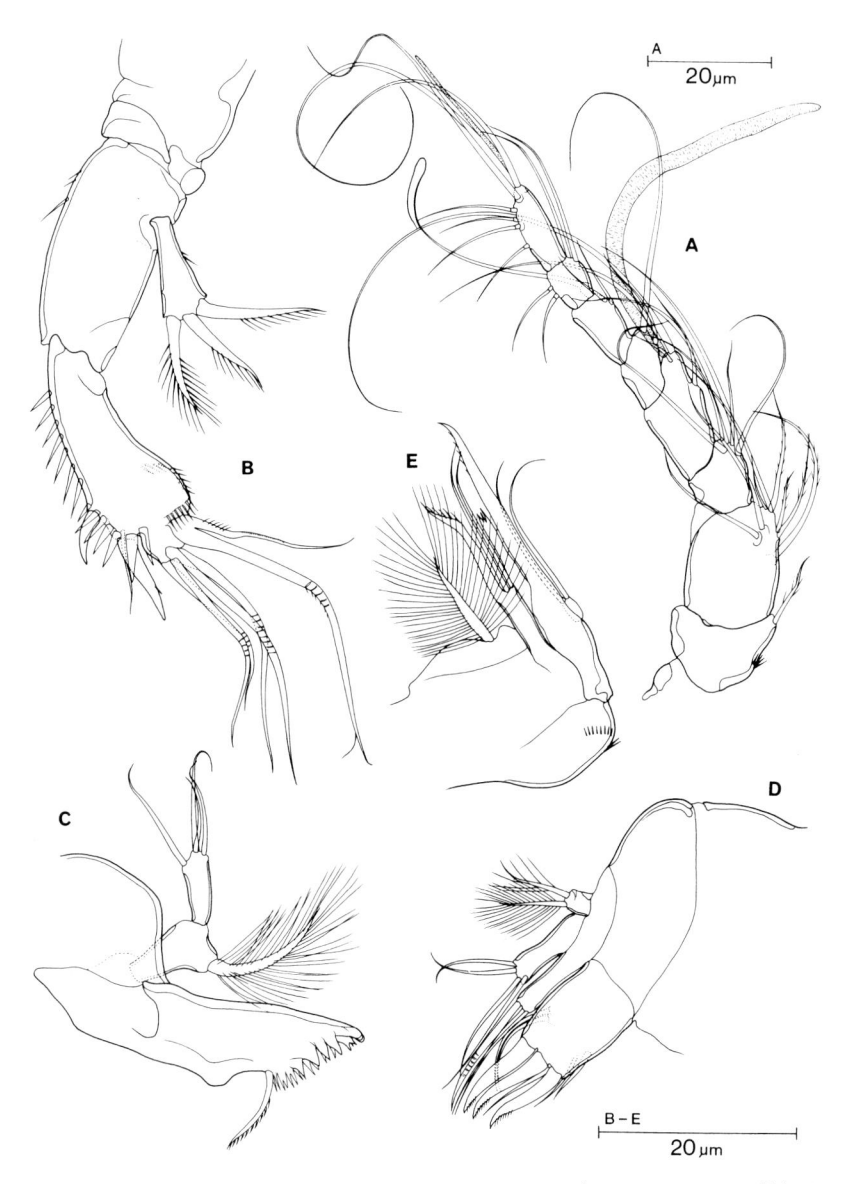

Fig. 5. *Nitocra galapagoensis* nov. spec. ♀. A. 1st Antenna. B. 2nd Antenna. C. Mandible. D. 1st Maxilla. E. 2nd Maxilla.

serts which is bipartite at its base. Outer margin and distal inner section of furca furnished with spinules. Egg sac plain, consisting of 6 – 14 eggs (Figs. 4A,B).

1st Antenna (Fig. 5A): 8 segments. 4th and last segment each with an aesthetasc.

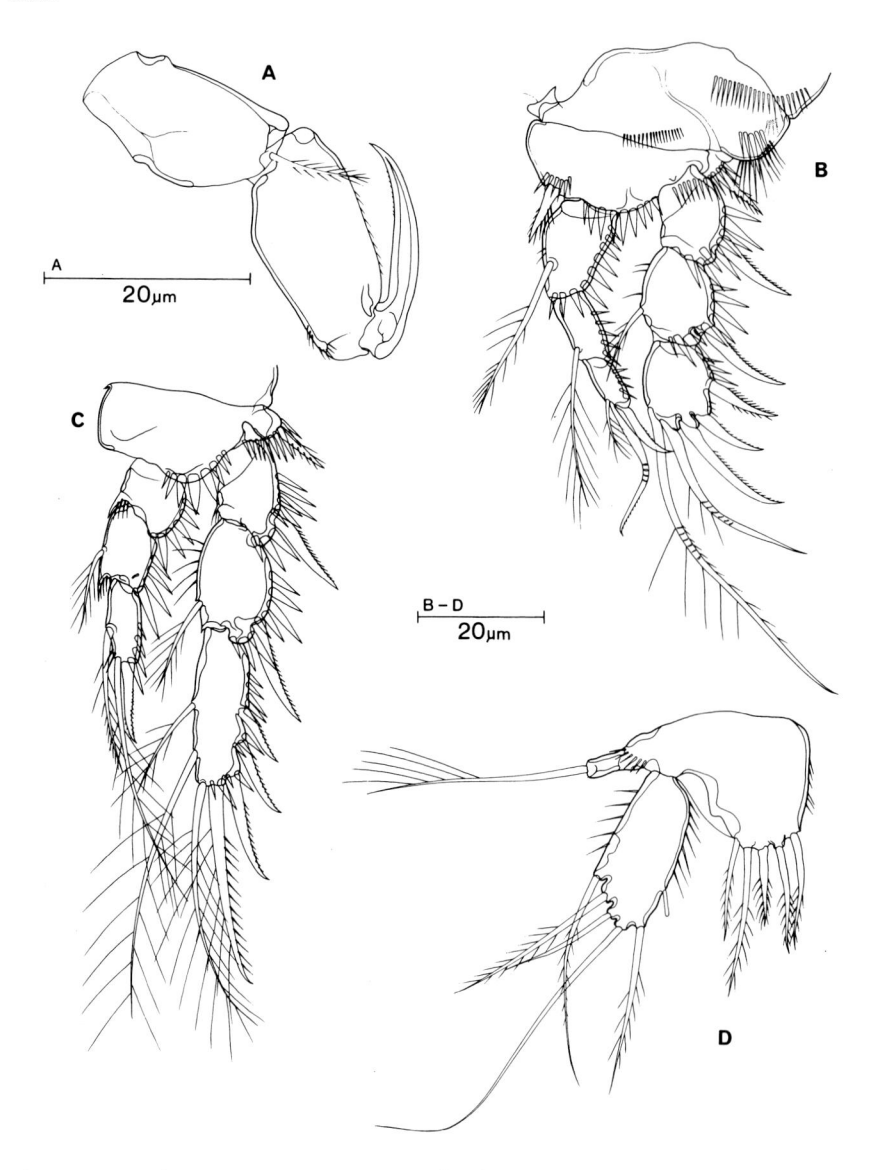

Fig. 6. *Nitocra galapagoensis* nov. spec. ! A. Maxilliped. B. P.1. C. P.2. D. P.5.

2nd Antenna (Fig. 5B): Allobasis with some spinules on inner margin. Free endopodite segment with a row of stout spinules, 2 spines and a hair-like seta on anterior margin. Apically 5 geniculate setae insert, longest one basally fused with a slender seta. Exopodite 1-segmented, furnished with 3 strong setae.

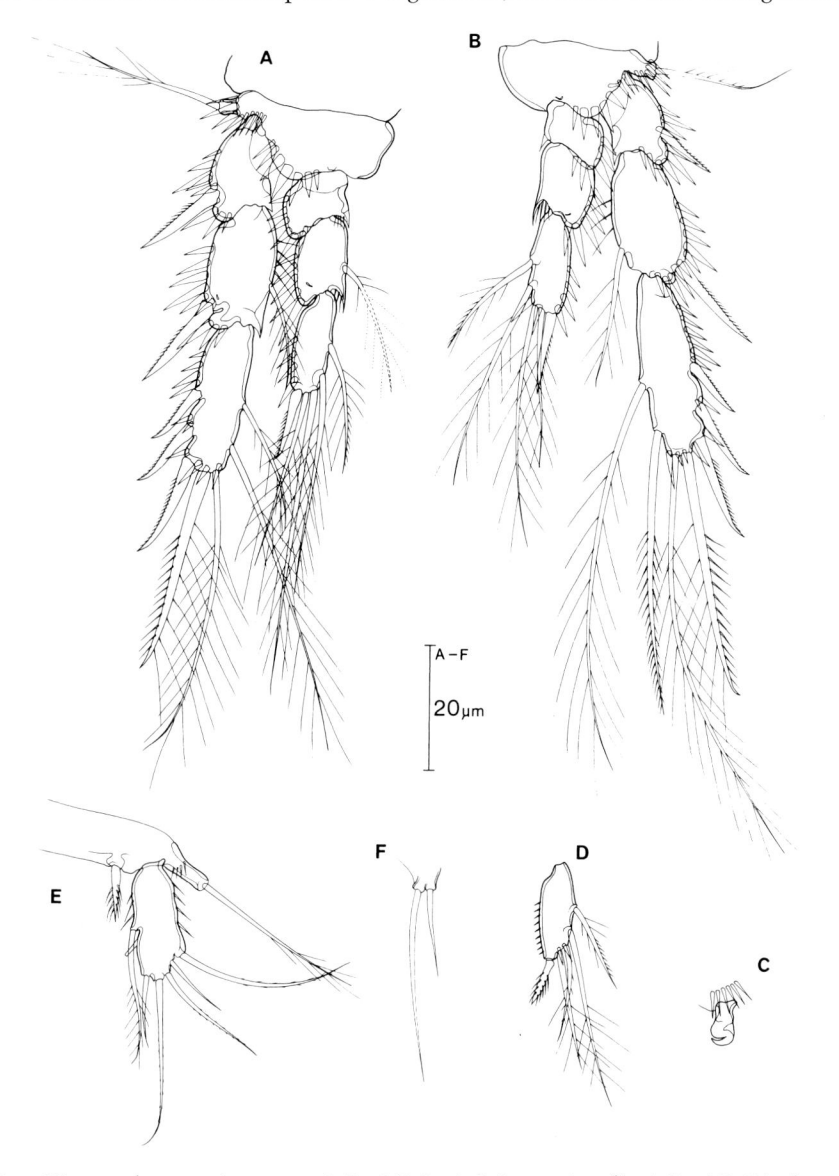

Fig. 7. *Nitocra galapagoensis* nov. spec. A. P.3 ♀. B. P.4 ♀. C. Inner spine of basis P.1 ♂. D. Distal segment of enp. P.3 ♂. E. P.5 ♂. F. P.6 ♂.

Mandible (Fig. 5C): Chewing edge of precoxa (coxa) with several teeth and a slightly bent seta. Palp 2-segmented. Basal segment with a hyaline plumose seta. Distal segment has 4 slender setae.

1st Maxilla (Fig. 5D): Arthrite of precoxa bears 3 spines and a seta on distal margin, 2 setae on inner side and 2 slender setae on surface. Coxa has 3 setae, basis with 4 setae (2 of them representing the endopodite?). Exopodite 1-segmented, carries 2 plumose setae.

2nd Maxilla (Fig. 5E): Syncoxa with 1 hyaline plumose seta and an endite, which bears 2 slender setae and a strong, distally bent appendage. Basis (allobasis) has a claw and a strong, terminally forked appendage. Endopodite 1-segmented, furnished with 2 slender setae.

Maxilliped (Fig. 6A): Basis (syncoxa) with 1 plumose seta; endopodite has 1 strong claw.

P.1 (Fig. 6B): Coxa with several rows of spinules. Basis with an inner and an outer spine, their insertion points set with spinules; distal margin likewise with spinules. Exopodite 3-segmented. Basal and middle segment with an outer spine and coarse spinules on outer margin; middle segment with an additional inner seta. Distal segment with 5 appendages and some spinules. Endopodite with 3 segments, basal segment slightly longer than middle one. Outer margin of segments spinulose. Basal and middle segment each with an inner plumose seta; distal segment bears 3 apical appendages.

P.2 – P.4 (Figs. 6C, 7A–B): Basis with an outer seta which is short in P.2; spinules extend near its insertion point and on distal margin. Exopodite 3-segmented. All segments are furnished with coarse spinules on outer edge. Basal and middle segment each with an outer spine; middle segment with an additional plumose seta on inner margin. Distal segment with 3 outer spines, 2 apical appendages (outer one comb-like) and 2 inner setae: in P.2 and P.3 proximal one shorter; in P.4 distal one shorter and densely pinnate. Endopodite 3-segmented, outer margins with coarse spinules. Basal segment short and asetose. Middle segment with an inner seta in P.2, occasionally with an inner seta in P.3 and without a seta in P.4. Distal segment with 3 appendages in P.2, and 5 appendages in P.3 and P.4.

Seta and spine formula:

	Exopodite	Endopodite
P.2	(0.1.223)	(0.1.120)
P.3	(0.1.223)	(0.0-1.221)
P.4	(0.1.223)	(0.0.221)

P.5 (Fig. 6D): Baseoendopodite with 1 plumose outer seta and 5 plumose setae on inner lobe. Inner and outer margin of exopodite with slender spinules;

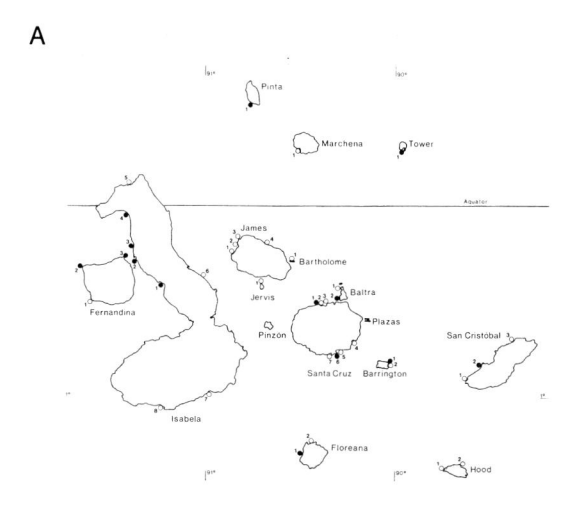

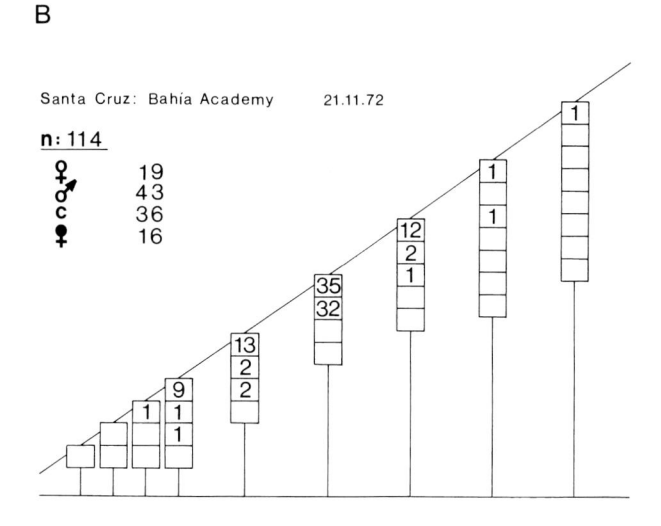

Fig. 8. *Nitocra galapagoensis* nov. spec. A. Distribution in the Galápagos Archipelago. The circles correspond to the sampling sites in Ax & Schmidt 1973; black circles mark the localities. B. Distribution in the beach of the Bahía Academy, Santa Cruz (n = number of animals on 21.11.72; ♀ = females, ♂ = males, c = copepodites, ♀ with black circles = ovigerous females; 0 – 15m = distance from the water line; rectangles = number of animals per 50 ccm).

distal margin with 5 slender setae. Furthermore, a short hyaline structure is to be seen on inner distal margin.

Male: Differs from the female in the following respects:
- Body length 0.25 – 0.36 mm.
- Abdominal somites with a stronger spinulation. In particular, the dorsal spinules and the ones inserting ventrally on anal somite between both furcal rami are distinctly longer than in female.
- 1st Antenna haplocer.
- Inner spine of basis P.1 modified (Fig. 7C).
- Length ratios of the setae of distal segment enp. P.3 different; furthermore, a row of short spinules extends on outer margin (Fig. 7D).
- Inner part of benp. P.5 with only 1 short plumose seta. Exopodite equipped with 6 setae; near insertion point of inner seta a blunt, hyaline structure can be observed (Fig. 7E).
- P.6 with 2 setae, outer one distinctly shorter than inner one (Fig. 7F).

Variability. The length ratios of the setae are somewhat variable. Occasionally, a ramus of a pereiopod is deformed. Middle segment of enp. P.3 with or without an inner seta. One and the same specimen may exhibit this difference. One distal segment of enp. P.4 of a female carries 6 appendages; its counterpart has 5 as usual. One benp. P.5 of a female has only 4 setae instead of the usual 5. Occasionally, the inner seta of exp. P.5 ♂ may be absent. The spinulation of anal operculum and lateral continuation is somewhat variable.

Distribution. *Nitocra galapagoensis* nov. spec. is widely distributed within the archipelago (Fig. 8A). The species occurs above all in the central area of the sandy beaches (Fig. 8B).

Etymology. The species name refers to the broad distribution within the Galápagos Archipelago.

Discussion. The new species is distinguished from the other members of this extensive genus by a unique combination of several features: (a) 2nd antenna with allobasis (other species according to text/drawings: *N. arctolongus* Shen & Tai, 1973, *N. reducta reducta* Schäfer, 1936, *N. reducta fluviatilis* Galhano, 1968, and the species incertae sedis *N. gracilimana* Giesbrecht, 1902); (b) P.1 enp. with very short basal segment (minority of *Nitocra*-species); (c) the specific seta and spine formula of P.2 – P.4; (d) P.5 ♀ exp. only with 5 setae (minority of *Nitocra*-species); P.5 ♂ benp. only with 1 seta (other species: *N. phreatica* Bozic, 1964). Closest relationship seems to exist with the cosmopolitan *N. lacustris*, to which five subspecies are currently attributed (*l. lacustris*

Schmankewitsch, 1875, *l. azorica* Kunz, 1983, *l. colombianus* Reid, 1988, l. *pazificus* Yeatman, 1983, *l. sinoi* Marcus & Por, 1961). (Sub-)populations of this species/species group occur in different freshwater, brackish, marine and high salinity water. Probably the population of *N. galapagoensis* nov. spec. can be traced back to passively transported specimens of *N. lacustris* s. l.

R e m a r k. Both *Nitocra*-species are no more closely related to each other than to species from other geographical regions. Therefore, it must be postulated that the ancestors of both species immigrated independently into the Galápagos area. However, after their dispersal within the archipelago, morphological alterations are obviously in the process of development.

Stenocopia Sars, 1907
Stenocopia sarsi nov. spec.

(Figs. 9 – 15)

L o c a l i t i e s a n d m a t e r i a l. **qn**: Santa Cruz: Bahía Academy (IX,6; **Locus typicus**. Monthly transects from March 1972 to February 1973); 62 ♀♀, 29 ovigerous ♀♀, 90 ♂♂, copepodites (number?; difficult to separate from those of *Stenocopia limicola*). Tower: Bahía Darwin (V,1; 22.1.73); 1 ♀.
11 specimens (7 ♀♀, 4 ♂♂) were dissected. Holotype female (partitioned upon 12 slides), reg. no. II Gal 81 a-l. Paratypes are 5 ♀♀ and 4 ♂♂, reg. no. I Gal 1090 – 1098. With the exception of habitus dorsal, caudal end dorsal and P.2 all drawings of the female are from holotype.

D e s c r i p t i o n

F e m a l e: Body length of dissected females from tip of rostrum to end of furcal rami 0.64 – 0.75 mm (holotype 0.68 mm). Rostrum broad, distal edge with 2 setules and an adjacent outer tooth, respectively; edge between setules slightly acute (Fig. 11A). Caudal dorsal margin of cephalothorax, pereiomeres and abdominal somites (except anal somite) with denticles, which are proximally accompanied by arched structures. Surface of cephalothorax and dorsal caudal part of somites (except penultimate and ultimate somite) furnished with fine setules. Genital double-somite laterally armed with spinules; dorsolaterally subdivided. Ventral caudal margin of genital double-somite and following two somites indented. Anal somite ventrally with spinules. Somite which follows genital double-somite has a subapical row of spinules; penultimate somite with a median row of spinules. Anal operculum seemingly bare but notched, laterally with a slender seta. Furcal rami about 5 – 6 times longer than broad. Proximally on outer margin a pore (?) can be seen. Subapically on outer margin a slender seta and a short setule arise; proximal to them on dorsal side another slender seta inserts; it is accompanied by a short seta. Subapi-

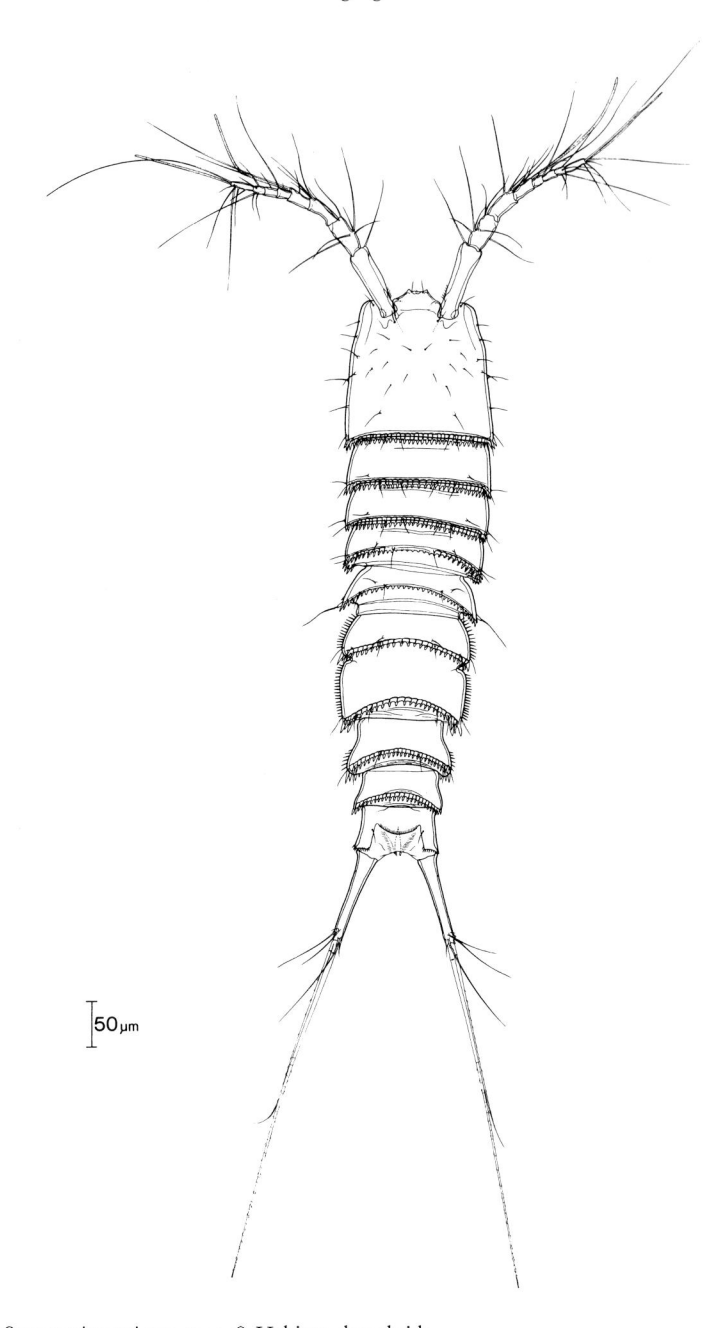

Fig. 9. *Stenocopia sarsi* nov. spec. ♀. Habitus, dorsal side.

cally on dorsal side is a long, slender seta which is bipartite at base. Apical edge with a short inner seta and 2 terminal hairy setae, the more interior of which is about 3 times longer than outer one (Figs. 9, 10A-B).

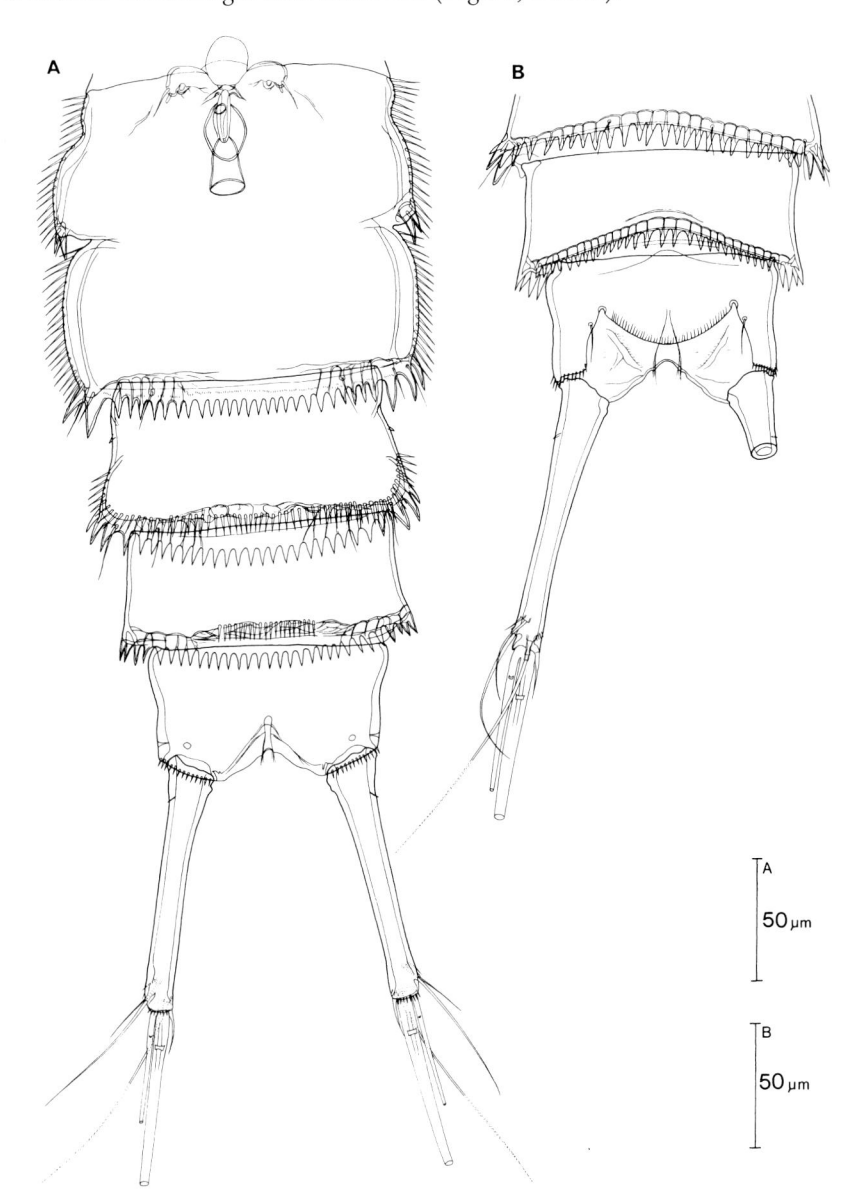

Fig. 10. *Stenocopia sarsi* nov. spec. ♀. A. Abdomen, ventral side. B. Caudal part of Abdomen, dorsal side.

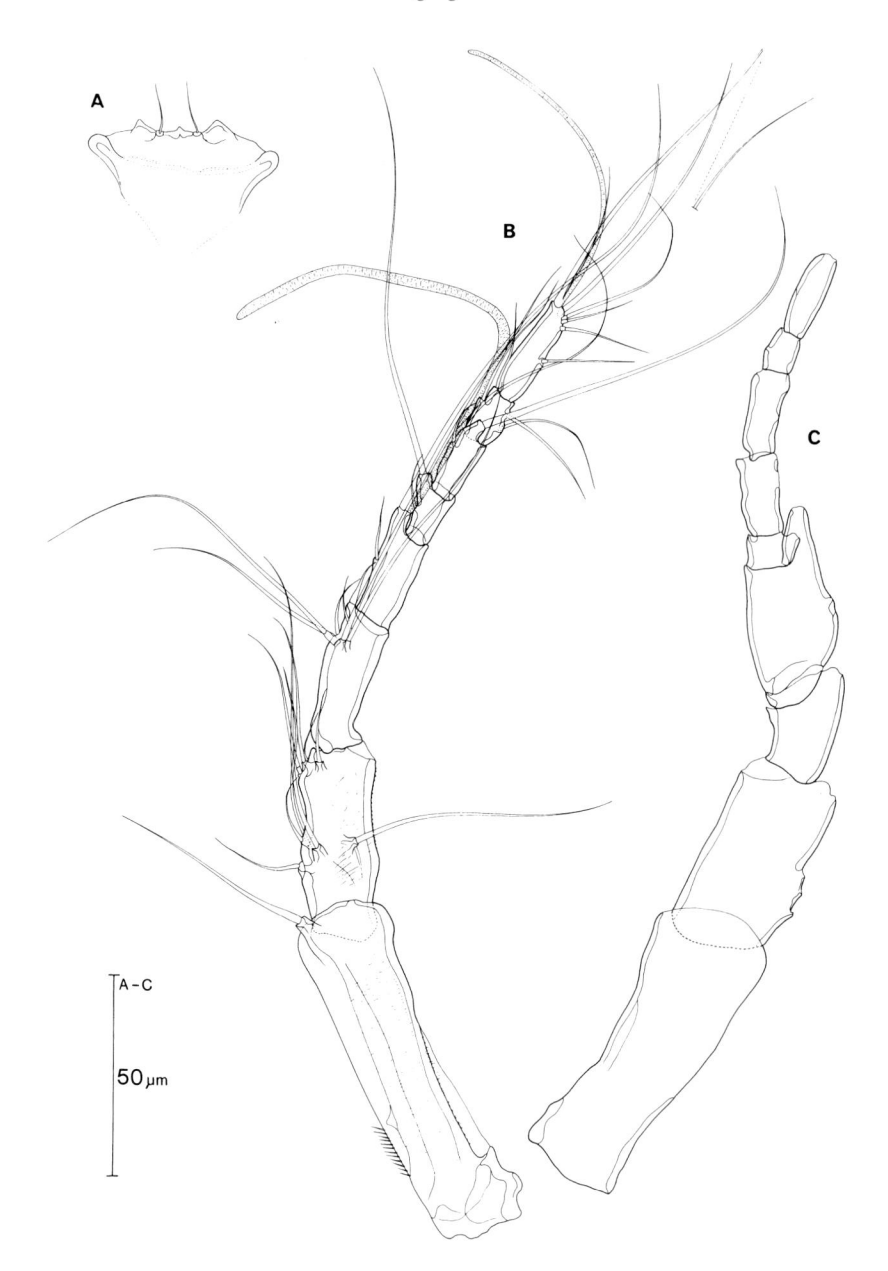

Fig. 11. *Stenocopia sarsi* nov. spec. A. Rostrum ♀. B. 1st Antenna ♀. C. 1st Antenna ♂ (setae omitted).

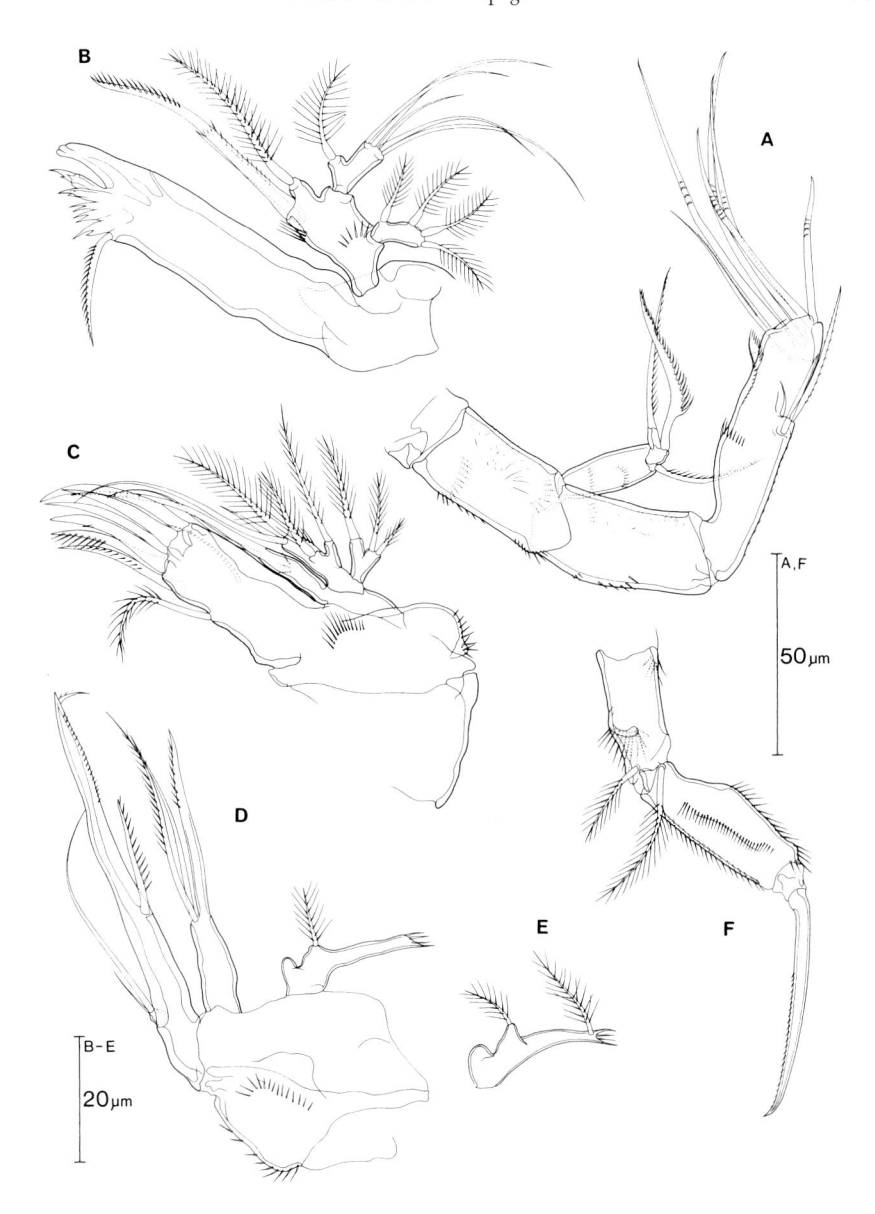

Fig. 12. *Stenocopia sarsi* nov. spec. ♀. A. 2nd Antenna. B. Mandible. C. 1st Maxilla. D. 2nd Maxilla. E. Proximal endite of 2nd Maxilla, other specimen. F. Maxilliped.

1st Antenna (Fig. 11B): 8 segments, last but one segment may have a line which suggests a former subdivision. 1st segment elongated. 4th and last segment each with an aesthetasc.

2nd Antenna (Fig. 12A): Basis with several spinules. Endopodite 2-segmented; 2nd segment subapically with 3 setae of different length and apically with 5 geniculate setae, the outermost of which is basally fused with one slender seta. Exopodite 2-segmented. Basal segment with one seta, distal segment short bearing 1 slightly curved seta and another seta sitting on a socle (3rd segment?).

Mandible (Fig. 12B): Chewing edge of precoxa (coxa) with several teeth and a seta. Coxa-basis (basis) with several spinules, 2 strong appendages and 1 plumose seta with weak contours. Endopodite has 1 plumose seta laterally and 5 slender setae apically. Exopodite bears 4 plumose setae.

1st Maxilla (Fig. 12C): Arthrite of precoxa with 6 spines apically, 1 plumose seta laterally and 1 seta on surface. Coxa bears 2 setae; basis with 3 setae of different lengths. Endopodite apparently fused with basis, has 2 plumose setae. Exopodite 1-segmented, with 3 plumose setae.

2nd Maxilla (Figs. 12D-E): Syncoxa with 2 endites. Proximal endite irregularly shaped, with 1 hyaline plumose seta and a strong appendage which is fused with the endite. Occasionally, this appendage is also furnished with a plumose seta (either a variable state or this seta is easily lost). Distal endite with 1 strong seta which is basally fused with the endite and 2 slender setae. Basis (allobasis) with 1 strong claw which terminates in a short and a longer, bent piece, and 1 seta which is half as long as the claw. Endopodite 1-segmented, bearing 2 setae of different length.

Maxilliped (Fig. 12F, somewhat distorted): Basis (syncoxa) with long spinules and 2 plumose setae. Proximal endopodite segment (basis) with 2 rows of spinules near inner margin and 1 row of spinules on dorsal margin. Distal endopodite segment (endopodite) carries 1 claw which has spinules on inner margin.

P.1 (Fig. 13A): Coxa with long, slender spinules. Basis with an inner and an outer plumose seta. Exopodite with 3 segments of about the same length. Middle segment with a seta on inner margin. Distal segment with 5 setae/spines, 2 inner setae geniculate. Endopodite 3-segmented. Basal segment longer than exp., inner and outer margins with fine setules, subapically with an inner plumose seta. Middle segment short, with an inner seta. Distal segment about twice as long as middle one, with 1 slender seta and 2 claw-like appendages.

P.2 – P.4 (Figs. 13C, 14A-B): Coxa with a few rows of spinules. Basis with an outer seta. Exopodite 3-segmented. All segments with spinules on outer margin; proximal segment of P.2 and P.3 has an additional row of spinules on caudal surface (not drawn). Proximal and middle segments each with an inner plumose seta. Distal segment with 3 slender outer spines and 2 apical appendages (1 spine-, 1 seta-like); inner margin with 2 plumose setae in P.2, 3 plumose setae in P.3 and 3 setae in P.4 (proximal and distal ones short, middle one elongated). Endopodite 3-segmented. Proximal segment short, carries 1 inner

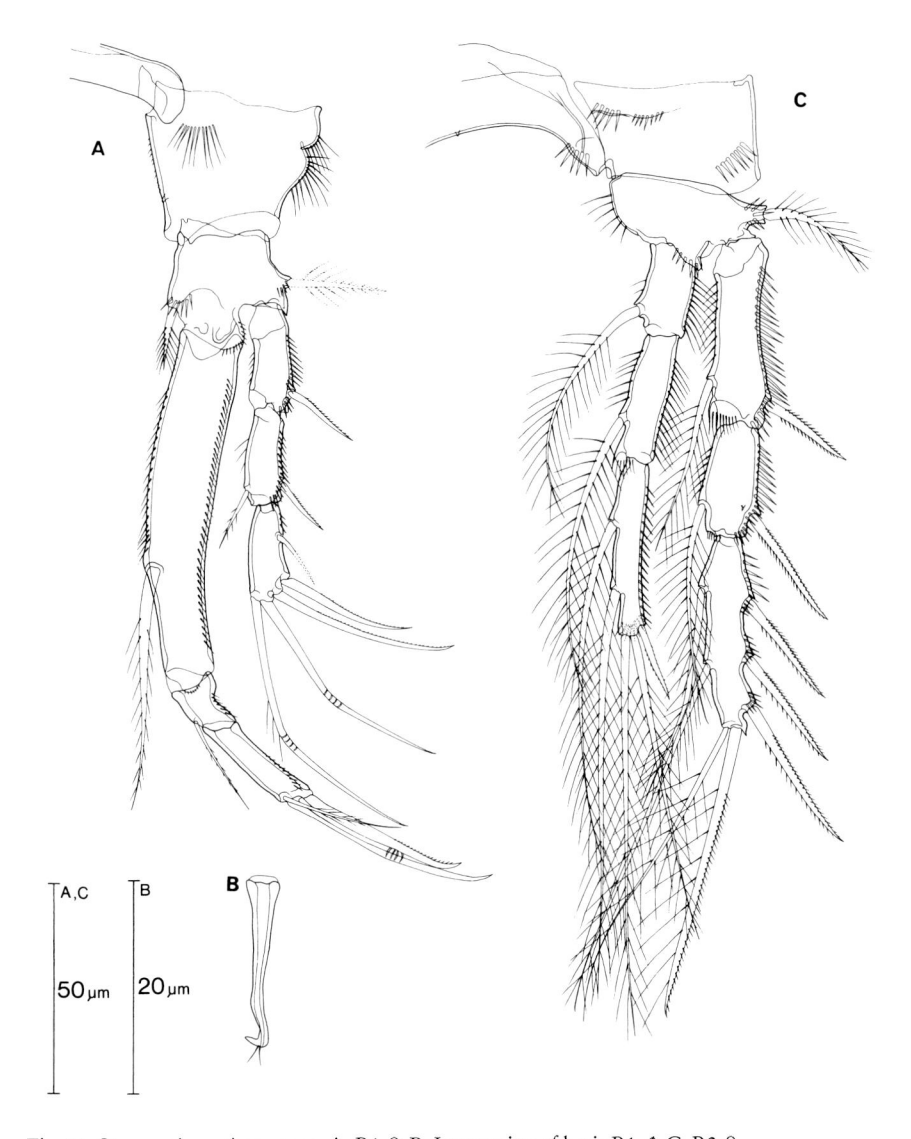

Fig. 13. *Stenocopia sarsi* nov. spec. A. P.1 ♀. B. Inner spine of basis P.1 ♂. C. P.2 ♀.

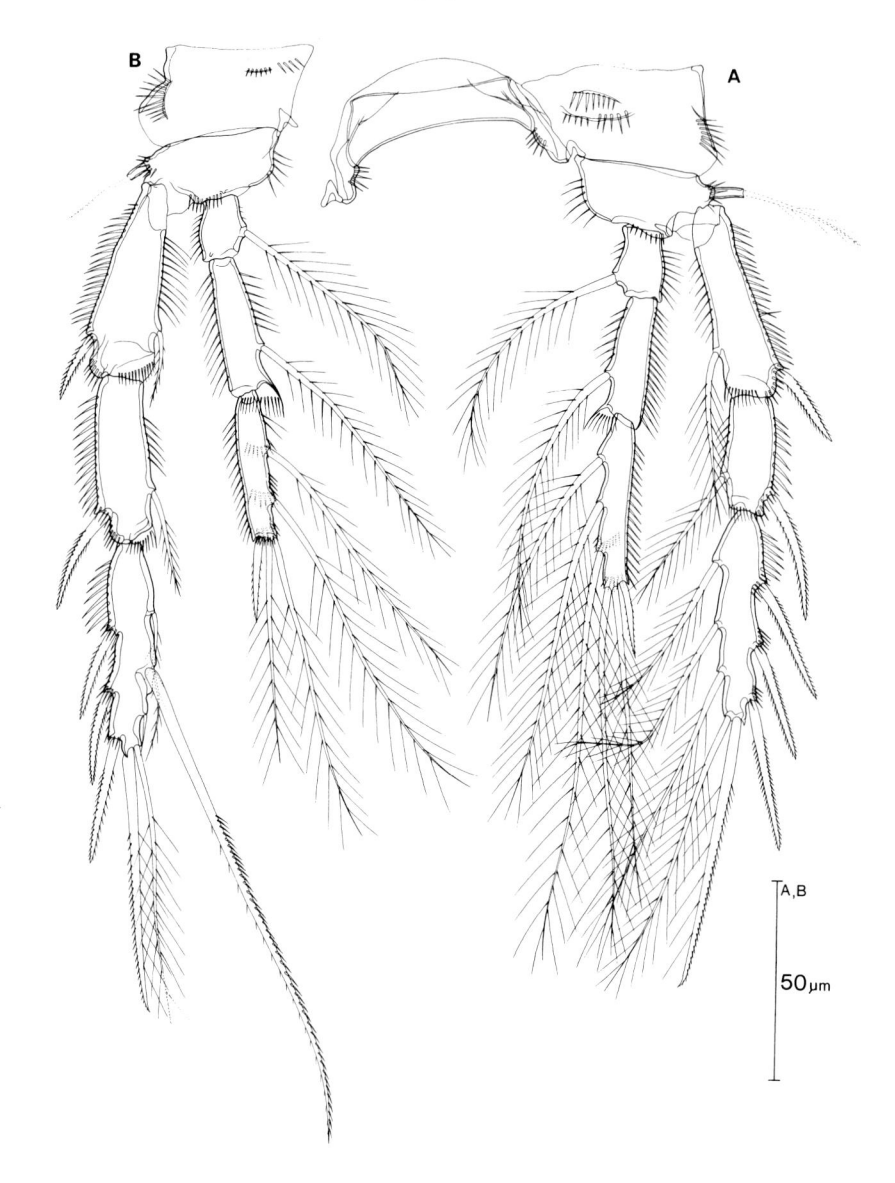

Fig. 14. *Stenocopia sarsi* nov. spec. ♀. A. P.3. B. P.4.

plumose seta. Middle segment with a tooth-like projection on distal inner edge, also furnished with 1 inner plumose seta. Distal segment with 5 appendages in P.2 and P.4 and 6 ones in P.3.

Seta and spine formula:

	Exopodite	Endopodite
P.2	(1.1.223)	(1.1.221)
P.3	(1.1.323)	(1.1.321)
P.4	(1.1.323)	(1.1.221)

P.5 (Fig.15A): Baseoendopodite with an outer seta standing on a socle, inner part spinulose on inner margin, with 5 plumose setae. Outer margin of exopodite with coarse spinules; inner margin with slender ones. Distal edge with 2 short, 2 plumose setae and a fifth seta which inserts on a long bulge.

Male: Differs from the female in the following respects:
- Body length of dissected ♂♂: 0.53 – 0.60 mm.
- 1st Antenna haplocer (Fig. 11C).
- Inner appendage of basis P.1 modified (Fig. 13B).
- Baseoendopodites of both P.5 fused. Outer seta inserts on a socle. Distal margin with 1 stout inner seta and 2 rudimentary setae. Exopodite with 2 slender setae on distal outer margin, 1 slender distal seta and a short setule nearby. All four dissected males possess a long, basally slightly bulbous, seta on the inner margin of the left exopodite. Right exopodite has a small setule (may be lacking) at the corresponding location. P.6 a transverse plate, furnished with 2 slender setae on distal outer part; proximally on outer margin a minute setule may occur (Fig. 15B).

Variability. Proximal endite of 2nd maxilla with 1 or 2 plumose setae. Segments of endopodites P.2 - P.4 occasionally with additional or fewer setae (e.g. enp. P.2, P.4 with formula 1.2.221 or 1.1.220; enp. P.3 with formula 1.1.221). The situation of P.5 ♂ is somewhat unclear but it appears that the difference between left and right exopodite is obligatory and not variable.

Note: Another ♀ was found at locality IX, 5c (Santa Cruz; 24.5.72), which had the same seta and spine formula as *Stenocopia sarsi* nov. spec. but revealed some morphological differences. The single specimen was slightly damaged. It could not be determined whether it belonged to *S. sarsi* nov. spec. or to a new, closely related species.

Etymology. The new species is dedicated to G. O. Sars (1837 – 1927) for his extraordinary life-work on Crustacea.

Discussion. According to LANG (1948) the taxon *Stenocopia* belongs to the "Subfam." Stenocopiinae, which together with the much more extensive Ameirinae constitutes the Ameiridae. At present *Stenocopia* consists of seven

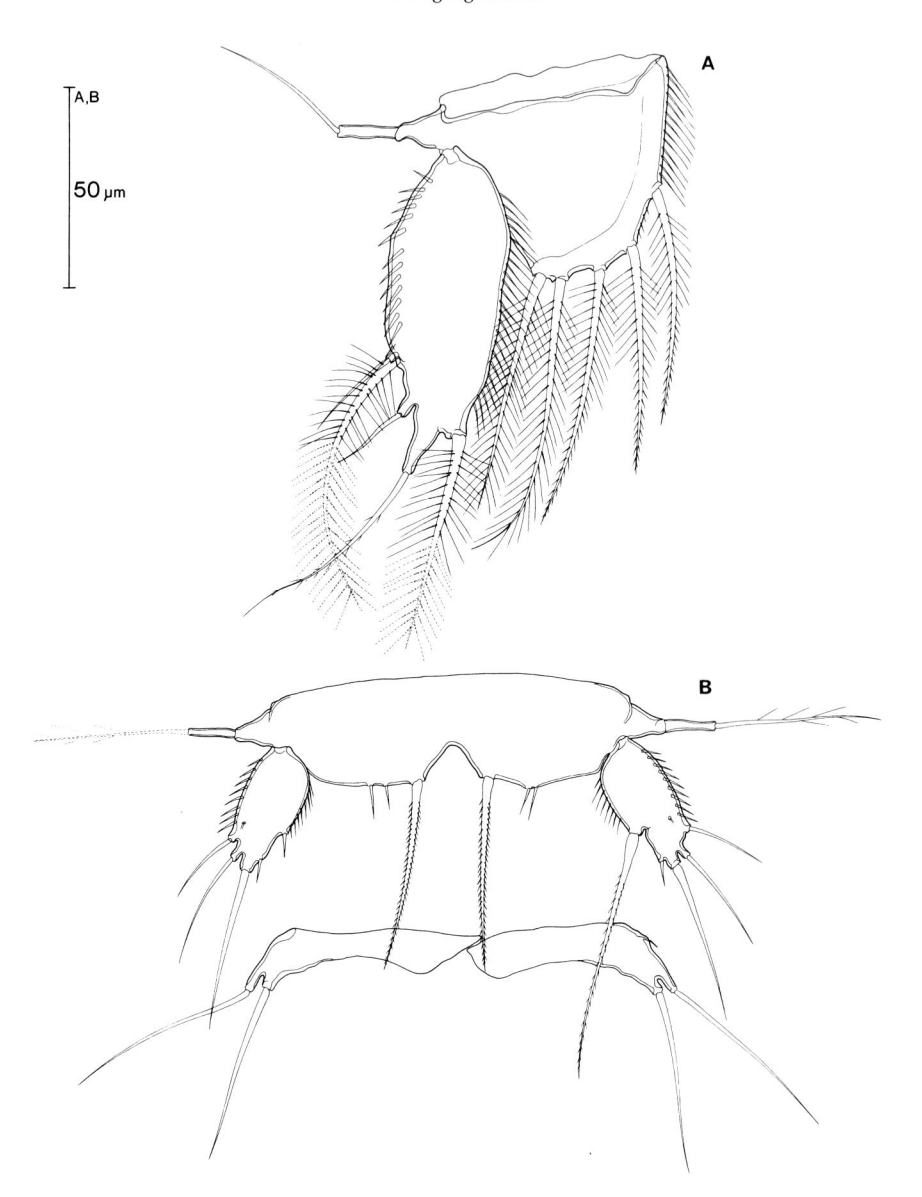

Fig. 15. *Stenocopia sarsi* nov. spec. A. P.5 ♀. B. P.5 and P.6 ♂.

species, which were subdivided by COTTARELLI et al (1985) into two groups on account of the segment number of enp. P.4.

The new species seems to be closely related to *S. longicaudata* (T. Scott, 1892) and *S. spinosa* (T. Scott, 1892) because of the 8-segmented 1st antenna ♀

and the seta and spine formula of P.1 – P.4. Differences between these species and *S. sarsi* nov. spec. exist above all in the length ratios of the inner setae of distal segment P.4 (compare SARS 1907 and 1911), the shape of exp. P.5 ♀ and the setation of P.5 ♂ (as to *S. spinosa* see POR 1964). Furthermore, the length - width ratio of the furca is worth mentioning. In contrast to the characteristic, laterally expanded epimeral plates of cephalothorax and the three following somites of *S. spinosa* in *S. sarsi* nov. spec., these epimeral plates are normal shaped.

The asymmetric setation of left and right P.5 as in the four dissected ♂♂ of Santa Cruz is neither reported for *S. longicaudata* nor for *S. spinosa*.

Stenocopia limicola Willey, 1935

(Figs. 16 – 19)

Localities and material. qn: Santa Cruz: Bahía Academy (IX,6; monthly transects from March 1972 to February 1973); 19 ♀♀, 20 ovigerous ♀♀, 25 ♂♂, copepodites (number?; difficult to separate from those of *Stenocopia sarsi* nov. spec.).

ql: Fernandina: Punta Espinosa (I,3; 25.9.72); 2 ♂♂. Santa Cruz: Bahía Academy (IX,5b; 16.12.72, rockpool); 1 ♀, 1 ovigerous ♀, 1 ♂. Santa Cruz: Bahía Academy (IX,5d; 15.2.72); 1 ♂.

11 specimens (4 ♀♀, 7 ♂♂) were dissected.

Description

Female: Body length of dissected females from tip of rostrum to end of furcal rami 0.48 – 0.55 mm. Rostrum more or less rectangular; distal margin with two vaults, inside accompanied by a slender setule (Fig. 17A). Genital double-somite dorsolaterally subdivided; dividing line indented. Dorsal caudal margin of hyaline frill of genital double-somite and following somite indented, subapically with several slender setules and delicate denticles. Hyaline frill of penultimate somite similar, but weaker and without setules. Margin of anal operculum smooth or slightly notched. At about middle of inner longitudinal margin of anal somite a slender seta inserts; distal margin with several spinules, inner ones coarse. Ventral caudal margin of genital double-somite and following two somites very weakly toothed. Subapically genital double-somite has some lateral spinules; next somite a continuous ventral row of slender spinules; penultimate somite a row of spinules on middle part. Last somite ventrally with some spinules on distal edge. Furcal rami about 4 times longer than broad. Proximal on outer margin a pore (?), on inner margin several fine hairs are to be seen. Subapically on the dorsal side a slender seta and an accompanying short seta are located; more distally the dorsal seta, which is bipartite at base. Subdistal on outer margin another slender seta inserts. Distal

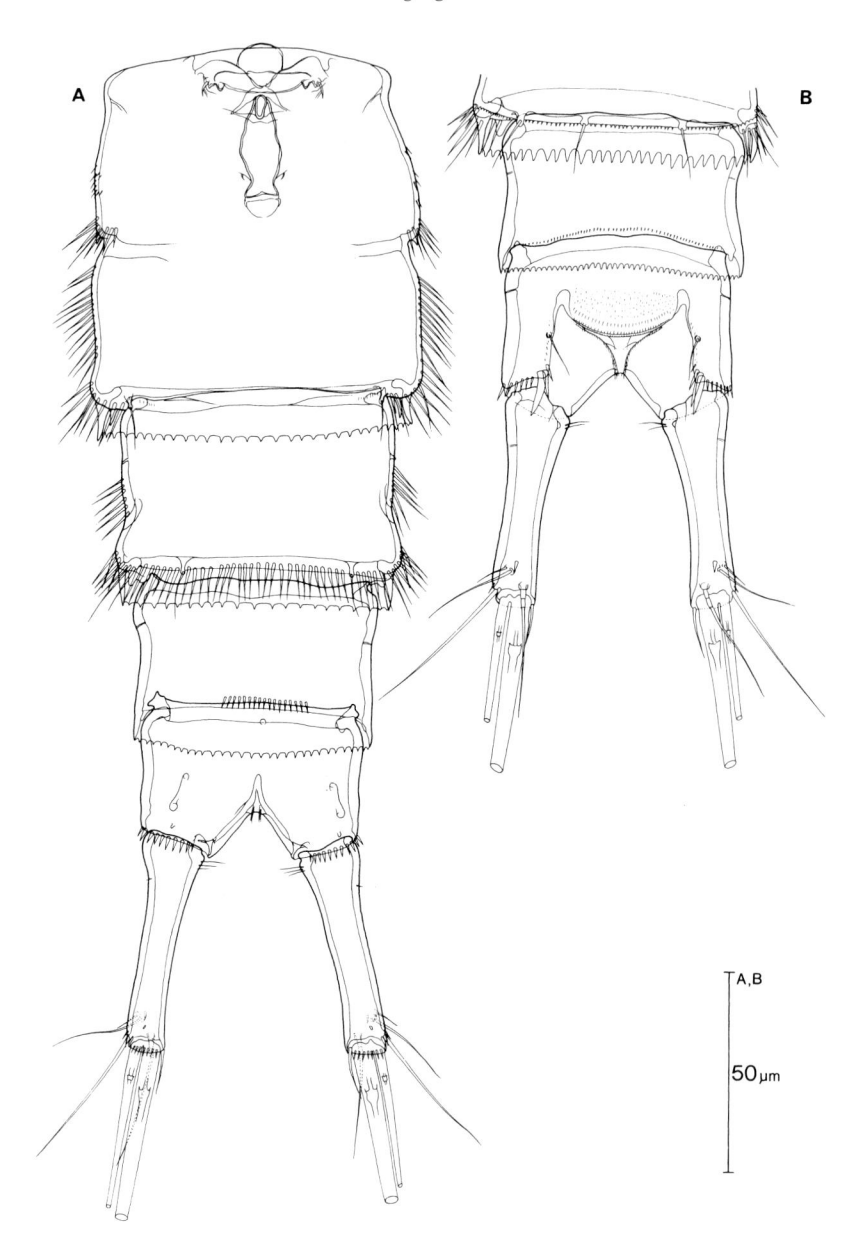

Fig. 16. *Stenocopia limicola* ♀. A. Abdomen, ventral side. B. Caudal part of Abdomen, dorsal side.

Fig. 17. *Stenocopia limicola* ♀. A. Rostrum and 1st Antenna. B. 2nd Antenna. C. Mandible. D. Chewing edge of the other Mandible. E. 1st Maxilla. F. 2nd Maxilla.

margin of furca spinulose; 3 terminal setae arise, inner one short, middle one longest (Figs. 16A-B).

1st Antenna (Fig. 17A): 9 segments. Proximal and following segment elongated. 4th and last segment each with an aesthetasc.

2nd Antenna (Fig. 17B): Basis has some spinules. Endopodite with 2 segments. Distal segment subapically with 3 setae on anterior margin. Terminally 5 geniculate setae insert, outer one basally fused with a slender seta. Exopodite 2-segmented; proximal segment carries an apical seta, distal short segment with 2 setae, posterior one basally swollen.

Mandible (Figs. 17C-D): Precoxa (coxa) with one distal main tooth on the palp side and some smaller prongs. Opposite part of chewing edge with 1 plumose seta and 1 or 2 strong tooth/teeth nearby. Each one of the 3 dissected females (as well as the 3 dissected males) has one gnathobase with 1 tooth, while its counterpart possesses 2 teeth. Proximal part of precoxa has another pointed structure. Coxa-basis (basis) with several spinules and 3 setae; one weakly contoured and plumose. Endopodite with 1 lateral seta and 4 apical setae. Exopodite has 4 short plumose setae.

1st Maxilla (Fig. 17E): Arthrite of precoxa with 6 appendages on distal edge, 2 setae laterally and 2 setae on surface. Coxa carries 2 slender setae and 1 strong appendage. Basis has 3 setae. Endopodite represented by 2 setae. Exopodite 1-segmented, has 3 setae.

2nd Maxilla (Fig. 17F): Syncoxa with 2 endites. Proximal endite with a strong, distally pronged appendage, which is basally fused with the endite and bears 2 short plumose setae. Distal endite with 3 setae. Basis (allobasis) with 1 claw and 1 seta. Endopodite 1-segmented, furnished with 3 setae.

Maxilliped (Fig. 18A): Basis (syncoxa) with several short spinules and a row of long spinules; distally 2 plumose setae insert. Proximal endopodite segment (basis) has two rows of spinules near inner margin, dorsal margin also spinulose. Distal endopodite segment (endopodite) bears 1 claw and 1 accompanying slender seta.

P.1 (Fig. 18B): Coxa with several rows of spinules. Basis with 1 inner and 1 outer plumose appendage; near point of insertion with some spinules; middle part of distal margin protrudes, furnished with a row of spinules. Exopodite 3-segmented. Middle segment has an inner seta; distal segment with 5 appendages. Endopodite 3-segmented. Proximal segment extends beyond exp.; inner and outer margin spinulose. Inner margin with 1 plumose seta somewhat distal from centre. Middle segment shortest, bearing 1 seta. Distal segment has 2 claw-like appendages and 1 seta.

P.2 – P.4 (Figs. 18C-D, 19A): Coxa has a proximal row of spinules and another row near outer part. Basis carries 1 outer seta and some rows of spinules on distal margin. Exopodite 3-segmented. Proximal segment has a slender spine and a row of spinules on outer margin and another one nearby on caudal surface (in P.2 and P.3, not drawn; in P.4 not observed). Inner margin with fine spinules and 1 plumose seta; distal inner part with a fan of spinules.

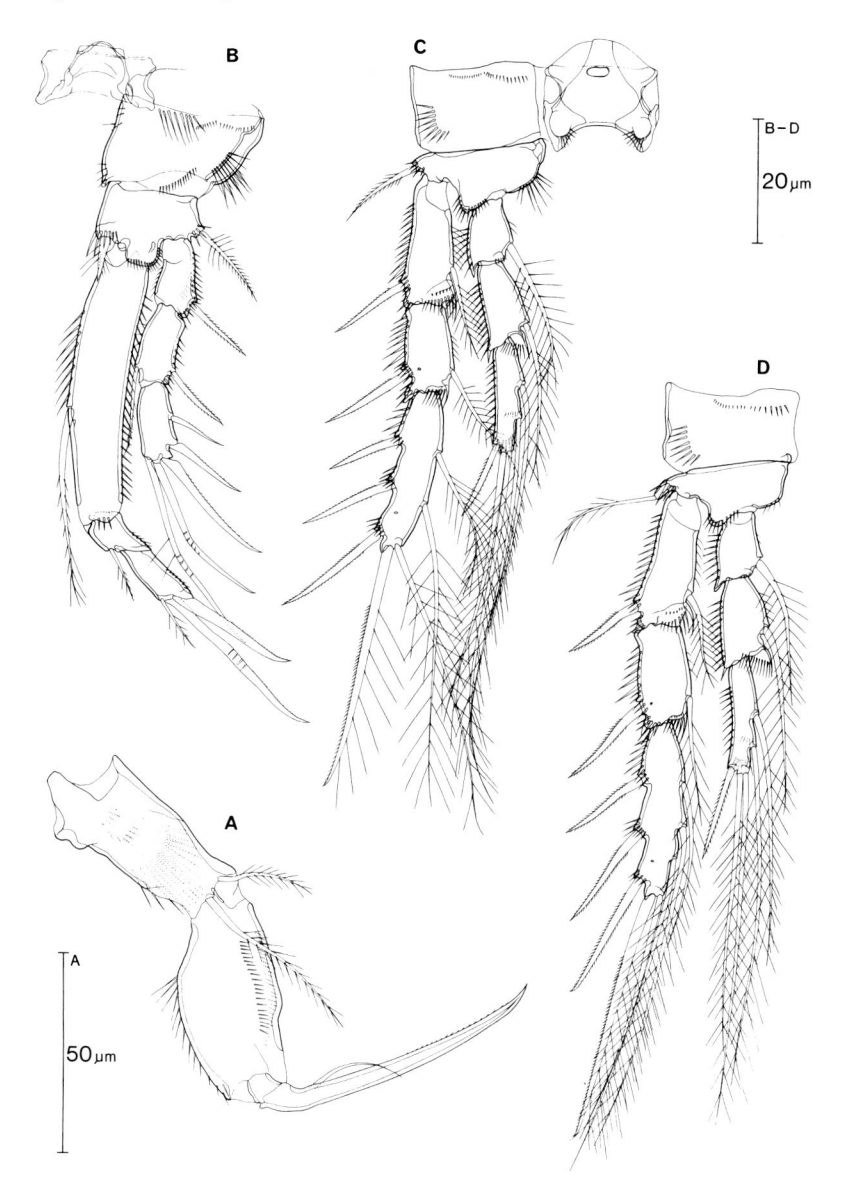

Fig. 18. *Stenocopia limicola* ♀. A. Maxilliped. B. P.1. C. P.2. D. P.3.

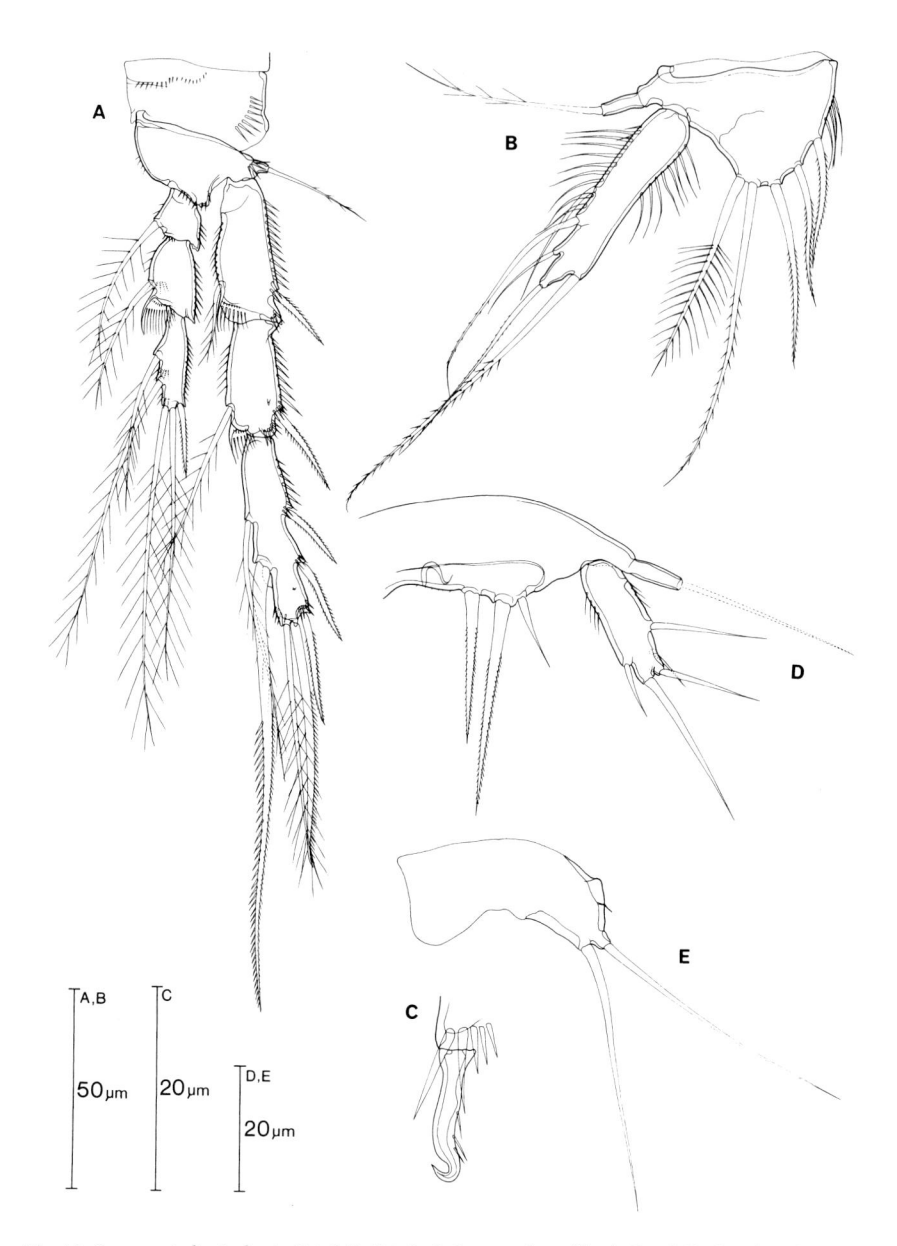

Fig. 19. *Stenocopia limicola*. A. P.4 ♀. B. P.5 ♀. C. Inner spine of basis P.1 ♂. D. P.5 ♂. E. P.6 ♂.

Middle segment ornamented as well except of the additional row of spinules. Distal segment with 3 slender outer spines and 2 appendages on distal edge. Inner margin carries 2 plumose setae in P.2 and 3 plumose setae in P.3. P.4 with 2 setae (distal one stout). Distal to the stout inner seta some spinules can be observed, but there is apparently no rudimentary seta present. Endopodite has 3 segments, each one spinulose on outer margin. Proximal segment shortest, with 1 inner plumose seta. Middle segment also with 1 inner plumose seta; distally with a fan of spinules. Distal segment carries 2 setae on inner margin, 2 apical setae and an outer appendage subapically.

Seta and spine formula:

	Exopodite	Endopodite
P.2	(1.1.223)	(1.1.221)
P.3	(1.1.323)	(1.1.221)
P.4	(1.1.223)	(1.1.221)

P.5 (Fig. 19B): Baseoendopodite with an outer seta which inserts on a socle; inner part with some spinules on inner margin and 5 setae, outermost but one longest. Exopodite slender, inner and outer margin spinulose, with 5 setae.

Male: Differs from the female in the following respects:
- Body length of dissected ♂♂: 0.40 – 0.48 mm.
- 2nd abdominal somite (i.e. somite following the one with P.6) ventrally with a continuous row of spinules as well as following somite (the latter corresponding to the female).
- 1st Antenna haplocer.
- Inner appendage of basis P.1 modified (Fig. 19C).
- Baseoendopodites of both P.5 fused. Outer seta stands on a socle. Inner part with 3 setae, middle one longest. Exopodite slender, bears 5 setae (Fig. 19D).
- P.6 with 2 setae of about the same length and a minute setule on outer margin (Fig. 19E).

Note: Both ♂♂ from Fernandina (length 0.42 and 0.45 mm) show some slight differences to the ones from Santa Cruz: for example, the length-breadth ratio of the furca is somewhat greater.

Discussion. Since no fundamental morphological difference can be ascertained between the present animals and the ones which were described from Bermuda by WILLEY (1935), both populations are interpreted to belong to one and the same species. WILLEY reports the benp. P.5 ♂ to have only 2 setae but nevertheless draws an outer minute setule. As can be seen from the present ♂♂ the length of this rudimentary seta is variable.

It is interesting that corresponding to *Nitocra bisetosa* also in the present dissected animals attributed to *Stenocopia limicola* both mandibulae show a different structure in their gnathobase, i.e., the cutting edge the near plumose seta has one or two strong tooth/teeth. However, whether or not this difference occurs constantly on right or left mandible was not checked, as it is obviously the case in *N. bisetosa* (see MIELKE 1994). This difference was not observed in *S. sarsi* nov. spec. At present nothing can be said about any phylogenetical consequences since too few ameirid species have been investigated concerning this phenomenon.

Remark: As in both *Nitocra*-species reported in this paper, the pair of the *Stenocopia*-species cannot be derived from a single hypothetic stem species. The ancestors of both species must have reached the Galapagos Archipelago independently.

Diosaccidae Sars, 1906
Balucopsylla Rao, 1972
Balucopsylla ? triarticulata Wells & Rao, 1987

(Figs. 3B, 20 – 22)

Localities and material. qn: Tower: Bahía Darwin (V,1; 22.1.73); 4 ♀♀, 5 ♂♂, 14 copepodites. Santa Cruz: Playa Borrero (IX,2; bimonthly transects from April 1972 to February 1973); 7 ♀♀, 1 ovigerous ♀, 5 ♂♂, 5 copepodites. Santa Cruz: Bahía Academy (IX,6; monthly transects from March 1972 to February 1973); 57 ♀♀, 15 ovigerous ♀♀, 74 ♂♂, 73 copepodites. Barrington: Exposed beach (XI,1; 1.4.72); 465 ♀♀, 25 ovigerous ♀♀, 377 ♂♂, 78 copepodites. Hood: Bahía Gardner (XIV,2; 30.6.72): 56 ♀♀, 1 ovigerous ♀, 34 ♂♂, 141 copepodites.

ql: Fernandina: Punta Espinosa (I,3; 25.9.72); 2 ♀♀, 3 ♂♂. Isabela: South Tagus (II,2; 25.10.72); 2 ♀♀, 1 ♂, 2 copepodites. Isabela: Punta Albemarle (II,5; 23.9.72); 3 ♀♀, 1 ♂. Tower: Lagoon (V,1b; 22.1.73); 6 ♀♀, 2 ovigerous ♀♀, 3 ♂♂. Santa Cruz: Puerto Nuñez (IX,4a; 3.3.72); 1 ♀. Santa Cruz: Private beach (IX,5d; 17.8.72); 10 ♀♀, 6 ♂♂. Santa Cruz: Private beach (IX,5d; 26.2.72); 10 ♀♀, 3 ovigerous ♀♀, 8 ♂♂. Baltra: Harbour (X,2a; 30.3.72); 2 ♀♀. Barrington : Exposed beach (XI,1; 22.2.72); 1 ♀, 1 ♂. San Cristóbal: Bahía Wreck (XIII,1; 20.11.72); 3 copepodites. San Cristóbal: Punta Bassa (XIII,2; 17.11.72); 11 ♀♀, 6 ovigerous ♀♀, 16 ♂♂, 14 copepodites.

Remark: 35 specimens (22 ♀♀, 13 ♂♂) were dissected. An unequivocal allocation of the present animals to *Balucopsylla triarticulata* is rather dubious. WELLS & RAO (1987) point out to a certain variability of their species (length, P.5 ♀, caudal ramus). Similar observations could be made within a sandy beach population of Punta Morales, Pacific coast of Costa Rica (MIELKE 1994). The same is true for the animals from the localities mentioned above. There are differences in the length ratios of setae; supernumerary setae occur (e.g., one ♀ has two well developed setae on the inner margin of the middle segment of one enp. P.4) or setae may be lacking (e.g., one exp. P.5 ♂ only with 3 setae).

However, these cases refer to single specimens, and each one can be clearly recognized as a genetic faux pas. On the other hand, the setation of the distal segments of enp. P.2 – P.4 of the Galápagos specimens is difficult to interpret. There are animals which show the same formula as given by WELLS & RAO:

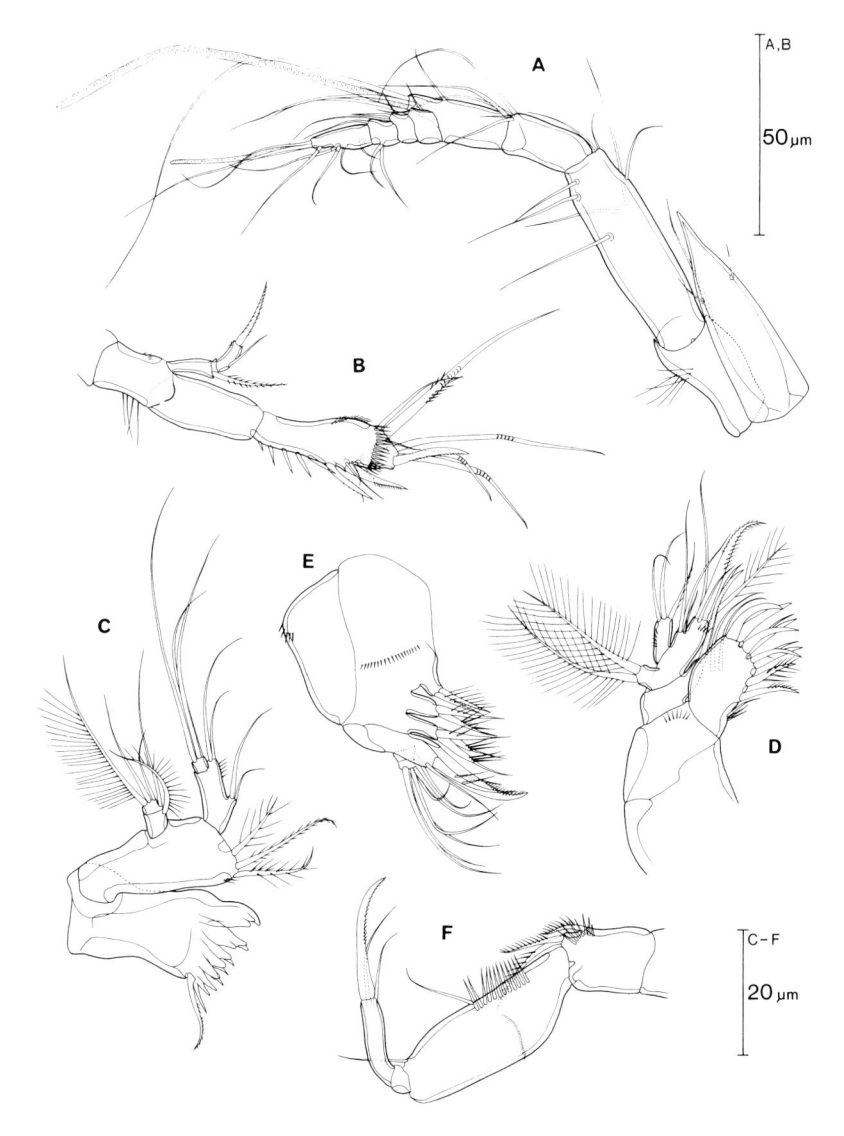

Fig. 20. *Balucopsylla ? triarticulata* ♀. A. Rostrum and 1st Antenna. B. 2nd Antenna. C. Mandible. D. 1st Maxilla. E. 2nd Maxilla. F. Maxilliped.

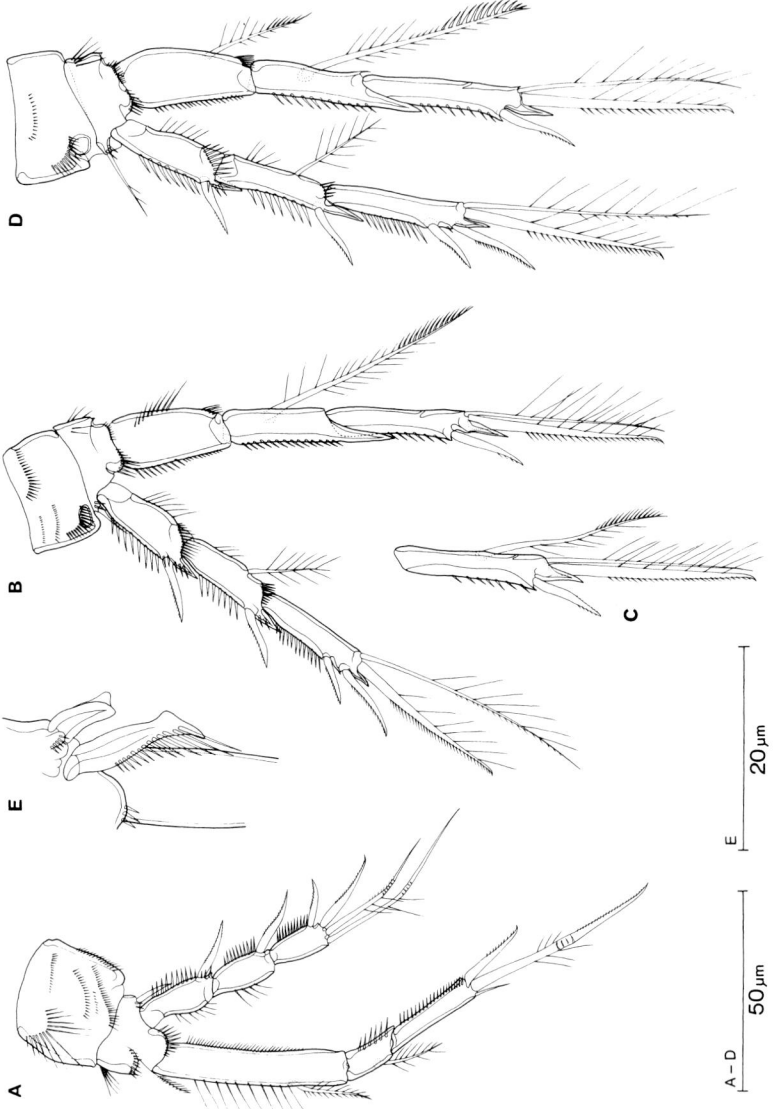

Fig. 21. *Balucopsylla ? triarticulata*. A. P.1 ♀. B. P.2 ♀. C. Distal segment enp. P.2, other specimen ♀. D. P.3 ♀. E. Inner spine of basis P.1 ♂.

021, whereas the majority of the dissected individuals possesses an additional inner seta: 121 (see Figs. 21B-C). However, the situation is complicated even more by the existence of intermediate forms. Several animals show both conditions on the endopodites of a single pair of legs. One ♀ has both enp. P.2 with 121 and enp. P.3 and P.4 with formula 021. Because of these transition sta-

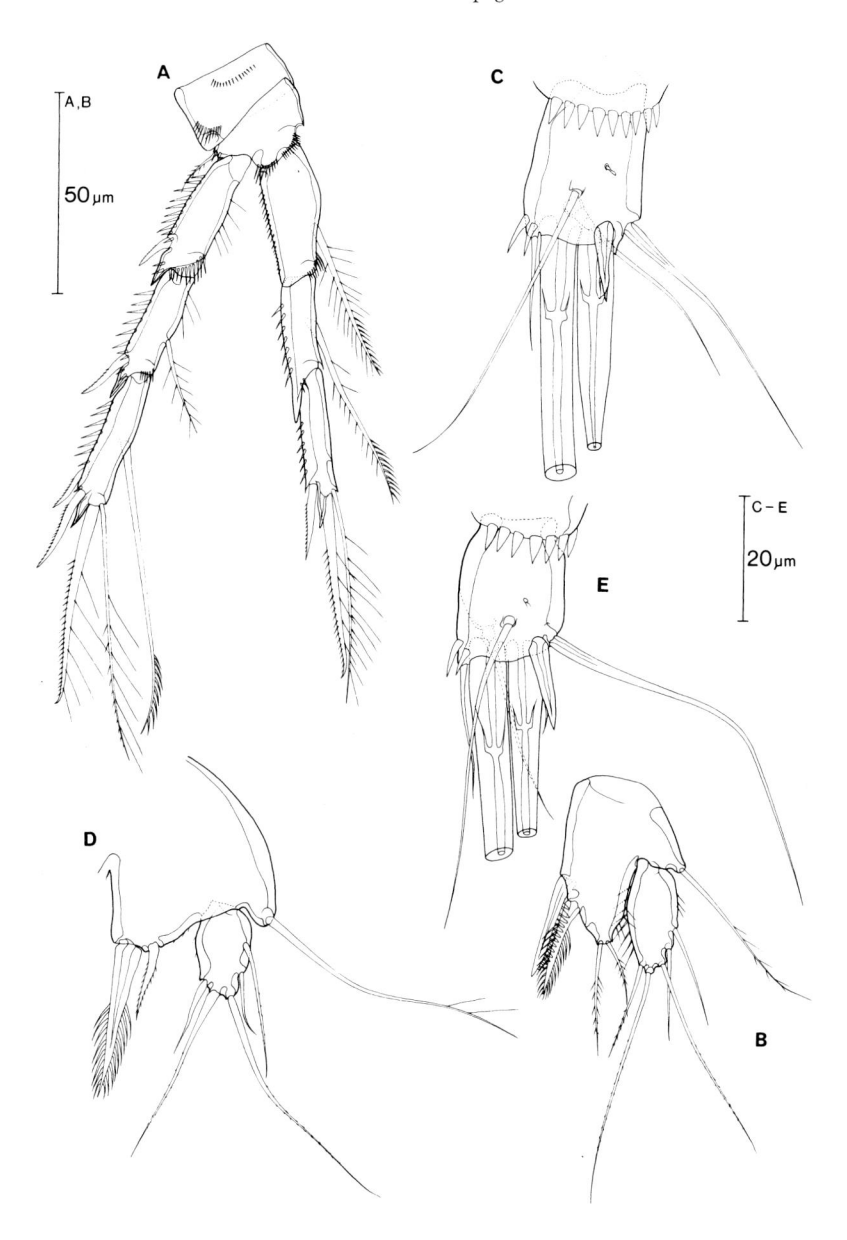

Fig. 22. *Balucopsylla ? triarticulata*. A. P.4 ♀. B. P.5 ♀. C. Furca, ventral side ♀. D. P.5 ♂. E. Furca, ventral side ♂.

ges and a more or less good morphological conformity (solely the specimens of Hood show some differences: enp. P.1 relatively shorter, length ratios of P.5) all Galápagos animals are considered, with reservations, to belong to one species.

Acknowledgements

The Galápagos-project was financially supported by Stiftung Volkswagenwerk.

Zusammenfassung

Von verschiedenen Sandstränden der Galápagos-Inseln werden fünf benthische Copepodenarten nachgewiesen. Davon gehören vier zu den Ameiridae, eine Art zu den Diosaccidae.

Nitocra bisetosa Mielke, 1993 wurde erstmals von Costa Rica gemeldet. Die Tiere von diversen Fundorten von Galápagos sind etwas variabel und werden nur mit Vorbehalt als eine Art betrachtet bzw. der Art von Costa Rica zugerechnet.

Eine andere, weit verbreitete *Nitocra*-Spezies erwies sich als neu für die Wissenschaft aufgrund einer besonderen Kombination von morphologischen Merkmalen: *N. galapagoensis*.

Stenocopia wird durch wenigstens zwei Arten repräsentiert. Die eine kann mit *S. limicola* identifiziert werden, die erstmals durch WILLEY (1935) von Bermuda nachgewiesen wurde. Die andere Stenocopia-Population wird als neue Art beschrieben: *S. sarsi*.

Weder die beiden *Nitocra*- noch die beiden *Stenocopia*-Arten können jeweils auf eine hypothetische Stammart zurückgeführt werden, die sich, nach ihrer Einwanderung, in die hier vorgestellten Spezies aufgespalten haben würde. Es ist zu postulieren, daß alle vier Arten unabhängig voneinander den Inselkomplex erreicht haben. Trotzdem gibt es klare Befunde für evolutive Abwandlungen, insbesondere bei *N. bisetosa*, die sich in der Folge der Ausbreitung im Archipel herausbildeten.

Die einzige hier vorgestellte Diosaccidenart erwies sich als eine recht variable Form. Ihre Bestimmung als *Balucopsylla triarticulata* Wells & Rao, 1987 muß als fraglich gelten.

References

Ax, P. & P. Schmidt (1973): Interstitielle Fauna von Galapagos. I. Einführung. Mikrofauna Meeresboden **20**, 1 – 38.

Cottarelli, V., P .E . Saporito & A. C. Puccetti (1985): Un interessante arpacticoide di acque interstiziali della Tailandia: *Stenocopia reducta* n. sp. (Crustacea, Copepoda). Boll. Mus. civ. St. nat. Verona **12**, 307 – 317.

Huys, R. & G. A. Boxshall (1991): Copepod Evolution. Ray Soc. London, 468 pp.

Lang, K. (1948): Monographie der Harpacticiden. Nordiska Bokh. Stockholm, 1682 pp.

– (1965): Copepoda Harpacticoidea from the Californian Pacific coast. Kungl. Svenska Vetenskaps. Handl. **10**, 1 – 566.

Mielke, W. (1979): Interstitielle Fauna von Galapagos. XXV. Longipediidae, Canuellidae, Ectinosomatidae (Harpacticoida). Mikrofauna Meeresboden **77**, 1 – 106.

– (1993): Species of the taxa *Orthopsyllus* and *Nitocra* (Copepoda) from Costa Rica. Microfauna Marina **8**, 247 – 266.

– (1994): New records of two copepod species (Crustacea) from the Pacific coast of Costa Rica. Microfauna Marina **9**, 55 – 60.

Por, F. D. (1964): Les Harpacticoides (Crustacea, Copepoda) des fonds meubles du Skagerak. Cah. Biol. Mar. **5**, 233 – 270.

Sars, G. O. (1907): An Account of the Crustacea of Norway. Vol. V. Copepoda, Harpacticoida. Parts XIX & XX. Canthocamptidae (concluded), Laophontidae (part). Bergen Museum, 221 – 240.

– (1911): An Account of the Crustacea of Norway. Vol. V. Copepoda, Harpacticoida. Parts XXXIII & XXXIV. Supplement (continued). Bergen Museum, 397 – 420.

Wells, J. B. J. & G. C. Rao (1987): Littoral Harpacticoida (Crustacea: Copepoda) from Andaman and Nicobar Islands. Mem. Zool. Surv. India **16**, 1 – 385.

Willey, A. (1935): Harpacticoid Copepoda from Bermuda.- Part II. Ann. Mag. Nat. Hist. **15**, 50 – 100.

Interstitielle Fauna von Galapagos

I.	Einführung P. Ax & P. Schmidt, Mikrofauna Meeresboden **20**, 1–38 (1973).
II.	Gnathostomulida. B. Ehlers & U. Ehlers, Mikrofauna Meeresboden **22**, 1–27 (1973).
III.	Promesostominae (Turbellaria, Typhloplanoida).P. Ax & U. Ehlers, Mikrofauna Meeresboden **23**, 1–16 (1973).
IV.	Gastrotricha. P. Schmidt, Mikrofauna Meeresboden **26**, 1–76 (1974).
V.	Otoplanidae (Turbellaria, Proseriata). P. Ax & R. Ax, Mikrofauna Meeresboden **27**, 1–28 (1974).
VI..	*Aeolosoma maritimum dubiosum* nov. sp. (Annelida Oligochaeta, W. Westheide & P. Schmidt, Mikrofauna Meeresboden **28**, 1–10 (1974).
VII.	Nematoplanidae, Polystyliphoridae, Coelogynoporidae (Turbellaria, Proseriata) P. Ax & R. Ax, Mikrofauna Meeresboden **29**, 1–28 (1974).
VIII.	Trigonostominae (Turbellaria, Typhloplanoida) U. Ehlers & P. Ax, Mikrofauna Meeresboden **30**, 1–33(1974).
IX.	Dolichomacrostomidae (Turbellaria, Macrostomida), B. Sopott-Ehlers & P. Schmidt, Mikrofauna Meeresboden **34**, 1–20 (1974).
X.	Kinorhyncha, P. Schmidt Mikrofauna Meeresboden **43**, 1–15 (1974).
XI.	Pisionidae, Hesionidae, Pilargidae, Syllidae (Ploychaeta). W. Westheide, Mikrofauna Meeresboden **44**, 1–146 (1974).
XII.	*Myozona* Marcus (Turbellaria, Macrostomida). B. Sopott-Ehlers & P. Schmidt, Mikrofauna Meeresboden **46**, 1–19 (1974).

XIII.   *Ototyphlonemertes* Diesing (Nemertini, Hoplonemertini). H. Mock & P. Schmidt, Mikrofauna Meeresboden **51**, 1–40 (1975).
XIV.   Polycladida (Turbellaria). B. Sopott-Ehlers & P. Schmidt, Mikrofauna Meeresboden **54**, 1–32 (1975).
XV.   *Macrostomum* O. Schmidt, 1848 und *Siccomacrostomum triviale* nov. gen. nov. spec. (Turbellaria, Macrostomida). P. Schmidt & B. Sopott-Ehlers, Mikrofauna Meeresboden **57**, 1–45 (1976).
XVI.   Tardigrada. D. McKirdy, P. Schmidt & M. McGinty-Bayly, Mikrofauna Meeresboden **58**, 1–43 (1976).
XVII.   Polygordiidae, Saccocirridae, Protodrilidae, Nerillidae, Dinophilidae (Polychaeta). P. Schmidt & W. Westheide, Mikrofauna Meeresboden **62**, 1–38 (1977).
XVIII.   Nereidae, Eunicidae, Dorvilleidae (Polychaeta). W. Westheide, Mikrofauna Meeresboden **63**, 1–40 (1977).
XIX.   Monocelididae (Turbellaria, Proseriata). P. Ax & R. Ax, Mikrofauna Meeresboden **64**, 1–44 (1977).
XX.   Halacaridae (Acari). I. Bartsch, Mikrofauna Meeresboden **65**, 1–108 (1977).
XXI.   Lebensraum, Umweltfaktoren, Gesamtfauna. P. Schmidt, Mikrofauna Meeresboden **68**, 1–52 (1978).
XXII.   Zur Ökologie der Halacaridae (Acari). I. Bartsch & P. Schmidt, Mikrofauna Meeresboden **69**, 1–38 (1978).
XXIII.   Acoela (Turbellaria). U. Ehlers & J. Dörjes, Mikrofauna Meeresboden **72**, 1–75 (1979).
XXIV.   Microparasellidae (Isopoda, Asellota). N. Coineau & P. Schmidt, Mikrofauna Meeresboden **73**, 1–19 (1979).
XXV.   Longipediidae, Canuellidae, Ectinosomatidae (Harpacticoida). W. Mielke, Mikrofauna Meeresboden **77**, 1–107 (1979).
XXVI.   Questidae, Cirratulidae, Acrocirridae, Ctenodrilidae (Polychaeta). W. Westheide, Mikrofauna Meeresboden **82**, 1–24 (1981).
XXVII.   Byrsophlebidae, Promesostomidae Brinkmanniellinae, Kytorhynchidae (Turbellaria, Typhloplanoida). U. Ehlers & B. Ehlers, Mikrofauna Meeresboden **83**, 1–35 (1981).
XXVIII.   Laophontinae (Laophontidae), Ancorabolidae (Harpacticoida). W. Mielke, Mikrofauna Meeresboden **84**, 1–106 (1981).
XXIX.   Darcythompsoniidae, Cylindropsyllidae (Harpacticoida). W. Mielke, Mikrofauna Meeresboden **87**, 1–52 (1982).
XXX.   Podocopida 1 (Ostracoda). J. Gottwald, Mikrofauna Meeresboden **90**, 1–187 (1983).
XXXI.   Paramesochridae (Harpacticoida). W. Mielke, Microfauna Marina **1**, 63–147 (1984).
XXXII.   Epsilonematidae (Nematodes). E. Clasing, Microfauna Marina **1**, 149–189 (1984).
XXXIII.   Tubificidae (Annelida, Oligochaeta). Ch. Erséus, Microfauna Marina **1**, 191–198 (1984).
XXXIV.   Schizorhynchia (Plathelminthes, Kalyptorhynchia). U. Noldt & S. Hoxhold, Microfauna Marina **1**, 199–256 (1984).
XXXV.   Chromadoridae (Nematoda): D. Blome, Microfauna Marina **2**, 271–329 (1985).
XXXVI.   Tetragonicipitidae (Harpacticoida). W. Mielke, Microfauna Marina **5**, 95–172 (1989).
XXXVII.   Metidae (Harpacticoida). W. Mielke, Microfauna Marina **5**, 173–188 (1989).
XXXVIII.   *Haloplanella* Luther und *Pratoplana* Ax (Typhloplanoida, Plathelminthes). U. Ehlers & B. Sopott-Ehlers, Microfauna Marina **5**, 189–206 (1989).

Dr. Wolfgang Mielke
II. Zoologisches Institut und Museum der Universität Göttingen,
Berliner Straße 28, D-37073 Göttingen

Interstitial Fauna of Galapagos. XL.
Copepoda, part 8

Wolfgang Mielke

Abstract

Four species of interstitial copepods belonging to different taxa and of uncertain systematics were found in the quantitatively investigated sandy beach of the Bahía Academy/Santa Cruz (IX,6).

Only two males of a new species of the Tetragonicipitidae are available. Although the long basal segment of 1st antenna lacks the characteristic outer tooth, the species is considered to be a member of *Tetragoniceps* Brady, 1880: *T. santacruzensis* nov. spec. There exists no sisterspecies-relationship with *T. galapagoensis* Mielke, 1989, which has already been described from the Galapagos.

A few animals have conspicuous "club-like" appendages on their mouthparts and are assigned to *Micropsammis* Mielke, 1975: *M. galapagoensis* nov. spec. This genus is now ranked with the Paranannopidae (see GEE & HUYS 1991). It is suggested that the species *secunda* from List/Island of Sylt (Germany) is left in *Micropsammis* as originally done; the establishment of a separate genus, *Telopsammis* Gee & Huys, 1991 does not seem to be necessary.

Two animals (1♀, 1♂) probably represent a second species of *Carolinicola* Huys & Thistle, 1989: *C. galapagoensis* nov. spec. The only known species so far, *C. (= Hemimesochra) trisetosa* (Coull, 1973) was originally classed with the Cletodidae, then ascribed to the Canthocamptidae and finally integrated into the Paranannopidae (see HUYS & THISTLE 1989).

Psammocamptus galapagoensis nov. spec., a representative of the Canthocamptidae, is regarded as being closely related to *P. axi* Mielke, 1975, which has been described from the sandy beach of the east coast of List/Island of Sylt (Germany). Since no fundamental morphological differences exist, both species of *Bathycamptus* Huys & Thistle, 1989 are included in *Psammocamptus* Mielke, 1975.

Finally, some critical remarks are given concerning the present systematical work on harpacticoid copepods.

Fauna intersticial de Galápagos. XL. Copepoda, parte 8

Resumen

En la playa arenosa de la Bahía Academy/Santa Cruz (IX,6) investigada cuantitativamente se encontraron cuatro especies de copépodos intersticiales, pertenecientes a diferentes taxa y cuya posición sistemática parece incierta.

Solamente se colectaron dos machos de una especie nueva de los Tetragonicipitidae. Aunque falta el diente al borde externo del artejo basal de la 1ª antena, la especie se interpreta como perteneciente a *Tetragoniceps* Brady, 1880: *T. santacruzensis* nov. spec. Sin embargo no parece estar muy relacionada con *T. galapagoensis* Mielke, 1989, ya descrita de los Galápagos.

Algunos animales con apéndices vistosos en forma de clava situados en la región bucal se agruparon bajo *M. galapagoensis* nov. spec. a *Micropsammis*, un género que ahora se ha incorporado a los Paranannopidae (ver GEE & HUYS 1991). Además se propone dejar la especie *secunda* de List/Isla Sylt (Alemania) bajo *Micropsammis* como se la describió originalmente; el establecimiento de un género propio, *Telopsammis* Gee & Huys, 1991, se juzga como superfluo.

Dos animales (1♀, 1♂) probablemente representan la segunda especie de *Carolinicola* Huys & Thistle, 1989: *C. galapagoensis* nov. spec. La única especie de este género conocida hasta ahora, *C. (= Hemimesochra) trisetosa* (Coull, 1973), originalmente fue integrada a los Cletodidae, entonces atribuida a los Canthocamptidae y finalmente añadida a los Paranannopidae (ver HUYS & THISTLE 1989).

Psammocamptus galapagoensis nov. spec., un representante de los Canthocamptidae, es considerado como pariente cercano de *P. axi* Mielke, 1975, que fue descrito de la playa arenosa en la costa oriental de List/Isla Sylt (Alemania). Las dos especies de *Bathycamptus* Huys & Thistle, 1989 deberán clasificarse bajo *Psammocamptus* Mielke, 1975, porque no existen diferencias morfológicas fundamentales entre los dos géneros.

Finalmente se presentan algunas notas respecto al trabajo sistemático actual sobre los copépodos harpacticoideos.

A. Introduction

This eighth contribution to the Galapagos benthic copepods deals with four species, poor in individuals and, thus far, exclusively known from the meiobenthically best investigated beach, the one at the Bahía Academy/Santa Cruz (see AX & SCHMIDT 1973); only *Psammastacus galapagoensis* nov. spec. was additionally collected at Punta Espinosa, Fernandina.

The interpretation of the systematic position and phylogenetic relationships of the introduced species is problematical. Therefore, all statements made here on the systematics of the new species should be regarded as preliminary. Above all, because their presumably closest known relatives occur far from the Galapagos Archipelago; this suggests the existence of many more species world-wide belonging to the respective supraspecific taxon.

The material has been deposited in the collections of the Zoological Museum of the University of Göttingen.

The interpretation of body, mouth parts and thoracopods is adopted from LANG (1948, 1965). With respect to the mouth parts, the interpretation of the components by HUYS & BOXSHALL (1991) is given in parentheses.

B. Results

Tetragonicipitidae Lang, 1948
Tetragoniceps Brady, 1880
Tetragoniceps santacruzensis nov. spec.

(Figs. 1 – 4)

Locality and material. qn: Santa Cruz: Bahía Academy (IX,6; **Locus typicus**. Monthly transects from March 1972 to February 1973); 2 ♂♂.

Both specimens were dissected. Holotype is 1 ♂, reg. no. I Gal 1122. Paratype is the other male, reg. no. I Gal 1121.

Description

Male: Body length from tip of rostrum to end of furcal rami 0.46 mm (holotype) and 0.45 mm (paratype). Rostrum basally with a distinct division line, slightly broader than long; subapically on both sides a slender setule inserts (Fig. 2A). Habitus slender, body sides extending more or less parallelly. Caudal margins of all somites obviously smooth. With the exception of penultimate somite the dorsal surface of all somites is equipped with two or more setules. Furthermore, one (occasionally two) median pore is to be seen. Ventral margin of 2nd and 3rd abdominal somites subdistally furnished with slender

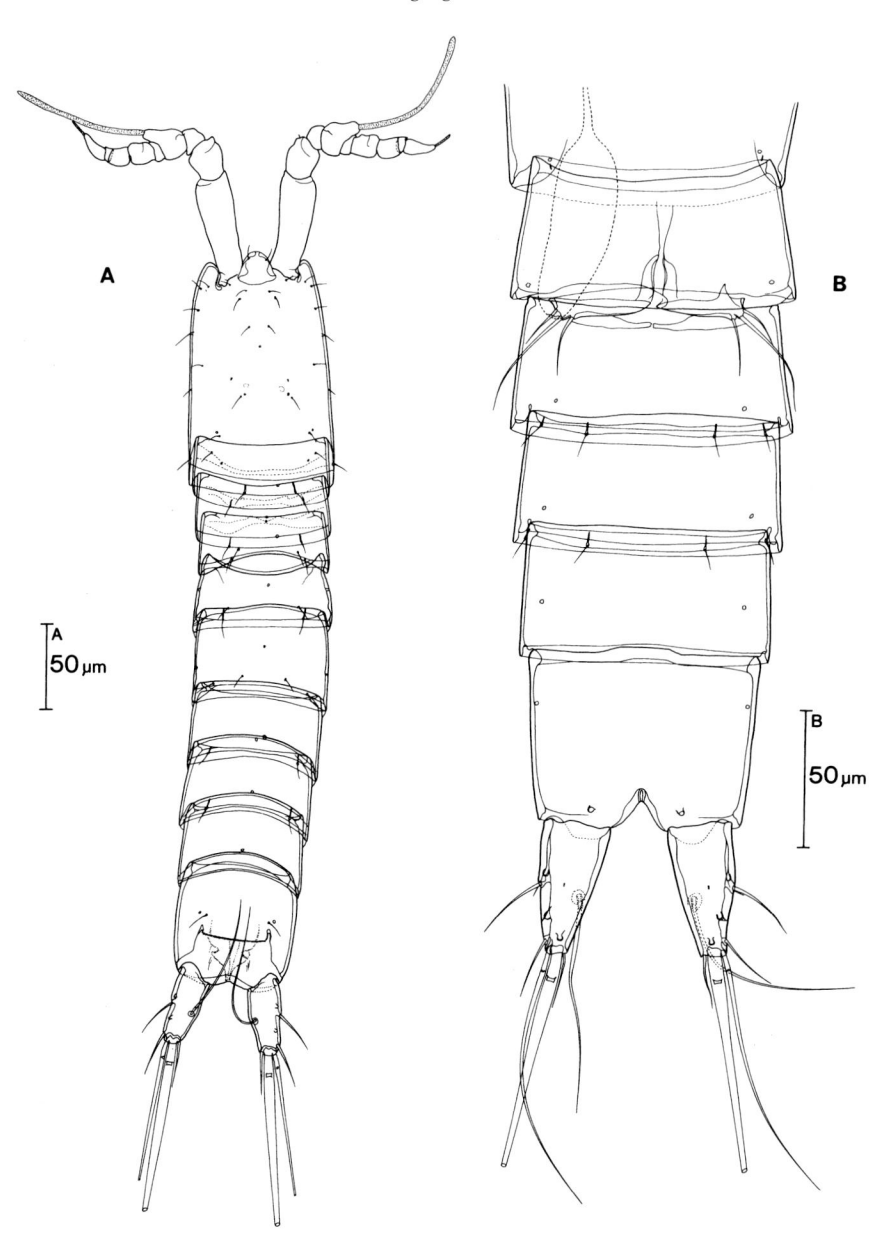

Fig. 1. *Tetragoniceps santacruzensis* nov. spec. ♂. A. Habitus, dorsal side. B. Abdomen, ventral side.

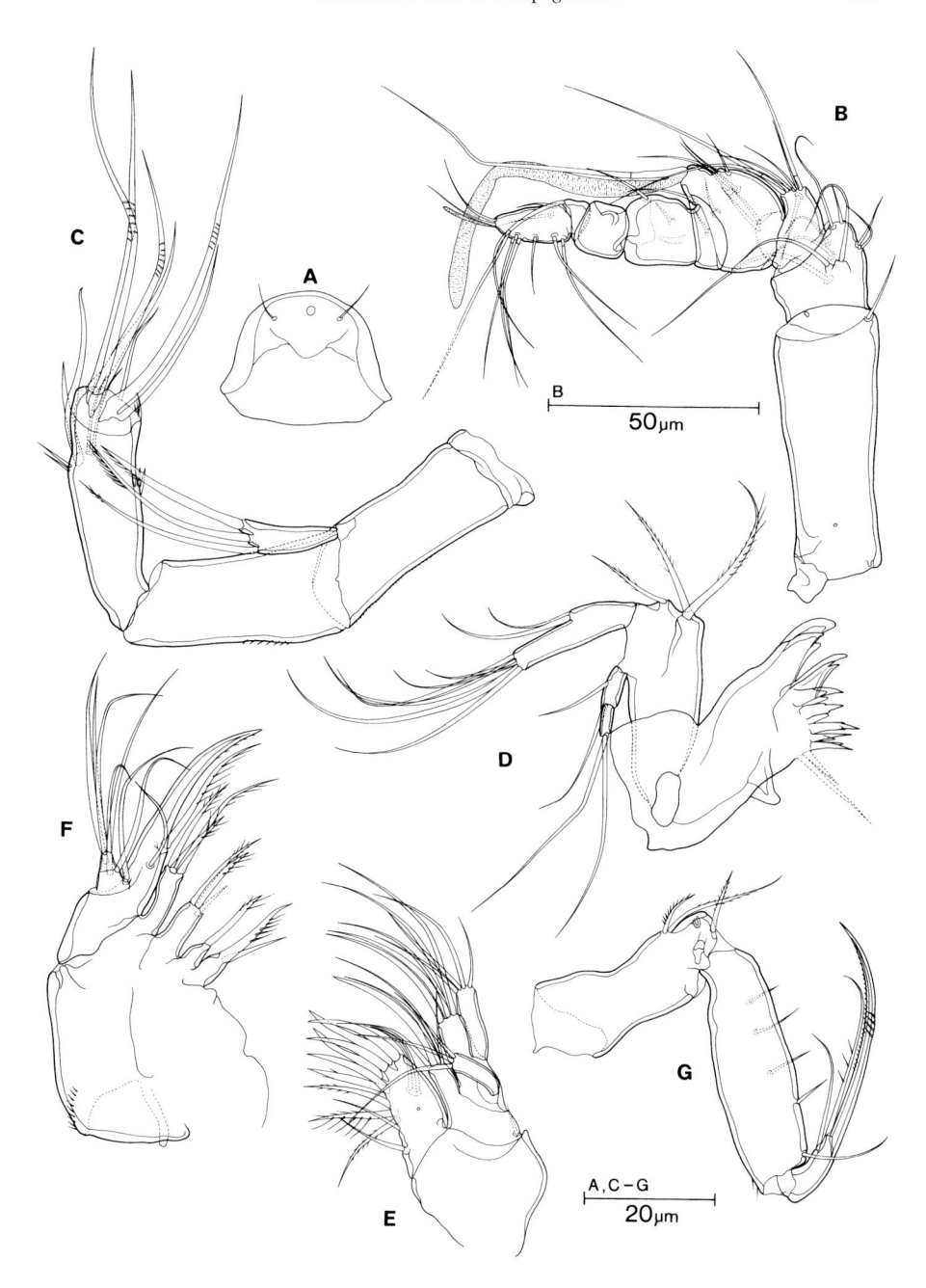

Fig. 2. *Tetragoniceps santacruzensis* nov. spec. ♂. A. Rostrum. B. 1st Antenna. C. 2nd Antenna. D. Mandible. E. 1st Maxilla. F. 2nd Maxilla. G. Maxilliped.

setules. Laterally, all abdominal somites have small pores. Anal operculum transverse, set with minute spinules. Furca tapering caudally, slightly more than twice as long as broad. About midlength on outer margin a slender seta arises, accompanied by a small setule; subapically on outer margin another slender seta inserts. Apical edge with 3 setae; inner one shortest, middle one longest. About middorsally, a long, slender seta inserts, which is bipartite at base (Figs. 1A,B).

1st Antenna (Fig. 2B): Haplocer, 9 segments. First segment elongated. 4th segment slightly swollen, distally on outer margin with an aesthetasc. Terminal segment has a small aesthetasc.

2nd Antenna (Fig. 2C): Basis and proximal endopodite segment delicately spinulose on anterior margin. Distal endopodite segment subapically with 2 slender setae and 2 spines of unequal length. Apically 4 geniculate setae insert, outer of which is basally fused with a slender seta; and 2 additional appendages. Exopodite 1-segmented, provided with 3 setae.

Mandible (Fig. 2D): Chewing edge of precoxa (coxa) with several acute teeth and a lateral seta. Coxa-basis (basis) has 3 weakly plumose setae. Endopodite with 2 setae on an insertion point at midlength and 6 terminal, naked setae. Exopodite 2-segmented. Basal segment bears 2, distal segment 2 (one of the four mandibles at hand has obviously 3) setae.

1st Maxilla (Fig. 2E): Arthrite of precoxa with 11 appendages distally and laterally, and 2 setae on surface. Coxa has 5 terminal setae and 1 short seta on exite. Basis carries 5 (?) setae. Endopodite and exopodite furnished with 4 and 3 setae, respectively.

2nd Maxilla (Fig. 2F): Syncoxa with 3 endites; proximal one deeply notched, with 3 setae altogether. Middle and distal endites as well with 3 setae (proximal appendage of middle endite dwarfed or deformed here). Basis (allobasis) has 1 claw and 2 (or 3?) setae. Endopodite weakly 2-segmented, bears obviously 7 setae. It is not clear whether 1 of them in reality belongs to the basis.

Maxilliped (Fig. 2G): Basis (syncoxa) with 3 setae subapically. Proximal endopodite segment (basis) has a few slender spinules and 2 setae. Distal endopodite segment (endopodite) armed with 1 claw and 2 accompanying slender setae.

P.1 (Fig. 3A): Outer and inner margins of coxa delicately spinulose. Basis with an inner spine; outer seta not observed. Exopodite 3-segmented; outer margins spinulose, inner margins hairy. Distal segment shortest, bears 4 appendages, both inner ones geniculate. Endopodite 2-segmented. Proximal segment reaching to end of exp.; outer and inner margins hairy, subterminally 1 plumose seta inserts. Distal segment slender, bears 2 long appendages, unequal in length.

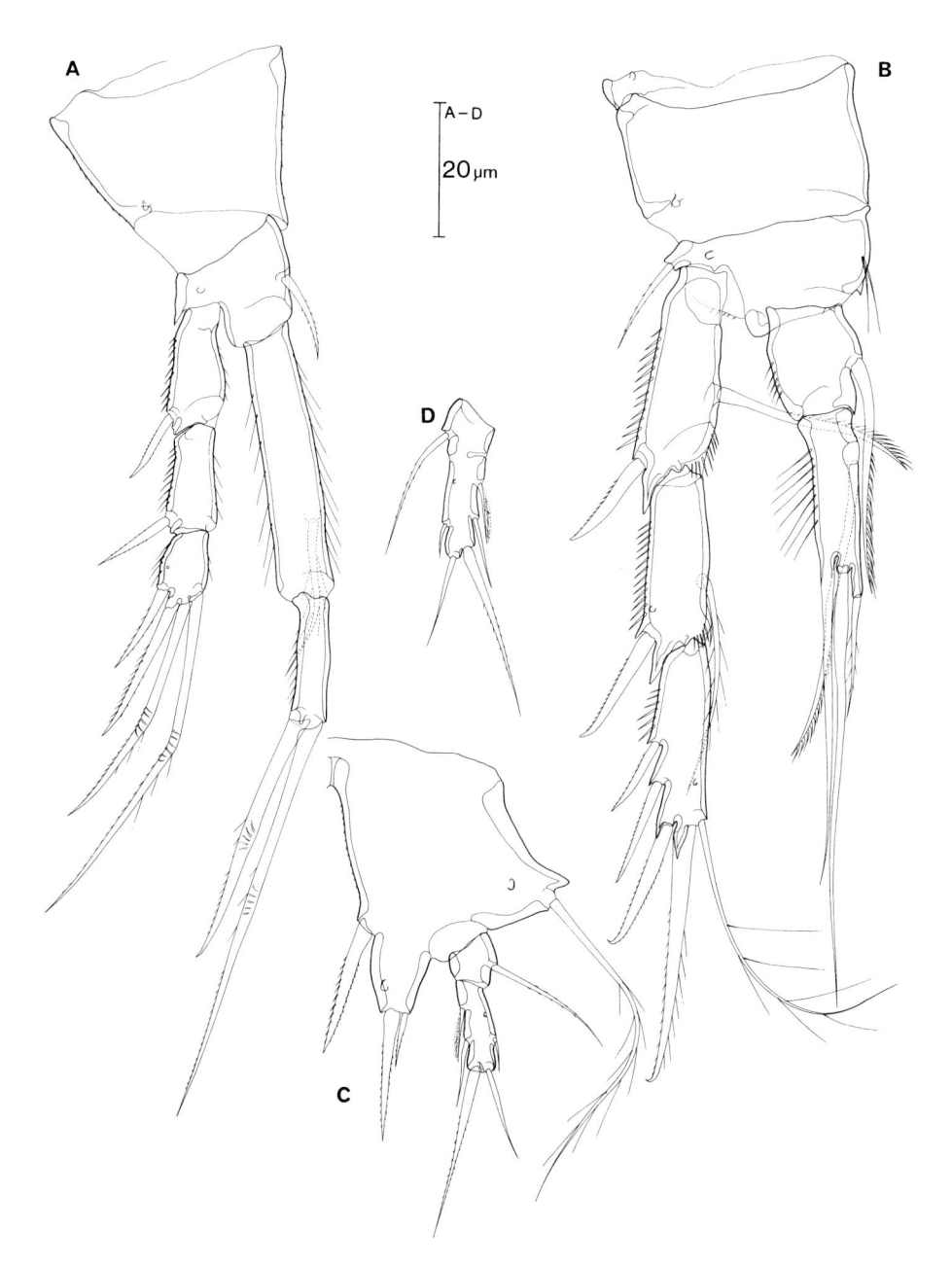

Fig. 3. *Tetragoniceps santacruzensis* nov. spec. ♂. A. P.1. B. P.2. C. P.5, left side. D. Exopodite P.5, right side.

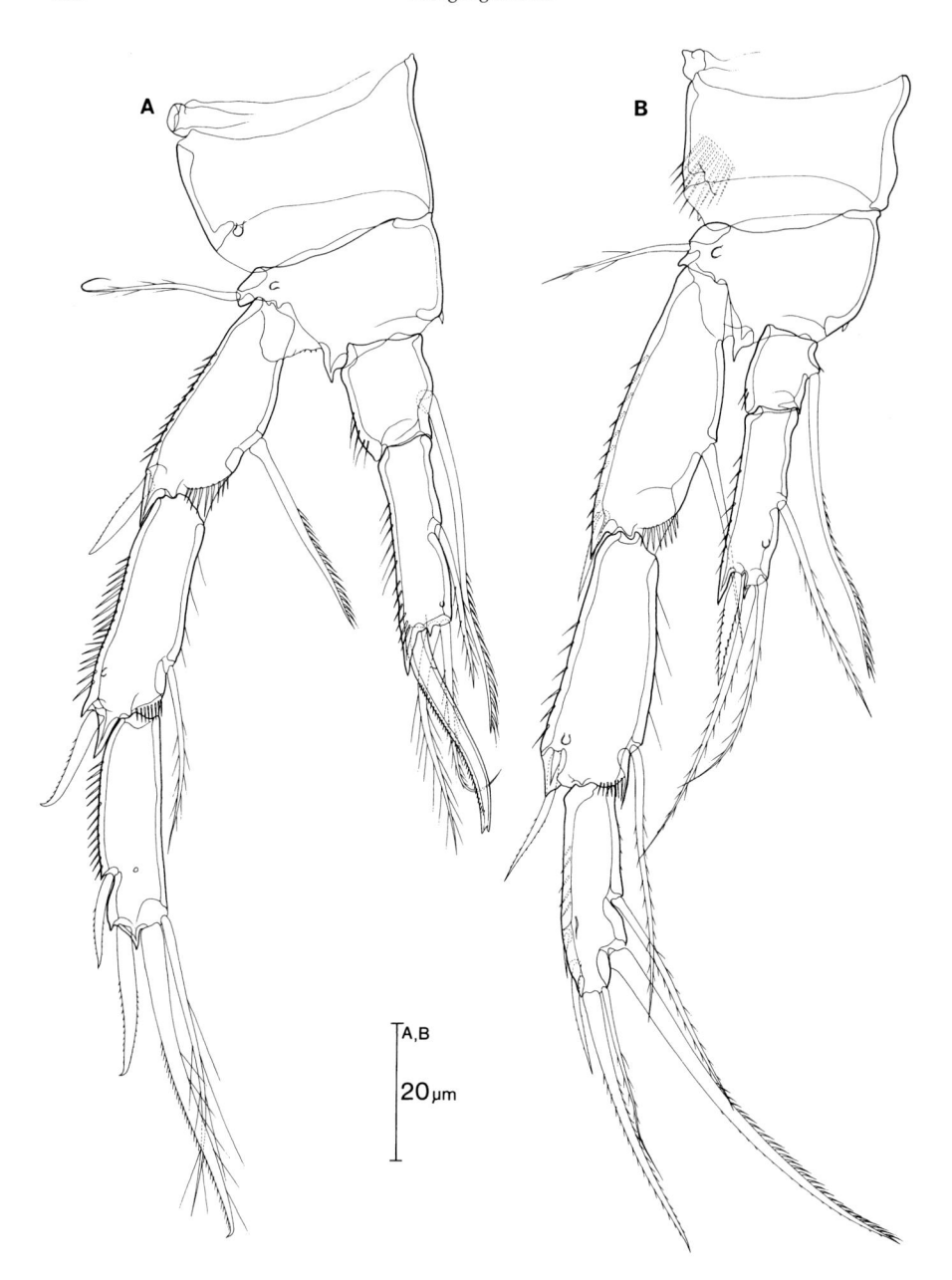

Fig. 4. *Tetragoniceps santacruzensis* nov. spec. ♂. A. P.3. B. P.4.

P.2 (Fig. 3B): Coxa seemingly without ornamentation. Basis has an outer seta, inner margin with a hook subdistally and 2 slender spinules. Exopodite consists of three segments of about the same length. Outer margins spinulose, distally tooth-like elongated. Proximal segment with an outer spine; subapically on inner side a row of spinules extends; inner margin carries a crest-like seta. Middle segment similar, but inner seta weakly plumose. Distal segment furnished with 3 outer spines, terminally with a long spine and a hairy seta. Endopodite 2-segmented. Proximal segment squarish, has an inner crest-like seta. Distal segment elongated, outer margin with long hairs, 1 crest-like seta on proximal inner margin and 3 setae of unequal length apically, outer of which seemingly fused with the segment.

P.3 (Fig. 4A): Resembles the preceding leg. Outer seta of basis longer, inner spinules not observed. Distal segment of exopodite only with 2 outer spines. Outer margin of distal segment of endopodite tooth-like elongated; apically 2 plumose setae and 1 stout, slightly bent and distally bifid appendage arise.

P.4 (Fig. 4B): Also resembles P.2. Coxa with a row of slender spinules on caudal surface. Outer spines of proximal and middle exopodite segments comparatively small. Distal exp. segment with 2 short outer spines, 2 apical and 2 inner appendages of different length. Outer margin of distal enp. segment tooth-like elongated, apically 2 plumose setae and a stout, outer spine insert.

Seta and spine formula:

	Exopodite	Endopodite
P.2	(1.1.023)	(1.121)
P.3	(1.1.022)	(1.121)
P.4	(1.1.222)	(1.121)

P.5 (Figs. 3C,D): Baseoendopodite with a slender, plumose outer seta; inner part with 1 stout spine on an inner subapical insertion point and 2 spines of unequal length apically. Exopodite of left P.5 distinctly 2-segmented, corresponding parts of right exp. only partly separated. Proximal segment bears an outer slender seta, distal segment furnished with 1 small outer seta, 2 apical and 2 inner setae (proximal inner seta apparently broadened?).

P.6 (Fig. 1B): Both legs display slight asymmetry. Each has 3 setae, middle one longest.

Female: Unknown.

Discussion. The Tetragonicipitidae of the Galapagos Islands have already been subject of an earlier publication (MIELKE 1989). Eight species were presented herein. Moreover, it was pointed out that the species stock of this family was much extensive. However, single specimens were purposely not described.

The two males from the Bahía Academy/Santa Cruz are attributed to *Tetragoniceps*, a species-group of about a dozen members world-wide, of which *T. galapagoensis* Mielke, 1989 has already been described in the paper mentioned above. The two Galapagos species obviously do not co-occur and are by no means sister species. *T. galapagoensis* is individual-rich and widely distributed in the archipelago; whereas *T. santacruzensis* nov. spec. apparently constitutes a numerically small and local population.

The integration of the new form into *Tetragoniceps* is questionable, particularly without knowledge of the female. The two males possess an elongated basal segment of A.1, but without a distal outer tooth, as well as a tooth-less 2nd segment of A.1. Since all other of the *Tetragoniceps* species exhibit such an outer tooth on the basal segment of A.1 a secondary loss of this structure must therefore be hypothesized. Other genera lacking teeth on the elongated 1st and on 2nd segments of A.1 cannot be taken into consideration, viz. *Pteropsyllus* (rudimentary exp. A.2, enp. P.1 3-segmented, structure of P.5 ♂) and *Phyllopodopsyllus* (part.; no species is known to have a 2-segmented exp. P.5 ♂; this plesiomorphic (?) feature is present at least in *Diagoniceps mexicana* Fiers, 1995; *Paraschizopera* (= *Nidiagoceps* = *Diagoniceps*) *menaiensis* (Geddes, 1968); *Tetragoniceps bergensis* Por, 1965; *Fearia* (= *Tetragoniceps*) *prima* Coull, 1971).

With regard to, e.g., the seta and spine formula of the pereiopods, structure of P.5 and furcal rami, the *Tetragoniceps* species are without doubt non-uniform: this suggests that "*Tetragoniceps* seems to be a polyphyletic grouping"(FIERS 1995). However, in systematics the primary objective should not be to search for reasons to erect new categories. It is more important to search for common derived characters. If the distal outer tooth of the basal segment A.1 can be verified as a common evolutionary novelty, then at least one autapomorphy for *Tetragoniceps* would exist: this would establish it as a monophyletic group. This taxon then would have undergone several evolutionary trends as indicated by KUNZ (1984), including the potential reduction of this tooth in *T. santacruzensis* nov. spec.

Paranannopidae Por, 1986
Micropsammis Mielke, 1975
Micropsammis galapagoensis nov. spec.

(Figs. 5 – 8)

Locality and material. qn: Santa Cruz: Bahía Academy (IX,6; **Locus typicus**. Monthly transects from March 1972 to February 1973); 2 ♀♀, 3 ♂♂.

2 ♀♀ and 2 ♂♂ were dissected. Holotype is 1 ♀, reg. no. I Gal 1123. Paratypes are 1 ♀ and 2 ♂♂, all dissected, reg. no. I Gal 1124 – 1126. R., A.1, A.2, Md., Mx.1, Mxp., P.2 – P.4, and Abdomen ventral are drawn from holotype.

Description

Female: Body length from tip of rostrum to end of furcal rami 0.28 mm (holotype) and 0.27 mm (paratype). Rostrum elongated, with 2 setules at about midlength and 2 setules subapically; furthermore, 2 hyaline elements are to be seen apically (Fig. 5C). Habitus slender. Dorsocaudal margin of cephalothorax and following three somites smooth. 3rd free somite dorsally with a double, half-circular ornamentation. Hyaline frill of following pereiomer, genital double-somite and following abdominal somite composed of rectangular, distally slightly indented lappets. Genital double-somite and following somite subdistally on ventral side with two rows of spinules. Penultimate somite dorsolaterally with a continuous row of spinules, dorsally forming a pseudoperculum; ventrally the hyaline frill is obviously slightly extended, probably also consisting of lappets. Ultimate somite short, subdivided. Furca more or less square, distal ventral margin spinulose. Outer margin subapically with 2 slender setae. Apically 3 setae insert, inner one short, middle one with spiny armature in its middle part, distal outer seta slightly plumose. About middorsally on inner margin a short plumose seta arises (Figs. 5A,B).

1st Antenna (Fig. 8A): 6-segmented, compact, furnished with several plumose setae; aesthetasc on 4th segment.

2nd Antenna (Fig. 6A): Allobasis has 1 plumose seta on anterior margin. Distal endopodite segment with two groups of spinules on anterior margin, distal margin with a row of spinules. Subapically 3 setae insert, apical edge with 6 setae, 4 of them geniculate. Exopodite 3-segmented. Basal segment obviously with 1 very weak seta laterally, distally with 1 plumose seta; middle segment has 1 plumose seta; distal segment with 1 seta laterally and 2 setae terminally.

Mandible (Fig. 6B): Chewing edge of precoxa (coxa) with several pointed teeth and 1 seta. Palp with 2 segments. Proximal segment carries 1 plumose seta. Distal part (enp.) with 2 setae and 1 stalked, apically vesica- or club-like ("claviform") appendage.

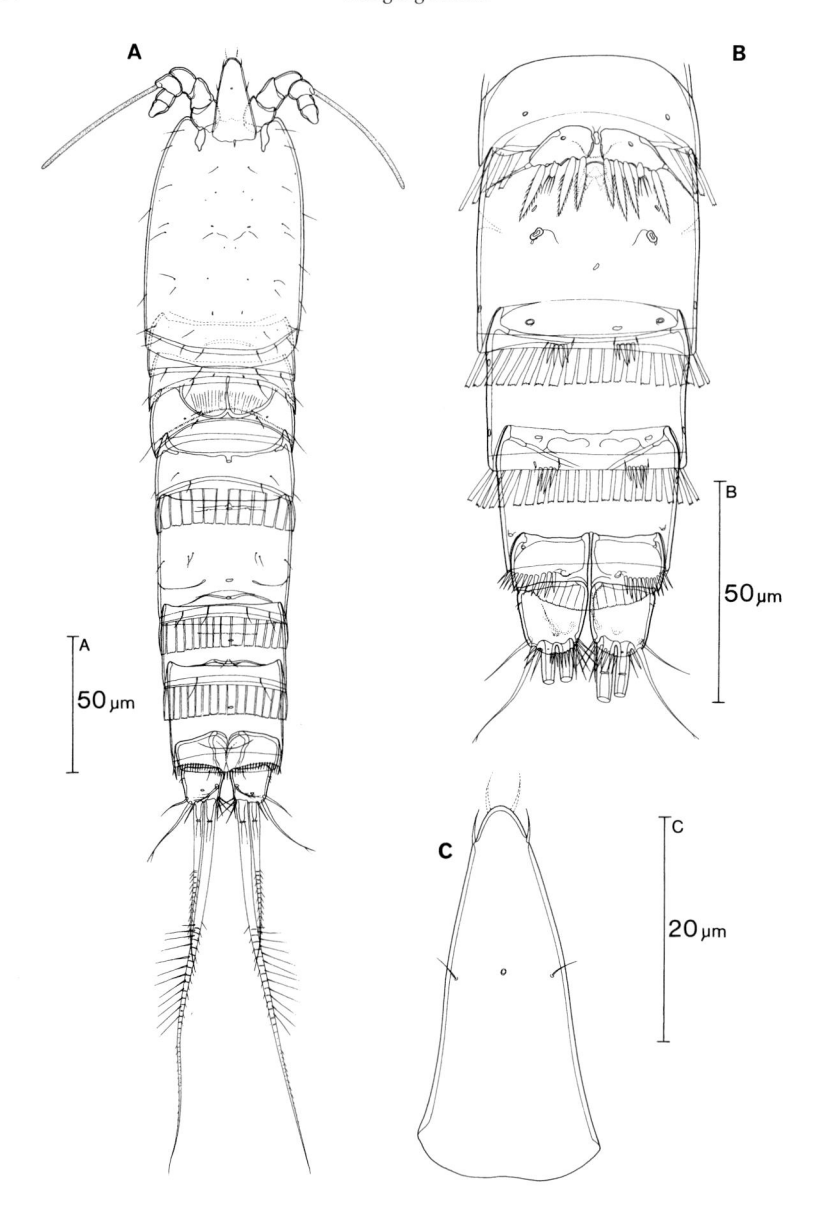

Fig. 5. *Micropsammis galapagoensis* nov. spec. ♀. A. Habitus, dorsal side. B. Abdomen, ventral side (note different armature of both P.5). C. Rostrum.

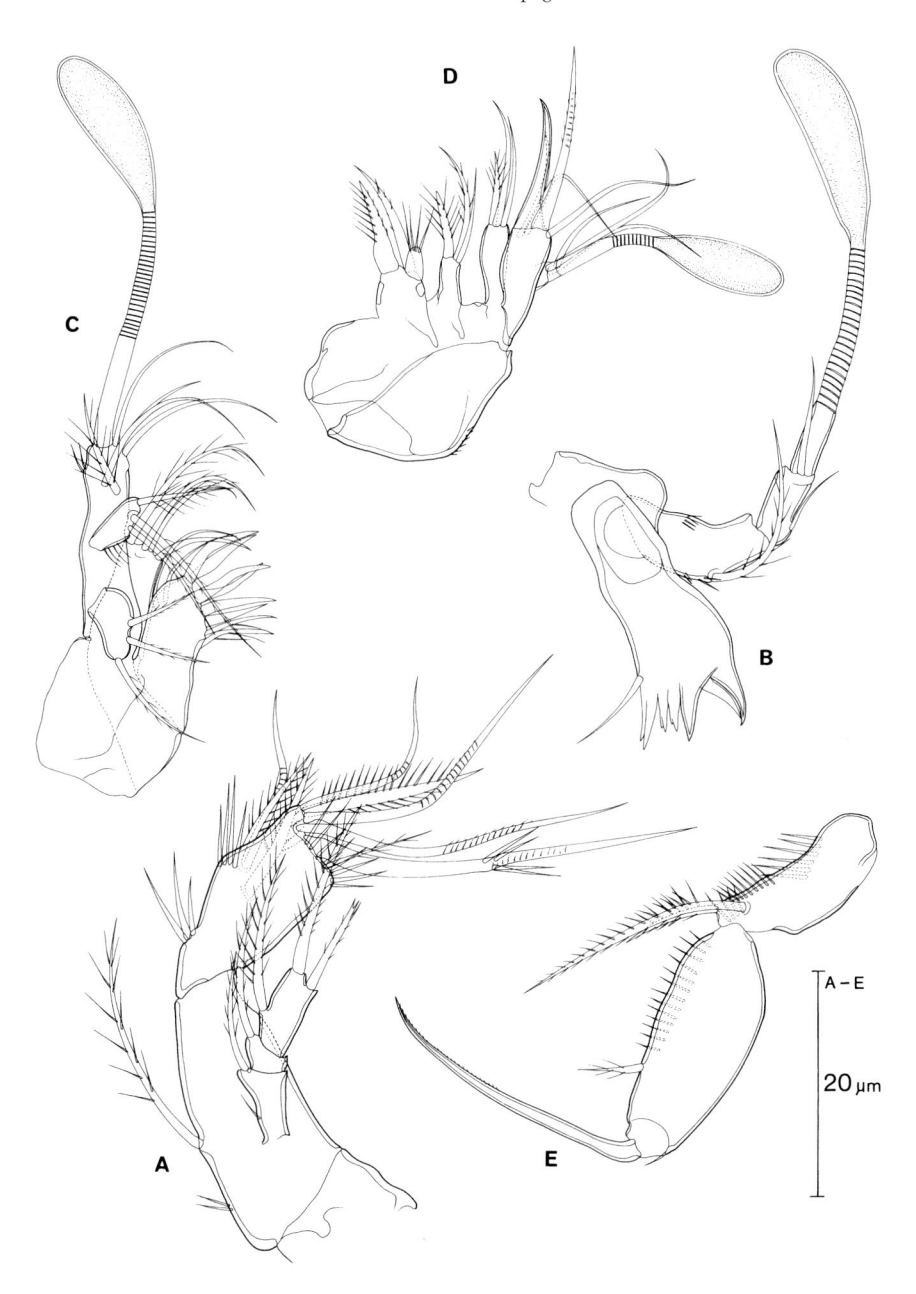

Fig. 6. *Micropsammis galapagoensis* nov. spec. A. 2nd Antenna ♀. B. Mandible ♀. C. 1st Maxilla ♀. D. 2nd Maxilla ♂. E. Maxilliped ♀.

1st Maxilla (Fig. 6C): Arthrite of precoxa with 3 pairs of spines and 2 setae on distal margin and 2 setae on surface. Coxa with 4 plumose setae. Basis elongated, bears 5 setae and 1 stalked, terminally vesica- or club-like appendage. Exopodite and endopodite each with 3 setae.

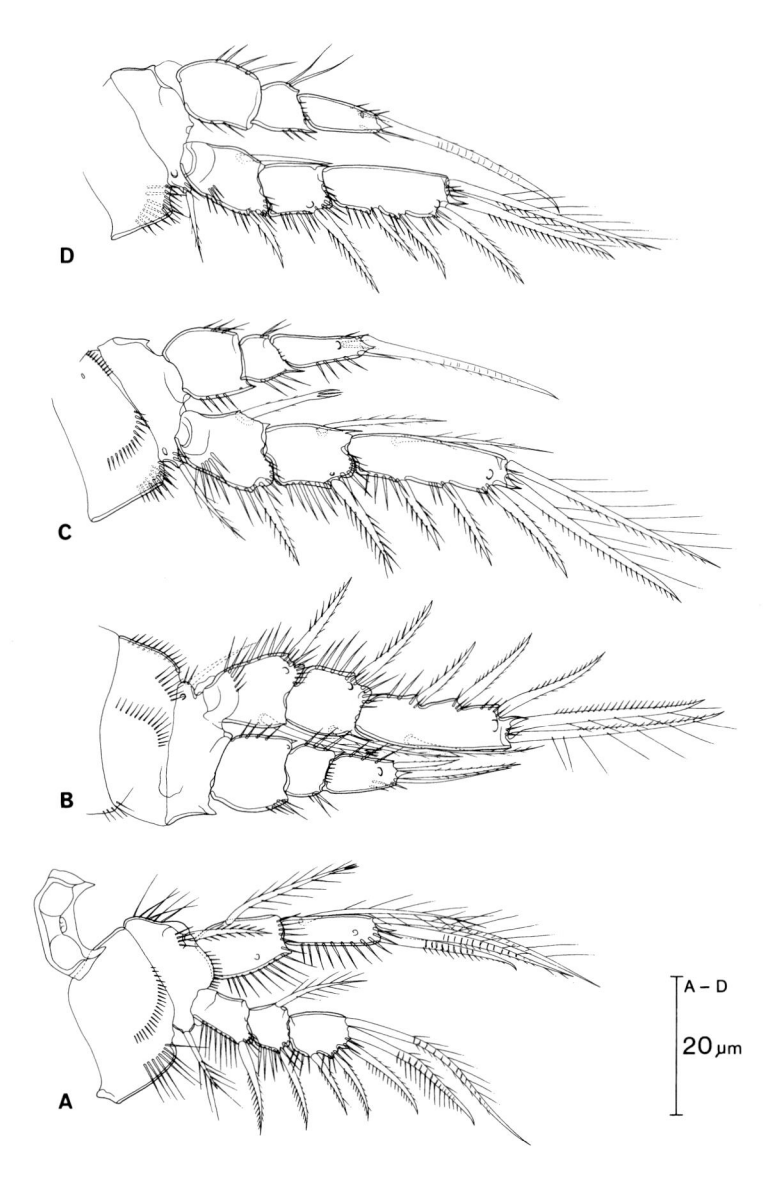

Fig. 7. *Micropsammis galapagoensis* nov. spec. A. P.1 ♂. B. P.2 ♀. C. P.3 ♀. D. P.4 ♀.

2nd Maxilla (as in male, Fig. 6D): Syncoxa with 3 endites, each with 3 appendages; the ones of proximal endite very different. Basis (allobasis) bears 1 claw, 1 strong seta and 3 slender setae. Endopodite weakly 2-segmented, furnished with 3 slender setae and a special appendage, which corresponds to the one of mandible and 1st maxilla.

Maxilliped (Fig. 6E): Basis (syncoxa) with rows of spinules and 2 plumose setae subdistally, unequal in length. Proximal endopodite segment (basis) furnished with a row of spinules near inner margin and 1 short plumose seta. Distal endopodite segment (enp.) bears 1 long, slender claw.

P.1 (as in male, Fig. 7A): Coxa with 2 rows of spinules of different length. Basis with long hairs on inner margin and a row of spinules distally. Inner appendage pinnate, outer seta ornamented with long spinules. Exopodite 3-segmented. Outer margins spinulose, outer spines slender. Middle segment has an inner seta. Distal segment with 5 appendages. Endopodite 2-segmented. Outer margins with long spinules. Proximal segment has 1 inner seta. Distal segment bears 1 seta basally on inner margin; apically with 3 setae, inner one longest.

P.2 – P.4 (Figs. 7B-D): Coxa ornamented with rows of spinules. Inner margin of basis ending in an acute tip. Exopodite with 3 segments; outer margins spinulose, outer spines pinnate. Proximal segment with an inner seta, which terminates in a prong-like manner. Middle and distal segment of P.2 and P.3 each with a slender plumose inner seta. Endopodite 3-segmented. Proximal and middle segment with spinules on outer and inner margin. Distal segment of enp. P.2 bears 3 appendages, the one of P.3 and P.4 has 1 long and robust seta, which is probably (not to ascertain unequivocally) accompanied by a vestigial seta.

Seta and spine formula:

	Exopodite	Endopodite
P.2	(1.1.123)	(0.0.021)
P.3	(1.1.123)	(0.0.0(1-2)0)
P.4	(1.0.023)	(0.0.0(1-2)0)

P.5 (Fig. 8B): Reduced to a plate, which carries 1 slender outer seta, 3 rootlike pinnate spines and 1 short appendage furnished with long filaments distally.

Male: Differs from the female in the following respects:
– Body length 0.24 – 0.27 mm.
– 1st Antenna subchirocer.
– Endopodite P.2 (Fig. 8C): Outer appendage of distal segment weakly separated from the segment. Middle element short. Inner seta longest.

- Endopodite P.3 (Fig. 8D): Middle segment with a distal hyaline projection.
- Both P.5 (Fig. 8E) fused; armature corresponds to that of the female.
- P.6 (Fig. 8E) has 1 root-like outer spine and 1 short inner seta.

Variability. The left P.5 of holotype has 1 additional root-like spine (see Fig. 5B).

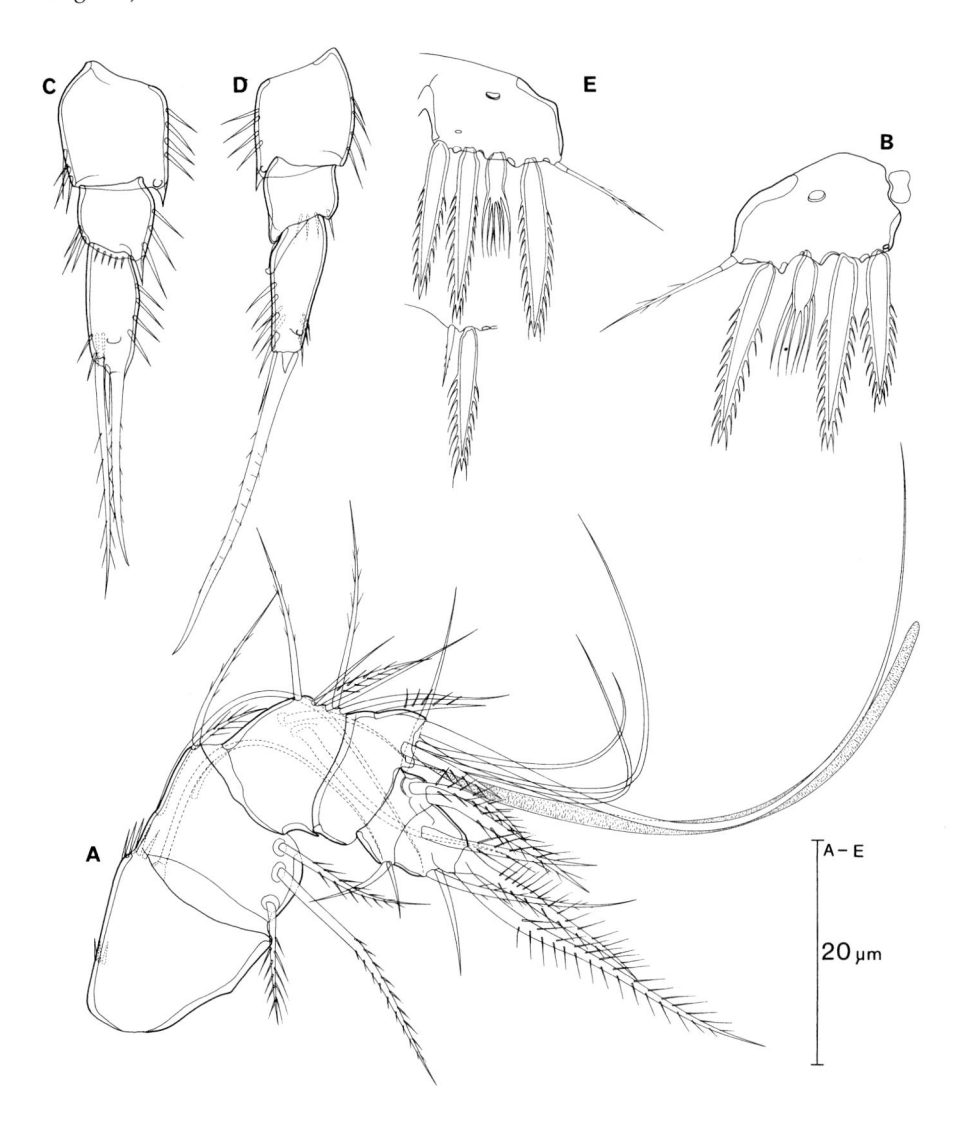

Fig. 8. *Micropsammis galapagoensis* nov. spec. A. 1st Antenna ♀. B. P.5 ♀. C. Endopodite P.2 ♂. D. Endopodite P.3 ♂. E. P.5 and P.6 ♂.

Discussion. The outstanding character of the Santa Cruz animals is the existence of vesica-/club-like appendages on the mouth parts. Species exhibiting such "claviform aesthetascs" are ranked nowadays among the Paranannopidae Por, 1986 and were comprehensively discussed by GEE & HUYS (1991). Within this "Verwandtschaftskreis" a monophyletic sub-group can be outlined which, thus far, comprises three monotypic genera previously classed with the Tachidiidae Sars, 1909, emend. Lang, 1948:

(1) *Leptotachidia iberica* Becker, 1974 from the Iberian deep-sea off Portugal (see BECKER 1974);

(2) *Micropsammis noodti* Mielke, 1975 and

(3) *Telopsammis secunda* (Mielke, 1975) (also belonging to *Micropsammis* according to the first description); both species were collected in List, Island of Sylt, Germany (see MIELKE 1975).

This species-group is characterized by several autapomorphies, e.g.:
– Hyaline frill of some somites deeply incised (palisadenartig).
– Structure of the mandibular palp.
– Fusion of benp. and exp. P.5 ♀ and ♂ (advanced state compared to the likely adelphotaxon *Paradanielssenia* Soyer, 1970).

GEE & HUYS (1991) established a new genus for *Micropsammis secunda*, viz. *Telopsammis*, and interpreted this species to be more closely related to *Leptotachidia iberica* than to *Micropsammis noodti*, enumerating more common features between *T. secunda* and *L. iberica* than between *T. secunda* and *M. noodti*, and referring to the principle of parsimony. However, the majority of these features are reductions, which **may** have developed convergently. On the other hand, *M. noodti* and *T. secunda* show some features, which can be interpreted as synapomorphies:

– Evolutionary starting-point of female A.1 with more than 6 segments. A first step led to a 6-segmented A.1 (*Micropsammis*). In a further step the 5-segmented A.1 developed, its terminal segment acquiring a specific bulbous appendage – "Brodskaya Organ" according to BECKER 1974 – (*Leptotachidia*).

– P.2 ♂ with a "special quality" of sexual dimorphism.

– General appearance of P.5. In the male the P.5 of *M. noodti* and *T. secunda* largely correspond. In the female the number of appendages is surely identical in *L. iberica* and *T. secunda*. However, *L. iberica* has 5 stout spines; whereas *T. secunda* exhibits 5 slender appendages. In contrast to the opinion of GEE & HUYS (1991) that benp. and exp. of P.5 ♀ are "indistinguishable", I think that they can be definitely distinguished. The former division of the inner part of benp. (3 setae) and the portion of exp. (2 setae) is slightly indicated by a small notch (*M. noodti* with a deep notch).

I therefore plead for the retention of the species *secunda* within the genus *Micropsammis*.

The new species is integrated into *Micropsammis*, because of:
– The 6-segmented A.1 ♀.
– The sexually dimorphic enp. P.2 ♂ (however, outer distal appendage not developed into an apophysis thus occupying a position between *Micropsammis* and *Leptotachidia*).

The following features can be stressed as being important for the new species:
– P.1 ♀ enp. distal segment, inner apical seta longest (plesiomorph?).
– Distal segment of exp. P.4 with 3 outer spines (plesiomorph); inner margin without seta (apomorph, convergently to *L. iberica* and *M. secunda*?).
– Reduced armature of enp. P.2 – P.4; the terminal segment has only 1 well developed seta (apomorph).
– Both P.5 ♀ and ♂ furnished with only 4 appendages, outer element but one distally set with filaments (apomorph); this character is also intermediate between *Micropsammis* and *Leptotachidia*.

Thus far, the species discussed above have only been found in widely separated habitats. This suggests the existence of many more species belonging to this taxon, whose level of classification (genus, sub-group, family) is absolutely insignificant.

? Paranannopidae Por, 1986
? Canthocamptidae Sars, 1906, emend. Monard, 1927, Lang, 1948
Carolinicola Huys & Thistle, 1989
Carolinicola galapagoensis nov. spec.

(Figs. 9 – 13)

Locality and material. qn: Santa Cruz: Bahía Academy (IX,6; **Locus typicus**. Monthly transects from March 1972 to February 1973); 1 ♀, 1 ♂.

Both animals were dissected. Holotype is the female, reg. no. I Gal 1119. Paratype is the male, reg. no. I Gal 1120.

Description

Female: Body length from tip of rostrum to end of furcal rami 0.40 mm. Rostrum basally with a distinct division line; near its middle a pore can be seen. Distal part tapering gradually. About distal 2/3 of its lateral margins a slender seta inserts, respectively. The rostrum seems to be enveloped by a hyaline cover (Fig.9B). Genital double-somite subdivided dorsolaterally. Ventral caudal margin of genital double-somite laterally with spinules. Following two

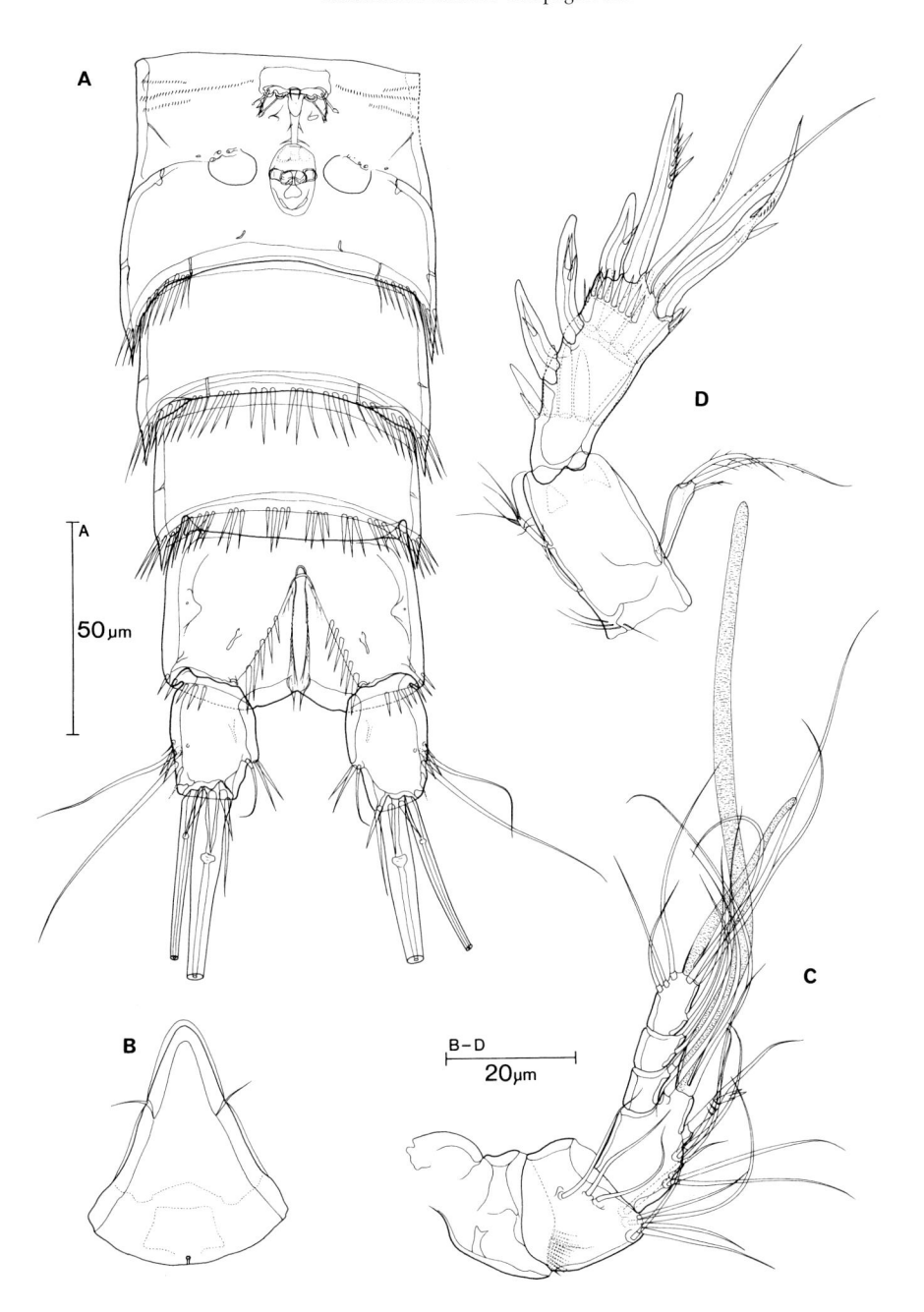

Fig. 9. *Carolinicola galapagoensis* nov. spec. ♀. A. Abdomen, ventral side. B. Rostrum. C. 1st Antenna. D. 2nd Antenna.

somites with groups of 3 – 4 spinules on ventral caudal margin. Anal somite ventrolaterally with short spinules on caudal margin; middle part of this somite with a triangular furrow, bordered by slender spinules. Anal operculum obviously smooth or slightly notched. Furca a little longer than broad. Laterally a minute setule, 2 slender setae and 2 – 3 spinules insert; moreover, a short hyaline structure is to be seen subdistally on a chitinous break. Apical margin furnished with some spinules and 3 setae, middle one longest and slightly bent. At about 2/3 of the inner margin a short basally bipartite seta and a few (2) accompanying slender spinules arise (Fig. 9A).

1st Antenna (Fig. 9C): 6 segments. 1st segment with a row of spinules and 1 short seta. 2nd and 3rd segments with several slender setae, 3rd segment furthermore with 1 spiny seta and an aesthetasc, which is basally fused with a seta. Distal three segments narrower. Last segment also furnished with an aesthetasc.

2nd Antenna (Fig. 9D): Allobasis proximally with a row of hair-like spinules; anterior margin has 2 setae, distal one terminating with several filaments. Free endopodite segment has 2 blunt spines on anterior margin, two rows of coarse spinules and another one with small spinules on surface. Apical edge furnished with 3 stout appendages and 2 slender setae. Exopodite consisting of one slender segment, distally bearing 3 setae of different length.

Mandible (Fig. 10A): Chewing edge of precoxa (coxa) with several teeth and a seta, which is furnished with setules on distal side. Coxa-basis (basis) with two rows of spinules and 1 distal seta, which is apically armed with long spinules. Endopodite 1-segmented, laterally with 1, apically with 3 slender setae. Exopodite obviously consisting of 1 small segment, bears 1 seta (bifid at tip?).

1st Maxilla (Fig. 10B): Arthrite of precoxa with several appendages on distal edge and 2 setae on surface. Coxa has 2 setae. Basis with 4(or 5?) setae. Endopodite represented by 3 setae (1 difficult to see because on the mouth part drawn it runs along the margin of the basis). Segment of exopodite seemingly fused with basis, bears 1 strong appendage and 1 short, thin seta.

2nd Maxilla (Fig. 10C): Syncoxa with some rows of spinules and two endites. Both endites are apically furnished with 3 appendages of different shape. Basis (allobasis) with 1 claw and 2 slender setae. Endopodite represented by 3 long setae and 1 accessory short setule.

Maxilliped (Fig. 10D): Basis (syncoxa) with long spinules on inner side and 1 seta subapically. Proximal endopodite segment (basis) furnished with long spinules. Distal endopodite segment (endopodite) with 1 strong claw, which bears some spinules and an accompanying setule basally.

P.1 (Fig. 11A): Precoxa with a distal row of spinules. Coxa with several rows of minute spinules and an outer row of coarse spinules. Basis also with several

Fig. 10. *Carolinicola galapagoensis* nov. spec. ♀. A. Mandible. B. 1st Maxilla. C. 2nd Maxilla. D. Maxilliped.

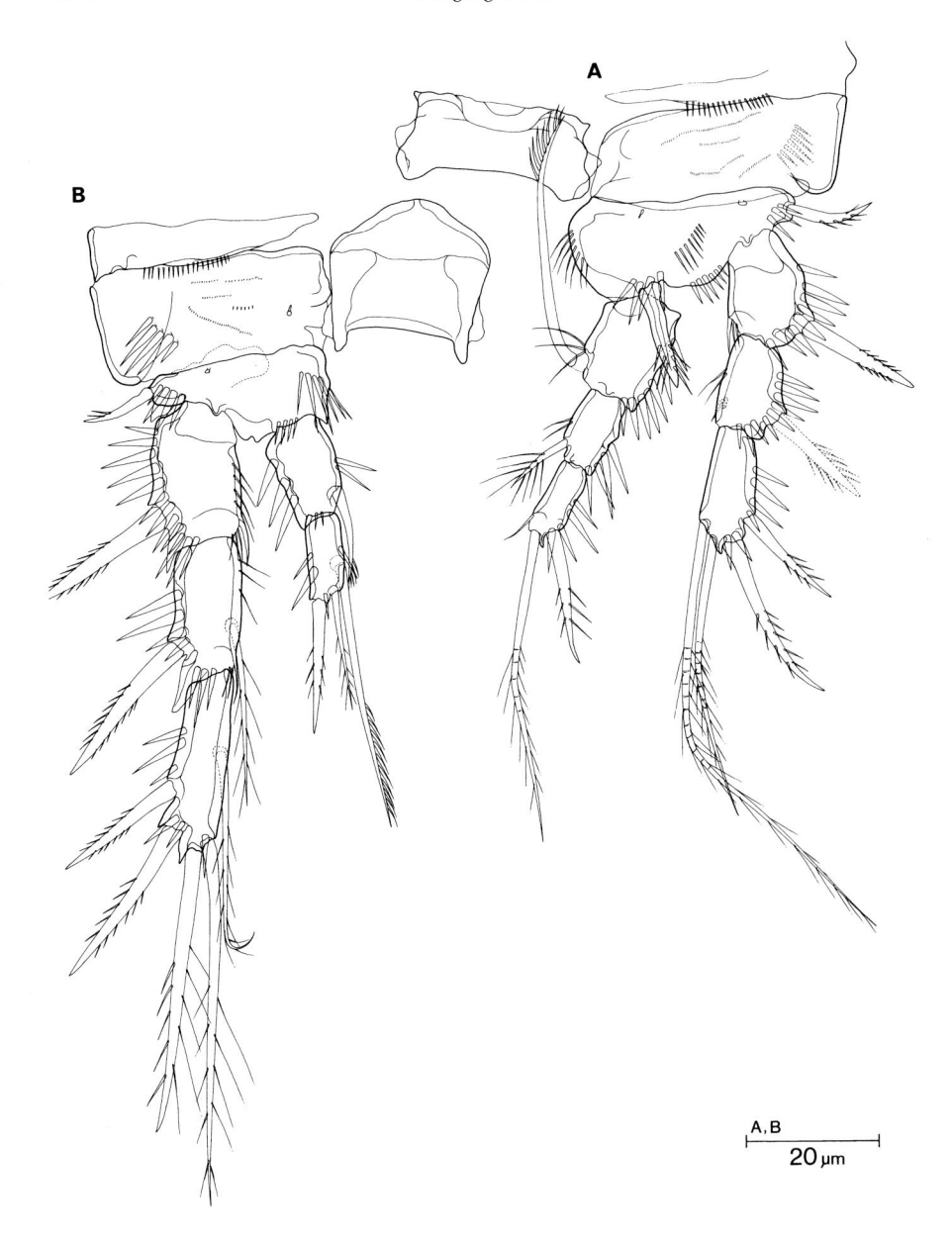

Fig. 11. *Carolinicola galapagoensis* nov. spec. ♀. A. P.1. B. P.2.

rows of spinules, an outer slender spine and an inner stout spine (here somewhat displaced). Exopodite 3-segmented, outer margins furnished with coarse spinules. Proximal segment has an outer spine, middle segment as well and an inner rudimentary seta; distal segment with 4 appendages. Endopodite with 3 segments, all with coarse spinules on outer margin. Proximal segment bears a long, distally armed seta. Middle and distal segments narrower. Middle segment has an inner plumose seta; distal segment bears a long, geniculate seta and a slender spine apically, subapically a rudimentary seta on inner margin.

P.2 – P.4 (Figs. 11B, 12A, 13A): Precoxa with a distal row of spinules. Coxa furnished with several rows of minute spinules on surface and an outer row of long spinules. Basis with an inner, outer and distal row of spinules and a short (P.2) or long (P.3 and P.4) outer seta, respectively. Exopodite with 3 segments, all with coarse spinules on outer margin. Outer margin of proximal and middle segment extended distally; each one furnished with a spine. Middle segment moreover with an inner plumose seta. Last segment bears 2 slender outer spines, 2 apical armed appendages and 1 (P.2) or 2 (P.3 and P.4) inner setae. These setae are plumose except distal one of P.4 which is saw-like. Endopodite 2-segmented. Proximal segment with an inner seta, which is short and brush-like in P.2 and P.3 and long in P.4. Distal segment with 3 setae in P.2 and 4 setae of different length in P.3 and P.4.

Seta and spine formula:

	Exopodite	Endopodite
P.2	(0.1.122)	(1.120)
P.3	(0.1.222)	(1.220)
P.4	(0.1.222)	(1.121)

P.5 (Fig. 13B): Baseoendopodite and exopodite fused. Part of benp. bears an outer plumose seta and 1 long plumose seta and 3 short armed setae medially. Part of exp. has 5 setae altogether.

Male: Differs from the female in the following respects:
- Body length of the only male found is 0.37 mm.
- 2nd, 3rd and penultimate somite each with groups of 3 – 4 spinules on ventral distal margin.
- 1st Antenna (Fig. 12B): More or less subchirocer, 8 segments. 1st segment with a row of spinules and a short seta. Small 3rd segment and 4th segment each with a spiny seta; 4th segment has a long aesthetasc. Segments 5 – 8 narrow. Anterior margin of segments 6 and 7 armed with structures forming a complicated pattern. Distal segment small, also bearing an aesthetasc.
- P.1: Setae as in female.

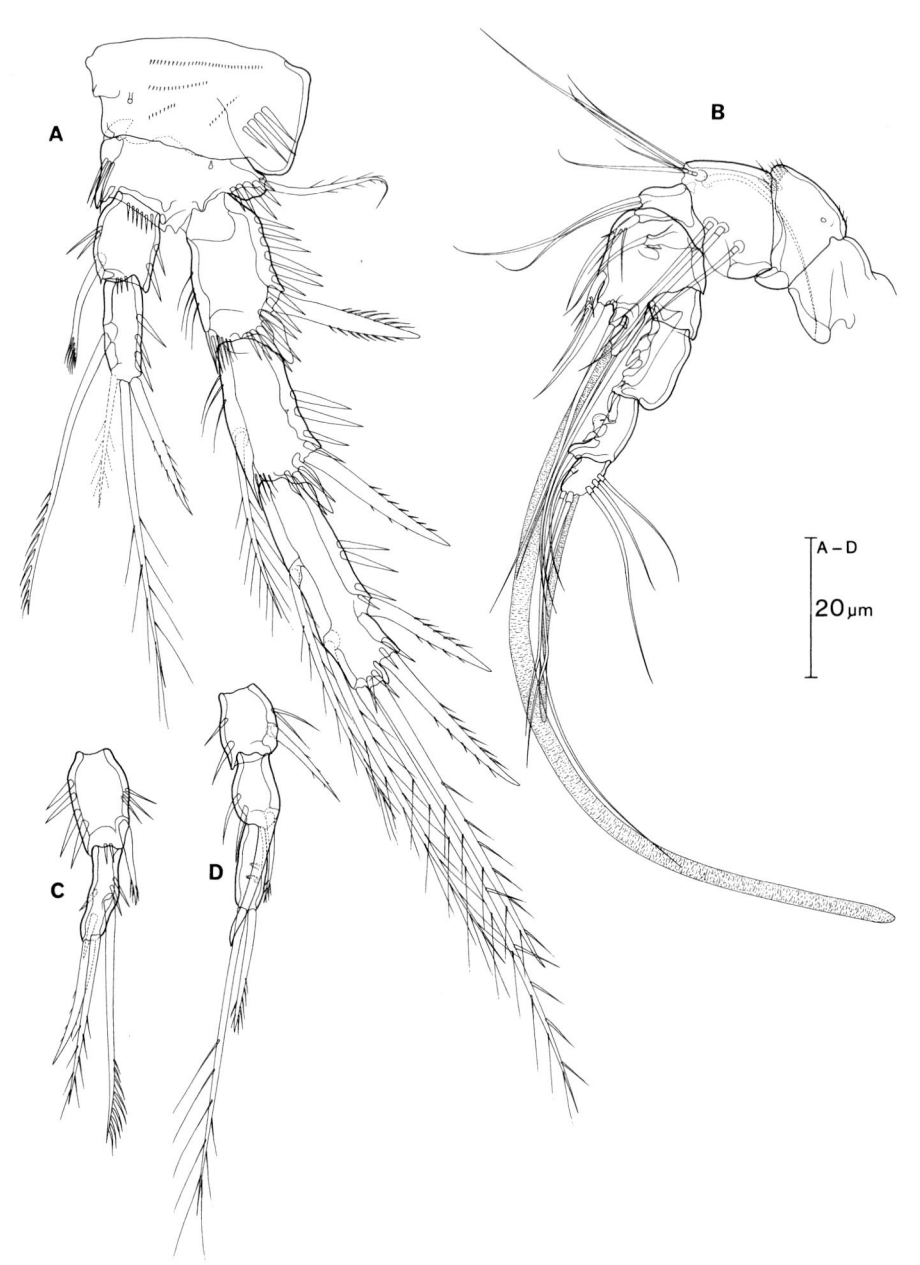

Fig. 12. *Carolinicola galapagoensis* nov. spec. A. P.3 ♀. B. 1st Antenna ♂. C. Endopodite P.2 ♂. D. Endopodite P.3 ♂.

Fig. 13. *Carolinicola galapagoensis* nov. spec. A. P.4 ♀. B. P.5 ♀. C P.5 (left) and both P.6 ♂.

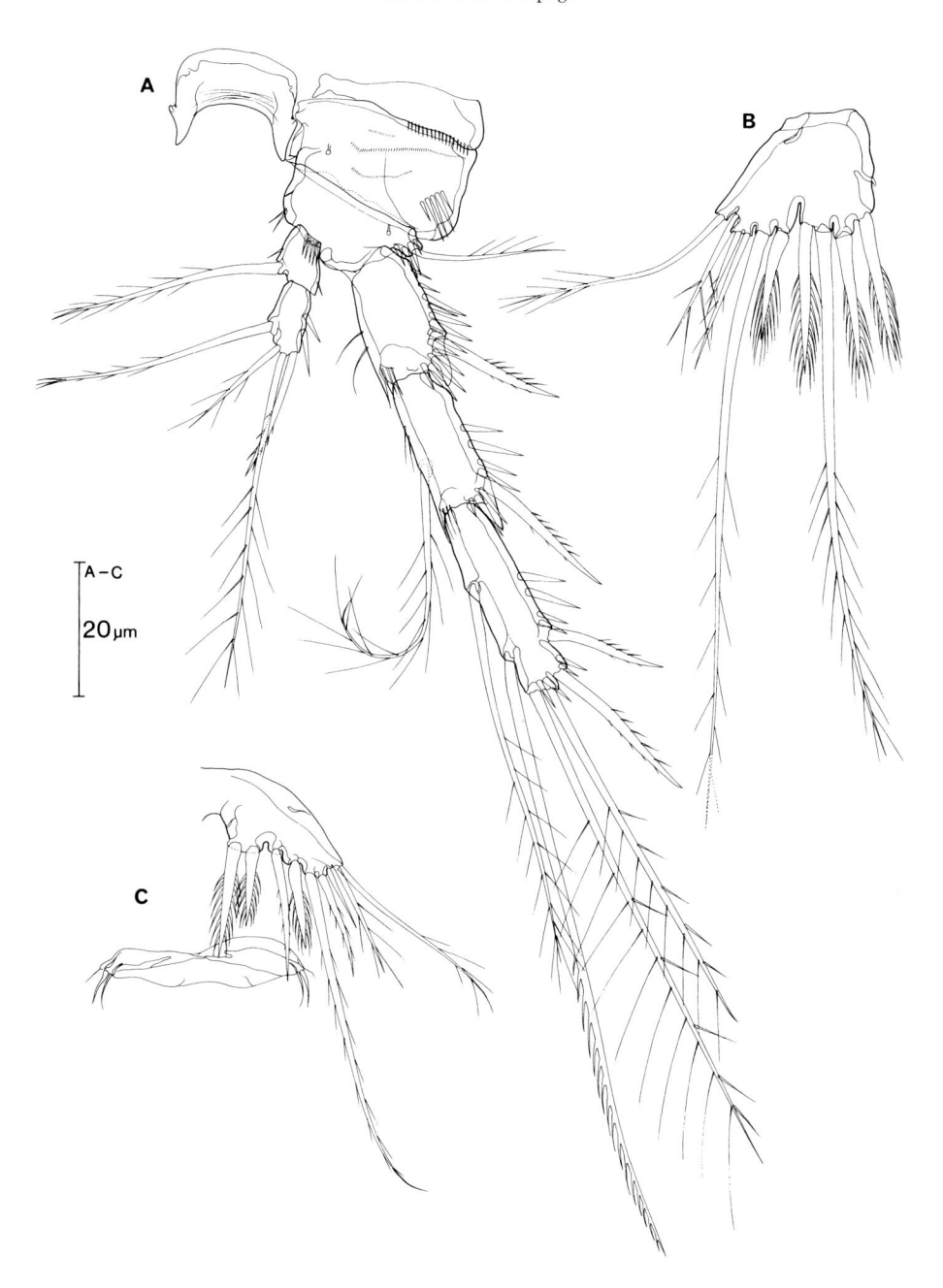

- P.2 (Fig. 12C): Middle seta of distal segment enp. longer.
- P.3 (Fig. 12D): Endopodite 3-segmented. Proximal segment has an inner seta. Middle segment with an inner seta and a terminal slender apophysis. Distal segment bears 2 setae, outer of which is more than twice as long as inner one.
- P.4: Distal outer seta of terminal segment enp. distinctly longer than in female.
- P.5 (Fig. 13C): Benp. and exp. fused. Benp. bears an outer slender seta and 2 armed setae on inner part. Portion of exopodite with 6 setae of different length.
- P.6 (Fig. 13C): Transverse plate with outer lobe that has 2 small setae.

Discussion. In the course of the disintegration of "*Hemimesochra* sensu lato" the new genus *Carolinicola* was established among others by HUYS & THISTLE (1989) in order to segregate *Hemimesochra trisetosa* Coull, 1973. This species was found off North Carolina, U.S.A., at four deep-sea locations (2400 – 5165 m); up to now only the female is known (COULL 1973). The new species from the beach of Bahía Academy/Santa Cruz largely corresponds to the Carolina species. Certainly, there exist some distinct differences. The most crucial point concerns the enp. P.1, which consists of three segments (plesiomorphic) in the Galapagos species instead of only two in *C. trisetosa* (middle and distal segments fused). Moreover, the A.1 is 6-segmented (plesiomorphic) (5-segmented in *C. trisetosa*, both distal segments obviously fused), the armature of the mandibular palp is reduced (apomorphic) compared to *C. trisetosa*. A few derived characters occur, but their significance – important synapomorphies or negligible convergences – is difficult to assess:

– Distal segment of exp. P.2 – P.4 has only 2 outer spines.
– Distal segment enp. P.2 with only 3 setae.
– Benp. and exp. P.5 fused.

The situation on anterior margin of allobasis A.2 is unclear. COULL (1973, Fig. 4) drew 1 hairy seta. The Galapagos animals exhibit 2 setae, the distal one is also furnished with long hairs.

HUYS & THISTLE regarded COULL's species as "an advanced member of the Paranannopidae", whereas the remaining former *Hemimesochra* species were left within the Canthocamptidae. The main reason concerned the "primitive" structure of the mandibular palp of *C. trisetosa*. Some of these species share some unequivocal, (syn-?) apomorphic characters with *Carolinicola*, e.g. *Boreolimella* Huys & Thistle, 1989; *Poria derketo* (Por, 1964) shows more or less the same structure of the P.1 as the Galapagos individuals suggesting an old phylogenetical connection.

The two specimens from Santa Cruz are (provisionally) assigned to *Carolinicola*. Probably the finding of the male of *C. trisetosa* will enable us to better clarify the relationship between both species. In any case they should not be separated from the former *Hemimesochra* species at the "family" level.

Canthocamptidae Sars, 1906, emend. Monard, 1927, Lang, 1948
Psammocamptus galapagoensis nov. spec.

(Figs. 14 – 18)

Localities and material. qn: Santa Cruz : Bahía Academy (IX,6; **Locus typicus**. Monthly transects from March 1972 to February 1973); 1 ♀, 3 ♂♂.
ql: Fernandina: Punta Espinosa (I,3; 25.9.72); 2 ♀♀, 1 ♂.
6 specimens were dissected. Holotype female, reg. no. I Gal 1111. Paratypes are 3 dissected ♂♂, reg. no. I Gal 1112 – 1114, all from Santa Cruz. The habitus is drawn from a female from Fernandina. All other drawings of the female are drawn from the holotype.

Description

Female: Body length from tip of rostrum to end of furcal rami 0.30 mm. Rostrum large, basally with separating line, subapically with 2 setules and apically with two small bulges (Fig. 15A). Cephalothorax with hairs on dorsolateral surface and on caudal hyaline frill. Following thoracic somites also with fine hairs on dorsal side. Genital double-somite partly subdivided; line of subdivision dorsolaterally with spinules. Caudal margins of genital double-somite and following somites – except ventral margin of anal segment – smooth; subapically the somites are furnished with spinules of different length, partly grouped together: genital double-somite on lateral side; following somite ventrally with a continuous row extending dorsolaterally; penultimate somite ventrally with an interrupted row of spinules; anal somite spinulose on ventral distal margin. Distal margin of anal operculum spinulose. Furca more than twice as long as broad. Distal margin spinulose. Subapically 1 long, slender seta inserts on outer edge and 1 short seta on dorsal surface nearby; both setae accompanied by spinules. Distal inner edge with 1 basally bipartite seta. Apically 3 setae arise: 1 small inner seta, 1 long middle and 1 shorter outer seta, both of latter ones furnished with spinules (Figs. 14A,B).

1st Antenna (Fig. 15A): Bent caudad. 7 segments. 4th and last segment each with an aesthetasc. Anterior edge with 5 thorny cones: 2nd segment with 1, 3rd segment with 2, 6th and 7th segment each with 1 cone.

2nd Antenna (Fig. 15B): Coxa with some spinules. Allobasis with some slender spinules; between them a slender seta inserts on a chitinuous gap. Free

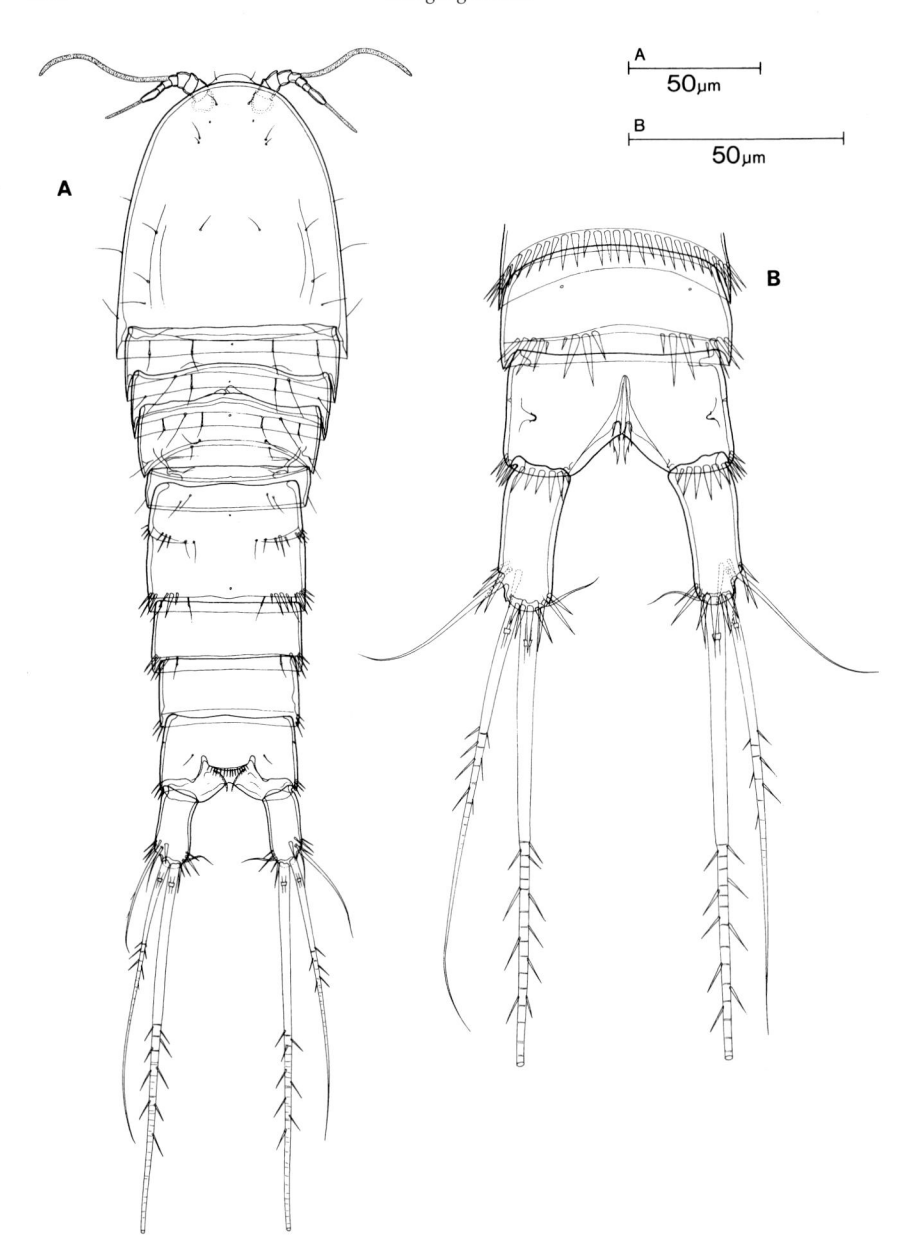

Fig. 14. *Psammocamptus galapagoensis* nov. spec. ♀. A. Habitus, dorsal side. B. Caudal part of Abdomen, dorsal side.

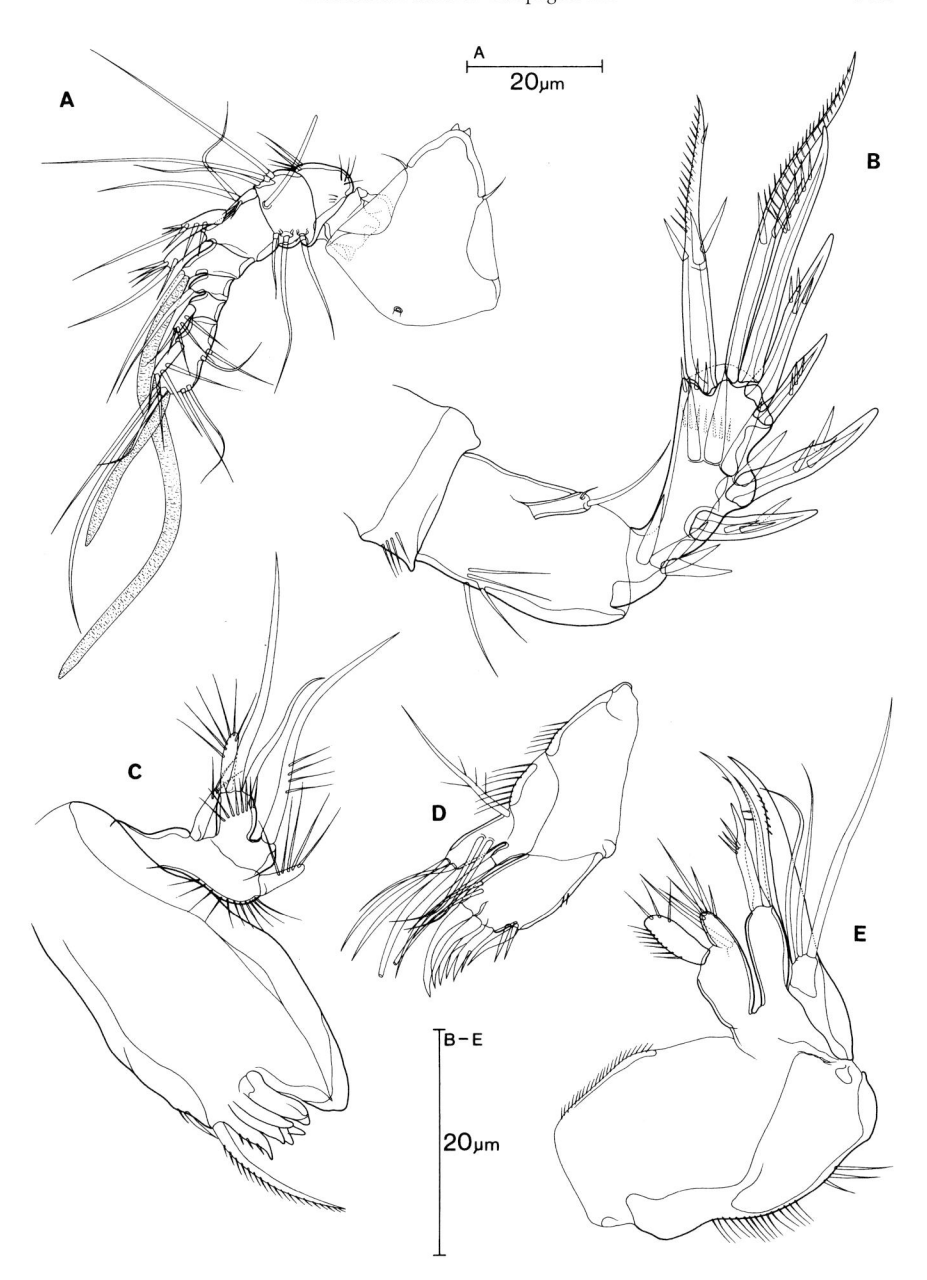

Fig. 15. *Psammocamptus galapagoensis* nov. spec. A. Rostrum and 1st Antenna ♀. B. 2nd Antenna ♀. C. Mandible ♀. D. 1st Maxilla ♂. E. 2nd Maxilla ♀.

endopodite segment furnished with coarse spinules. Anterior edge with 2 stout spines. Apical margin bears 5 spiny appendages. Exopodite 1-segmented, distally with 1 long and 1 minute seta.

Mandible (Fig. 15C): Chewing edge of corpus mandibulae with several teeth and 1 seta armed with one row of spinules. Coxa-basis (basis) with spinules on inner margin and 1 spinulose cone. Endopodite bears 1 lateral seta which has some spinules on proximal part, and 3 terminal appendages; subterminally there is a row of spinules. Exopodite obviously without a segment, represented only by 1 spinulose cone.

1st Maxilla (as in male, Fig. 15D): Arthrite of precoxa with several appendages on distal margin and 2 setae on surface. Coxa with 2, basis with 4 setae. Endopodite represented by 2 (or 3?) setae, exopodite consisting of 1 slender seta.

2nd Maxilla (Fig. 15E): Syncoxa spinulose on inner and outer margin; with two endites. Proximal endite carries 1 broad, spinulose appendage, another spinulose appendage which is fused with endite, and 1 slender seta. Distal endite with 3 setae of different length. Basis (allobasis) with 1 claw and probably 1 seta, or this seta belongs to the 1-segmented endopodite, which would then have 3 setae.

Maxilliped (as in male, Fig. 16A): Basis (syncoxa) with some spinules and 1 inner seta standing on a lobe. Proximal endopodite segment (basis) has a row of spinules near inner margin and some slender spinules on outer margin. Distal endopodite segment (endopodite) with 1 strong claw and an accompanying setule.

P.1 (Fig. 16B): Coxa with two rows of spinules on surface. Basis also with spinules near distal margin, an outer slender seta and an inner strong appendage. Exopodite with 3 segments which have coarse spinules on outer and distal margins. Proximal and middle segment each with a pinnate outer spine, middle segment moreover with a rudimentary inner seta. Distal segment with 4 appendages. Endopodite with 3 segments which are furnished with coarse spinules; proximal and middle segment each with a rudimentary inner seta, distal segment carries 1 inner short seta and 2 pinnate appendages.

P.2 – P.4 (Figs. 16C, 17A,B): Coxa with a row of spinules on surface near outer margin. Basis with spinules on inner, distal and outer margin; outer seta short in P.2, well developed and slender in P.3 and P.4. Exopodite 3-segmented. Proximal and middle segment with spinules and a slender spine on outer margin; inner margin with slender spinules and a short seta on middle segment. Distal outer edge tooth-like elongated. Distal segment has some spinules on outer and distal margin, 2 outer spines, 2 plumose distal setae and 1 slender seta on inner margin. Endopodite 2-segmented. Proximal segment with some spinules and a rudimentary seta on inner margin. Distal segment with some spinules on outer and inner margin; furthermore, this segment bears 3 setae in P.2, and 4 setae in P.3 and P.4. Distal segment of P.3 has an indentation on outer margin.

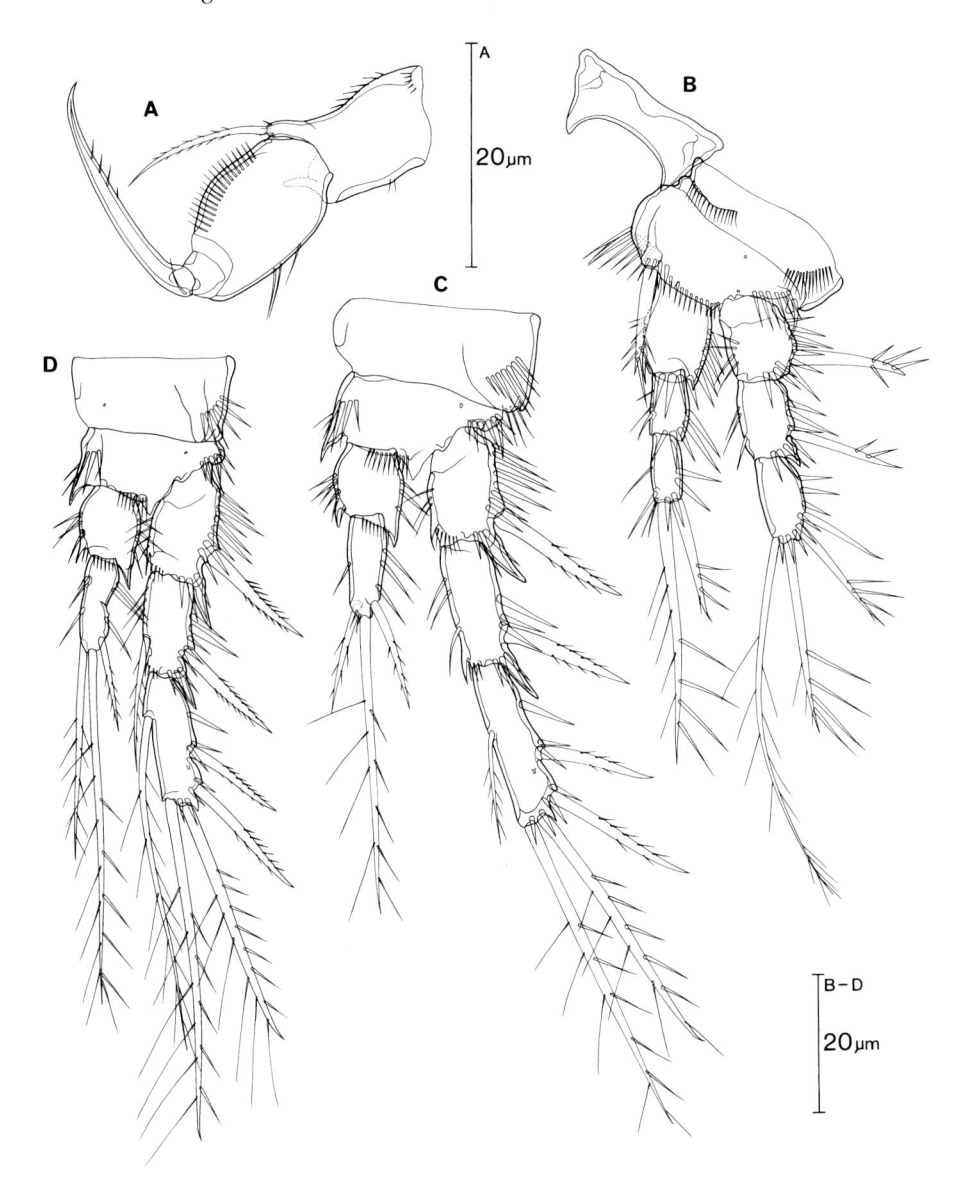

Fig. 16. *Psammocamptus galapagoensis* nov. spec. A. Maxilliped ♂. B. P.1 ♀. C. P.2 ♀. D. P.2 ♂.

Seta and spine formula:

	Exopodite	Endopodite
P.2	(0.1.122)	(1.111)
P.3	(0.1.122)	(1.220)
P.4	(0.1.122)	(1.121)

P.5 (Fig. 17C): Baseoendopodite and exopodite fused. Benp. with an outer seta; inner part with 2 inner plumose setae, 1 long pinnate seta and 1 short ou-

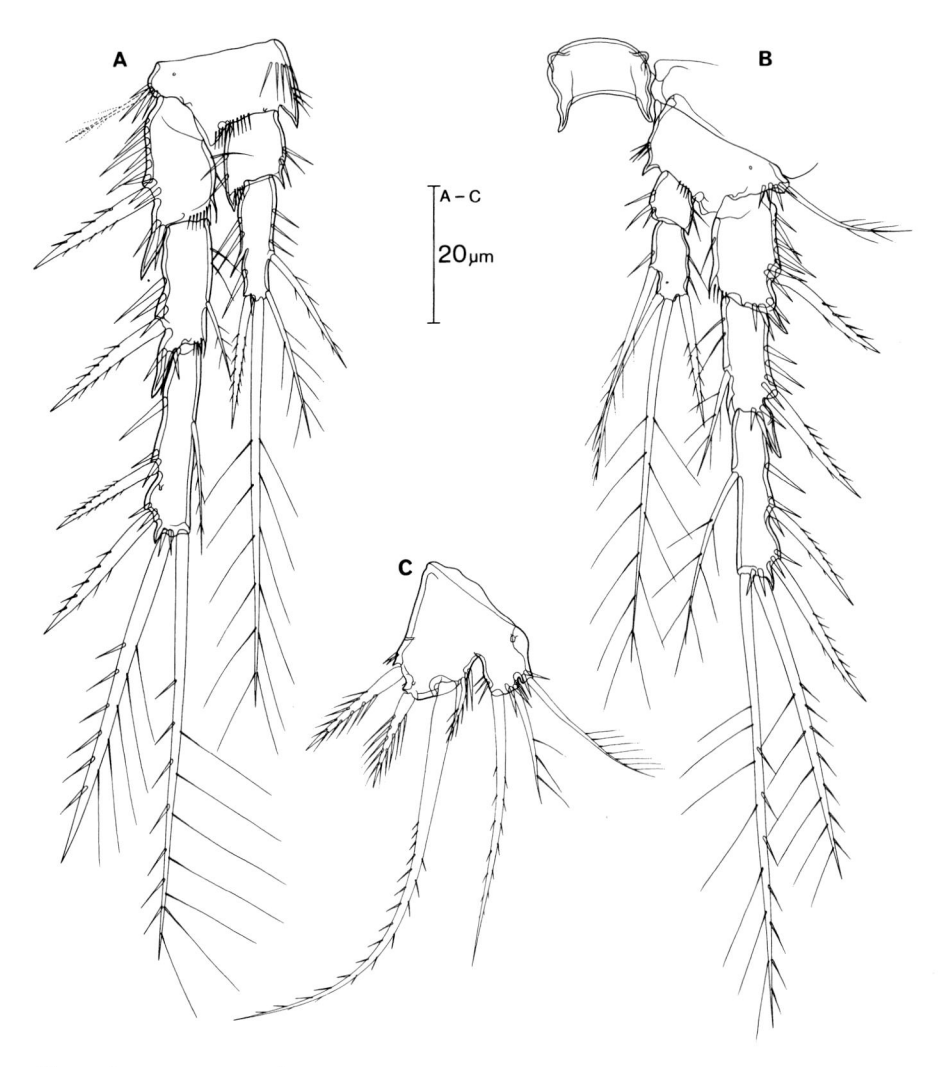

Fig. 17. *Psammocamptus galapagoensis* nov. spec. ♀. A. P.3. B. P.4. C. P.5.

ter seta. Exopodite with 1 long inner seta, 1 middle seta furnished with long hairs and 1 rudimentary outer setule.

Male: Differs from the female in the following respects:
– Body length of dissected ♂♂: 0.25 – 0.30 mm.
– 2nd, 3rd and 4th (penultimate) somite each with a continuous ventral row of spinules.
– 1st Antenna (Fig. 18A): Haplocer, 8 segments. 2nd segment with 2 small

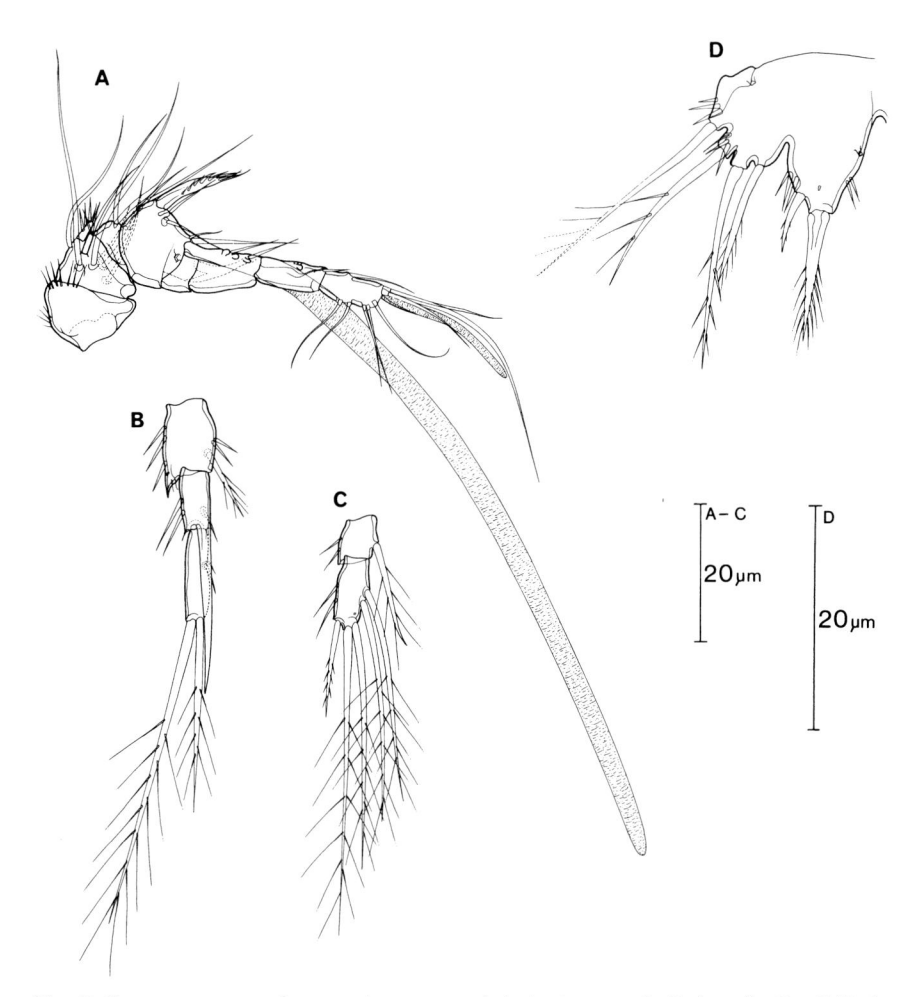

Fig. 18. *Psammocamptus galapagoensis* nov. spec. ♂. A. 1st Antenna. B. Endopodite P.3. C. Endopodite P.4. D. P.5.

thorny cones. 4th segment with 1 prominent thorny cone and 1 long aesthetasc. Distal segment with 1 short aesthetasc and several (9) slender setae.
- P.1: The rudimentary setae on proximal and middle segment enp. and middle segment exp. are slightly stouter. Inner seta of distal segment enp. well developed.
- P.2 (Fig. 16D): The inner setae of middle and distal segment exp. and of distal segment enp. are distinctly longer than in female. Inner margin of distal segment enp. carries a rudimentary seta. This seta may be longer than in Fig. 16D, or may be lacking. One ♂ has a distal segment with such a seta, its counterpart not.
- P.3 (Fig. 18B): As in P.2 the inner setae of middle and distal segment exp. are distinctly longer than in female. Endopodite 3-segmented. Proximal segment with a well developed inner seta. Inner edge of middle segment elongated into a long slender tooth; furthermore, a rudimentary seta inserts near base of this tooth. Distal segment has 2 apical plumose setae; inner margin with a rudimentary seta (or a chitinous hook).
- P.4 (Fig. 18C): Inner setae of middle and distal segment exp. and of proximal segment enp. are distinctly longer than in female. Distal segment enp. carries 5 setae (occasionally only 4 as in female).
- P.5 (Fig. 18D): Baseoendopodite and exopodite fused. Benp. with an outer seta; inner part bears 1 strong terminal seta and 1 small outer seta. Exopodite with 3 well developed setae and 1 outer rudimentary setule.
- P.6? No seta observed.

Variability. Apart from the items already mentioned above, the following divergences should be mentioned:
- Distal segment of right exp. P.4 of holotype has 3 outer spines. Its counterpart bears 2 as usual.
- Distal segment of left exp. P.4 of the ♂ from Fernandina has 2 setae on inner margin. The right exp. has 1 as usual.
- Basal segment of left enp. P.4 of one ♂ without inner seta; this seta, however, is present on the basis (!).

Note: 4 ♂♂ (2 of them were dissected) were found in a beach of Barrington (XI,2; 22.2.72) which largely resemble the species described above. However, there are some slight differences and, more important, they possess 2 setae on inner margin of distal segment both of P.3 and P.4. Probably these animals belong to another species.

Discussion. The animals of Santa Cruz and Fernandina closely resemble *Psammocamptus axi* Mielke, 1975, a species which was described from the eulittoral area of the east coast of List/Island of Sylt, Germany (MIELKE 1975). A re-examination of some characters of *P. axi* resulted in some different interpretations than in my description of 1975:

– The long distal segment of A.1 ♀ seems to be weakly subdivided.
– Anterior margin of allobasis A.2 with a slender seta, which is accompanied by long setules.
– Claw of maxilliped basally with a dwarfed seta.
– P.1 middle segment exp. and all three enp. segments apparently have a rudimentary inner setule. However, this is difficult to observe because of the presence of spinules nearby.
– Proximal, middle and probably distal segment of enp. P.3 ♂ also furnished with a rudimentary inner setule.

Further rudimentary setules were given with reservation as "(1)" in the seta and spine formula in my description (1975).

As to this aspect only minor or no differences between the two species exist. On the other hand, the following discriminations can be emphasized (*P. axi* in parenthesis):

– Exp. A.2 has only 1 well developed seta (2).
– P.2 enp. distal segment with inner seta (0).
– P.3 ♀ exp. middle segment with well developed inner seta (0).
– P.4 ♀ formula of exp. as in *P. axi* but in *P. galapagoensis* nov. spec. the seta on inner margin of middle segment is stronger, the seta on inner margin of distal segment is weaker.
– P.1 and P.2 exhibit sexual dimorphism (not observed in *P. axi*) as to structure and length of certain setae.
– P.3 ♂ enp. has longer inner setae on proximal and distal segment.
– P.4 with sexual dimorphism as in *P. axi* but terminal outer appendage of distal segment enp. ♂ normal (modified).
– Furca subdistally with 1 short and 1 long outer seta (2 long setae).

In 1989 HUYS & THISTLE established the genus *Bathycamptus* for a new species, *B. eckmani*, collected from "bathyal muds in San Diego Trough, off California", at three locations in a depth of more than 1000 m. Furthermore, they incorporated *Heteropsyllus minutus* Wells, 1965, a species founded solely on one male and of unclear systematic position (WELLS 1965). The animal was found at the Fladen ground on "sub-littoral muds off the coast of Scotland", at a depth of 146 m.

In accordance with HUYS & THISTLE (1989, p. 119) the species of *Bathycamptus* and *Psammocamptus* can be considered as being most closely related. In addition, both *Psammocamptus* species live in sandy habitats (AX & SCHMIDT 1973, MIELKE 1975), i.e. are not "mud-dwelling copepods" as HUYS

& THISTLE allege. The morphological differences between both genera are not impressive. For example, the seta and spine formula of the pereiopods is more extensive in *Bathycamptus*. However, as can be seen from one male from Fernandina (see "variability" above) and the animals from Barrington (see "note" above) the occurrence of two inner setae on distal segment exp. P.4 is latently extant in *Psammocamptus* thus stressing the close relationship. On the other hand, the differences in sexual dimorphism is somewhat perplexing. *P. galapagoensis* nov. spec. exhibits the greatest range of dimorphism of all four (*B. minutus*?) species since endopodites **and** exopodites P.1 – P.4 show differences between female and male. However, the modified outer appendage on distal segment enp. P.4 ♂ in the other three species is merely a normal seta in *P. galapagoensis* nov. spec. Therefore, the sexual dimorphism seems to be of restricted significance for the interpretation of the phylogenetic relationship of this species-group. It does not appear to be advisable to introduce a new genus for each deviation. Bearing in mind that many more species are to be expected world-wide by reason of the scattered distribution of the species known at present, it would be reasonable to unite them in one supra-specific taxon, namely *Psammocamptus* (on grounds of its priority).

C. General considerations

According to the current trend in harpacticoid systematics, it would not have caused problems to establish a new genus for each one of the four species presented above. However, arguments for separation should not be the primary consideration in systematics. On the contrary, the search for common derived characters should be given priority. If at least one synapomorphy with another species/species-group cannot be evidenced or hypothesized and if the addition of a species to a taxon would turn this into a paraphylum then the erection of an own monotypic genus may be reasonable.

The systematic position of a number of harpacticoid groups/genera is still uncertain, their respective adelphotaxon unknown and their incorporation into one of the existing "families" is debatable. In recent years, above all, R. HUYS has offered several meritorious contributions to harpacticoid systematics. However, statements like "deserves family status" or "to make the boundaries more narrow" are arbitrary and subjective assessments that may questioned by others. If, for example, *Orthopsyllus* Brady & Robertson, 1873 is unequivocally a monophylum, its classificatorial designation as a genus or a family (diagnosis of Orthopsyllidae announced for Sarsia by HUYS 1990 but still not available; therefore ignored by me in MIELKE 1993. The same is true for Idyanthidae and Zosimidae in HUYS et al. 1992) is absolutely unimportant. To separate genera from probably paraphyletic families (such as *Orthopsyllus* from the Canthocamptidae, *Karllangia* Noodt, 1964 from the Ameiridae, *Psammotopa* Pennak, 1942 from the Diosaccidae etc.) is nevertheless unsatisfactorily if their adelphotaxon relationship and the very difficult but unavoidable clarification of the remainder of the "family" has not been worked out. As AX (1995) pointed out in his new book (a) classificatorial terms are not obligatory and (b) evident paraphyletic assemblages should indeed not be tolerated but can be provisionally kept as groups written in quotation-marks (like e.g. "Reptilia"). This procedure could also reduce the inflation of new categorical terms.

Acknowledgements

The Galapagos-project was financially supported by Stiftung Volkswagenwerk.

Zusammenfassung

Im quantitativ untersuchten Sandstrand der Bahía Academy/Santa Cruz (IX,6) wurden vier Arten von interstitiellen Copepoden gefunden, die zu verschiedenen Taxa gehören und deren systematische Zuordnung unsicher erscheint.

Lediglich zwei Männchen liegen von einer neuen Art der Tetragonicipitidae vor. Obwohl der charakteristische distale Außenrandzahn am Grundglied der 1.Antenne fehlt, wird die Art als zu *Tetragoniceps* Brady, 1880 gehörig interpretiert: *T. santacruzensis* nov. spec. Ein Schwesterart-Verhältnis zu der schon von Galapagos beschriebenen Spezies *T. galapagoensis* Mielke, 1989, besteht nicht.

Einige Tiere mit auffälligen „Klöppelborsten" an den Mundwerkzeugen werden als *M. galapagoensis* nov. spec. zu *Micropsammis* Mielke, 1975, gestellt, ein Genus, welches nunmehr bei den Paranannopidae eingereiht wird (s. GEE & HUYS 1991). Es wird ferner vorgeschlagen, die Art *secunda* von List/Insel Sylt (Deutschland) wie ursprünglich bei *Micropsammis* zu belassen; die Errichtung eines eigenen Genus, *Telopsammis* Gee & Huys, 1991, wird für unnötig erachtet.

Zwei Tiere (1 ♀, 1 ♂) repräsentieren möglicherweise die zweite Art von *Carolinicola* Huys & Thistle, 1989: *C. galapagoensis* nov. spec. Die bislang einzig bekannte Art dieses Genus, *C. (= Hemimesochra) trisetosa* (Coull, 1973), wurde ursprünglich zu den Cletodidae gestellt, dann den Canthocamptidae zugeschrieben und letztendlich bei den Paranannopidae eingereiht (s. HUYS & THISTLE 1989).

Psammocamptus galapagoensis nov. spec., ein Vertreter der Canthocamptidae, wird als nah verwandt mit *P. axi* Mielke, 1975 betrachtet, welcher vom Sandstrand an der Ostküste von List/Insel Sylt (Deutschland) beschrieben wurde. Die beiden Arten von *Bathycamptus* Huys & Thistle, 1989 werden zu *Psammocamptus* Mielke, 1975 gestellt, da keine fundamentalen morphologischen Unterschiede zwischen beiden Gattungen bestehen.

Abschließend werden noch einige kritische Bemerkungen bezüglich der gegenwärtigen systematischen Arbeit über harpacticoide Copepoden gemacht.

References

AX, P. (1995): Das System der Metazoa I. Ein Lehrbuch der phylogenetischen Systematik. Gustav Fischer Verlag, 226 pp.

AX, P. & P. SCHMIDT (1973): Interstitielle Fauna von Galapagos. I. Einführung. Mikrofauna Meeresboden **20**, 1 – 38.

BECKER, K.-H. (1974): Eidonomie und Taxonomie abyssaler Harpacticoidea (Crustacea, Copepoda). Teil I. Cerviniidae – Ameiridae. "Meteor" Forsch.-Ergebnisse **18**, 1 – 28.

COULL, B. C. (1973): Meiobenthic Harpacticoida (Crustacea, Copepoda) from the deep sea off North Carolina. I. The genera *Hemimesochra* Sars, *Paranannopus* Lang, and *Cylindronannopus* n. g. Trans. Amer. Micros. Soc. **92**, 185 – 198.

FIERS, F. (1995): New Tetragonicipitidae (Copepoda, Harpacticoida) from the Yucatecan continental shelf (Mexico), including a revision of the genus *Diagoniceps* Willey. Bull. Inst. r. Sc. Nat. Belgique (Biologie) **65**, 151 – 236.

GEE, J. M. & R. HUYS (1991): A review of Paranannopidae (Copepoda: Harpacticoida) with claviform aesthetascs on oral appendages. Journ. Nat. Hist. **25**, 1135 – 1169.

HUYS, R. (1990): A new family of harpacticoid copepods and an analysis of the phylogenetic relationships within the Laophontoidea T. Scott. Bijdr. Dierk. **60**, 79 – 120.

HUYS, R. & D. THISTLE (1989): *Bathycamptus eckmani* gen. et spec. nov. (Copepoda, Harpacticoida) with a review of the taxonomic status of certain other deepwater harpacticoids. Hydrobiologia **185**, 101 – 126.

HUYS, R. & G. A. BOXSHALL (1991): Copepod Evolution. Ray Soc. London, 468 pp.

HUYS, R., P. M. J. HERMAN, C. H. R. HEIP & K. SOETAERT (1992): The meiobenthos of the North Sea: density, biomass trends and distribution of copepod communities. ICES J. mar. Sci. **49**, 23 – 44.

KUNZ, H. (1984): Systematik der Familie Tetragonicipitidae LANG (Crustacea, Harpacticoida). Mitt. zool. Mus. Univ. Kiel **2**, 33 – 48.

LANG, K. (1948): Monographie der Harpacticiden. Nordiska Bokh. Stockholm, 1682 pp.

LANG, K. (1965): Copepoda Harpacticoidea from the Californian Pacific coast. Kungl. Svenska Vetenskaps. Handl. **10**, 1 – 566.

MIELKE, W. (1975): Systematik der Copepoda eines Sandstrandes der Nordseeinsel Sylt. Mikrofauna Meeresboden **52**, 1 – 134.

MIELKE, W. (1989): Interstitielle Fauna von Galapagos. XXXVI. Tetragonicipitidae (Harpacticoida). Microfauna Marina **5**, 95 – 172.

MIELKE, W. (1993): Species of the taxa *Orthopsyllus* and *Nitocra* (Copepoda) from Costa Rica. Microfauna Marina **8**, 247 – 266.
WELLS, J. B. J. (1965): Copepoda (Crustacea) from the Meiobenthos of Some Scottish Marine Sublittoral Muds. Proc. Roy. Soc. Edinb. **69**, 1 – 33.

Interstitielle Fauna von Galapagos

I.	Einführung P. AX & P. SCHMIDT, Mikrofauna Meeresboden **20**, 1–38 (1973).
II.	Gnathostomulida. B. EHLERS & U. EHLERS, Mikrofauna Meeresboden **22**, 1–27 (1973).
III.	Promesostominae (Turbellaria, Typhloplanoida). P. AX & U. EHLERS, Mikrofauna Meeresboden **23**, 1–16 (1973).
IV.	Gastrotricha. P. SCHMIDT, Mikrofauna Meeresboden **26**, 1–76 (1974).
V.	Otoplanidae (Turbellaria, Proseriata). P. AX & R. AX, Mikrofauna Meeresboden **27**, 1–28 (1974).
VI..	*Aeolosoma maritimum dubiosum* nov. sp. (Annelida Oligochaeta, W. WESTHEIDE & P. SCHMIDT, Mikrofauna Meeresboden **28**, 1–10 (1974).
VII.	Nematoplanidae, Polystyliphoridae, Coelogynoporidae (Turbellaria, Proseriata) P. AX & R. AX, Mikrofauna Meeresboden **29**, 1–28 (1974).
VIII.	Trigonostominae (Turbellaria, Typhloplanoida) U. EHLERS & P. AX, Mikrofauna Meeresboden **30**, 1–33(1974).
IX.	Dolichomacrostomidae (Turbellaria, Macrostomida), B. SOPOTT-EHLERS & P. SCHMIDT, Mikrofauna Meeresboden **34**, 1–20 (1974).
X.	Kinorhyncha, P. SCHMIDT Mikrofauna Meeresboden **43**, 1–15 (1974).
XI.	Pisionidae, Hesionidae, Pilargidae, Syllidae (Ployachaeta). W. WESTHEIDE, Mikrofauna Meeresboden **44**, 1–146 (1974).
XII.	*Myozona* Marcus (Turbellaria, Macrostomida). B. SOPOTT-EHLERS & P. SCHMIDT, Mikrofauna Meeresboden **46**, 1–19 (1974).
XIII.	*Ototyphlonemertes* Diesing (Nemertini, Hoplonemertini). H. MOCK & P. SCHMIDT, Mikrofauna Meeresboden **51**, 1–40 (1975).
XIV.	Polycladida (Turbellaria). B. SOPOTT-EHLERS & P. SCHMIDT, Mikrofauna Meeresboden **54**, 1–32 (1975).
XV.	*Macrostomum* O. Schmidt, 1848 und *Siccomacrostomum triviale* nov. gen. nov. spec. (Turbellaria, Macrostomida). P. SCHMIDT & B. SOPOTT-EHLERS, Mikrofauna Meeresboden **57**, 1–45 (1976).
XVI.	Tardigrada. D. MCKIRDY, P. SCHMIDT & M. MCGINTY-BAYLY, Mikrofauna Meeresboden **58**, 1–43 (1976).
XVII.	Polygordiidae, Saccocirridae, Protodrilidae, Nerillidae, Dinophilidae (Polychaeta). P. SCHMIDT & W. WESTHEIDE, Mikrofauna Meeresboden **62**, 1–38 (1977).
XVIII.	Nereidae, Eunicidae, Dorvilleidae (Polychaeta). W. WESTHEIDE, Mikrofauna Meeresboden **63**, 1–40 (1977).
XIX.	Monocelididae (Turbellaria, Proseriata). P. AX & R. AX, Mikrofauna Meeresboden **64**, 1–44 (1977).
XX.	Halacaridae (Acari). I. BARTSCH, Mikrofauna Meeresboden **65**, 1–108 (1977).
XXI.	Lebensraum, Umweltfaktoren, Gesamtfauna. P. SCHMIDT, Mikrofauna Meeresboden **68**, 1–52 (1978).
XXII.	Zur Ökologie der Halacaridae (Acari). I. BARTSCH & P. SCHMIDT, Mikrofauna Meeresboden **69**, 1–38 (1978).
XXIII.	Acoela (Turbellaria). U. EHLERS & J. DÖRJES, Mikrofauna Meeresboden **72**, 1–75 (1979).
XXIV.	Microparasellidae (Isopoda, Asellota). N. COINEAU & P. SCHMIDT, Mikrofauna Meeresboden **73**, 1–19 (1979).

XXV. Longipediidae, Canuellidae, Ectinosomatidae (Harpacticoida). W. MIELKE, Mikrofauna Meeresboden **77**, 1–107 (1979).
XXVI. Questidae, Cirratulidae, Acrocirridae, Ctenodrilidae (Polychaeta). W. WESTHEIDE, Mikrofauna Meeresboden **82**, 1–24 (1981).
XXVII. Byrsophlebidae, Promesostomidae Brinkmanniellinae, Kytorhynchidae (Turbellaria, Typhloplanoida). U. EHLERS & B. EHLERS, Mikrofauna Meeresboden **83**, 1–35 (1981).
XXVIII. Laophontinae (Laophontidae), Ancorabolidae (Harpacticoida). W. MIELKE, Mikrofauna Meeresboden **84**, 1–106 (1981).
XXIX. Darcythompsoniidae, Cylindropsyllidae (Harpacticoida). W. MIELKE, Mikrofauna Meeresboden **87**, 1–52 (1982).
XXX. Podocopida 1 (Ostracoda). J. GOTTWALD, Mikrofauna Meeresboden **90**, 1–187 (1983).
XXXI. Paramesochridae (Harpacticoida). W. MIELKE, Microfauna Marina **1**, 63–147 (1984).
XXXII. Epsilonematidae (Nematodes). E. CLASING, Microfauna Marina **1**, 149–189 (1984).
XXXIII. Tubificidae (Annelida, Oligochaeta). CH. ERSÉUS, Microfauna Marina **1**, 191–198 (1984).
XXXIV. Schizorhynchia (Plathelminthes, Kalyptorhynchia). U. NOLDT & S. HOXHOLD, Microfauna Marina **1**, 199–256 (1984).
XXXV. Chromadoridae (Nematoda): D.BLOME, Microfauna Marina **2**, 271–329 (1985).
XXXVI. Tetragonicipitidae (Harpacticoida). W. MIELKE, Microfauna Marina **5**, 95–172 (1989).
XXXVII. Metidae (Harpacticoida). W. MIELKE, Microfauna Marina **5**, 173–188 (1989).
XXXVIII. *Haloplanella* Luther und *Pratoplana* Ax (Typhloplanoida, Plathelminthes). U. EHLERS & B. SOPOTT-EHLERS, Microfauna Marina **5**, 189–206 (1989).
XXXIX. Copepoda, part 7. W. Mielke, Microfauna Marina **11**, 115–152 (1997).

Dr. Wolfgang Mielke
II. Zoologisches Institut und Museum der Universität Göttingen,
Berliner Straße 28, D-37073 Göttingen

Electronmicroscopical investigations of male gametes in *Ptychopera westbladi* (Plathelminthes, Rhabdocoela, "Typhloplanoida")

Beate Sopott-Ehlers and Ulrich Ehlers

Abstract

Steps of spermiogenesis and the ultrastructure of the aciliary male gametes of *Ptychopera westbladi* are described. During maturation of male gametes two differentiations of special interest appear: (1) transitory centrioles one of which elongates considerably and (2) microvilli-like evaginations of the surface membrane of outgrowing spermatids. The mature sperm cell is filiform in shape and characterized by cortical microtubules, a single row of mitochondria, dense bodies, "minimitochondria", a nucleus with twisted strings of fibrous chromatin and the absence of cilia. "Minimitochondria", dense bodies and microtubules intermingle with each other in the presumed functional fore-end of the cell. Mikrovilli-like evaginations, "minimitochondria" and their special arrangement with other cell organelles are hypothesized to be autapomorphies of *P. westbladi* or of a supraspecific taxon (e.g. *Ptychopera*) including this species. An aciliarity of male gametes is also known for *Proxenetes deltoides* and might represent an autapomorphic characteristic of the Trigonostominae.

A. Introduction

Male gametes of Plathelmimthes show modified organisations compared to the plesiomorphic type present in many aquatic Metazoa with an external fertilization of oocytes (see i.al. FRANZEN 1977). For species of the Trepaxonemata, spermatozoa in general are filiform in shape and bear two locomotory free cilia or flagella of a 9x2+"1" microtubular pattern. The sperm shaft is enclosed by cortical microtubules, and its cytoplasm includes glycogen, dense bodies, mitochondria and a nucleus with fibrous chromatin (see EHLERS 1985).

However, many variations from this basic pattern of the Trepaxonemata exist. These differences concern the number and the existence or the lack of cilia, as well as the equipment with organelles of the sperm body. Beyond this, events during spermiogenesis have shown divergent features in different taxa.

Characteristics of mature spermatozoa and differentiations appearing during gametogenesis have provided many data, suitable to elucidate phylogenetic relationships within the taxon Plathelminthes (see JUSTINE 1995, 1996; WATSON & ROHDE 1995).

Live sperm cells of *Ptychopera westbladi* (Luther, 1943), a representative of the Trigonostominae within the Trigonostomidae, appear aflagellate, when viewed under the lightmicroscope. However, the lightmicroscope does not provide enough information to clarify with certainty, whether cilia are definitely missing or whether these locomotory organelles are incorporated.

So, the aim of our electronmicroscopical studies was, to find out, whether cilia are really lacking or whether they are incorporated. Beyond this, the question arose, whether special features exist, which could help to come to a better understanding concerning the relationships within the nonmonophyletic taxon "Typhloplanoida" and the Rhabdocoela in general.

B. Material and Methods

Specimens of *Ptychopera westbladi* (Luther, 1943) derive from sand samples gathered on mud flats of the island of Sylt (North Sea). Extraction from the sediment, fixation, dehydration, embedding, sectioning and staining followed conventional processing steps. Series of sections were studied using a Zeiss EM 900 electronmicroscope.

C. Results

Cytogenesis of male gametes

Within the testes of *Ptychopera westbladi* all stages of gametogenesis can be observed. In the lateral segments late spermatids and residual bodies occur,

Fig. 1. A. Segment of a testis with spermatogonia, spermatocytes and spermatids. The arrows mark ▶ synaptonemal complexes. Scale = 2 µm. B, C. Nucleus in primary spermatocytes with early stages of synaptonemal complexes. Scale in B = 1 µm. Scale in C = 0.5 µm.

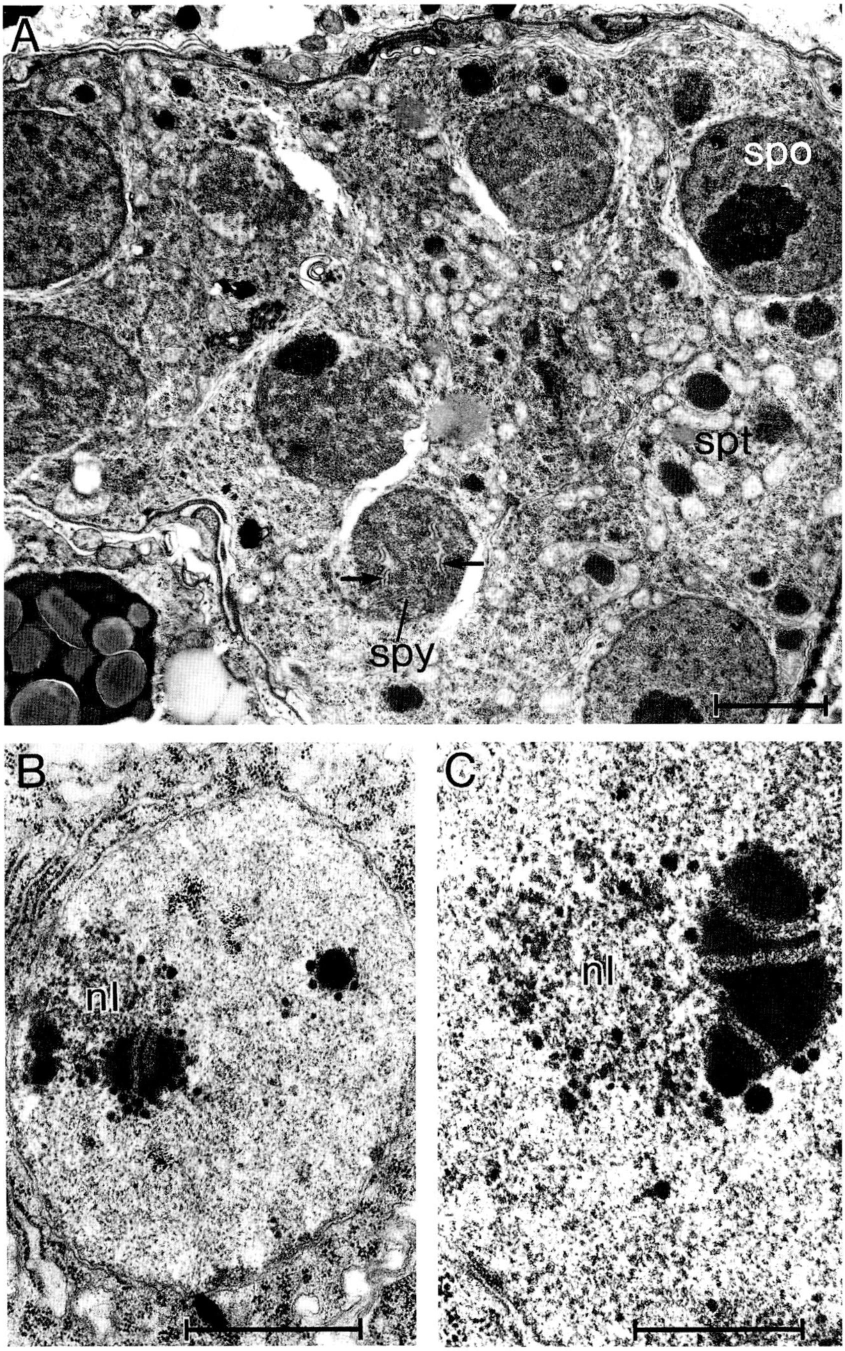
A
spo
spt
spy
B
nl
C
nl

while stages of spermatogenesis and spermiogenesis are found more centrally (Fig. 1A).

Spermatogenesis and spermiogenesis have been analyzed at the ultrastructural level for many species of the Plathelminthes. Spermatogenesis in *P. westbladi* is similar to that of other representatives of the Trepaxonemata. Spermatogonia exhibit a large nucleus containing a distinct nucleolus. The cytoplasm is electron-dark and shows a few mitochondria with weakly developed cristae.

Mitotic divisions of spermatogonia result in groups of spermatocytes. These are characterized by the existence of centrioles, chromatoid bodies or nuclear extrusions, differentiations which are typical of neoblasts and early stages of gametogenesis in general. Primary spermatocytes in *P. westbladi* can be distinguished by a loss of demarcation of the nucleolus and a simultaneously occuring assamblage of chromatin near synaptonemal complexes (Fig. 1B, C). Furthermore, well developed rough ER is especially found near the nucleus. Centrioles still have a diplosomal configuration (Fig. 2B, C).

With proceeding maturation the centrioles undergo dramatical changes. One of the centrioles elongates considerably. At the same time satellite structures are differentiated. These satellite structures consist of microtubules embedded in dark-staining material and have a somehow star-shaped appearance (Fig. 2D-G). Because of the dark substances the microtubules are difficult to document in micrographs.

Spermatids are arranged in rosettes, the nuclei being located in peripheral pouches. The nuclei are often surrounded by dense bodies and small roundish mitochondria. The cytoplasm of the cytophore is scattered with glycogen islets and contains voluminous complexes of rough ER and dictyosomes. Furthermore, small electron-lucent ovoid bodies enclosed by a bordering membrane are seen. These bodies were never found near Golgi complexes and possess a substructure which appeared in some instances similar to christae (Fig. 8A-C). Although it could not be definitely clarified, whether these bodies are minimitochondria or a second type of dense bodies, these organelles will be referred to as "minimitochondria" in the following.

As spermatids mature, chromatin commences to condense. At the same time a row of microtubules enclosed by dark substances appears beneath the cell membrane distal to the nucleus. The next step of development is the differentiation of a triangular cytoplasmic cap lined by electron-dense material at this side, which faces the surface membrane (Fig. 3A-C). In a few instances,

---

Fig. 2. A - C. Centrioles. B, C. Centrioles in diplosomal configurations. D - G. Different stages of ▶ outgrowing centrioles with satellite structures. Scale in all figures = 0.5 μm.

Male gametes in *P. westbladi* 197

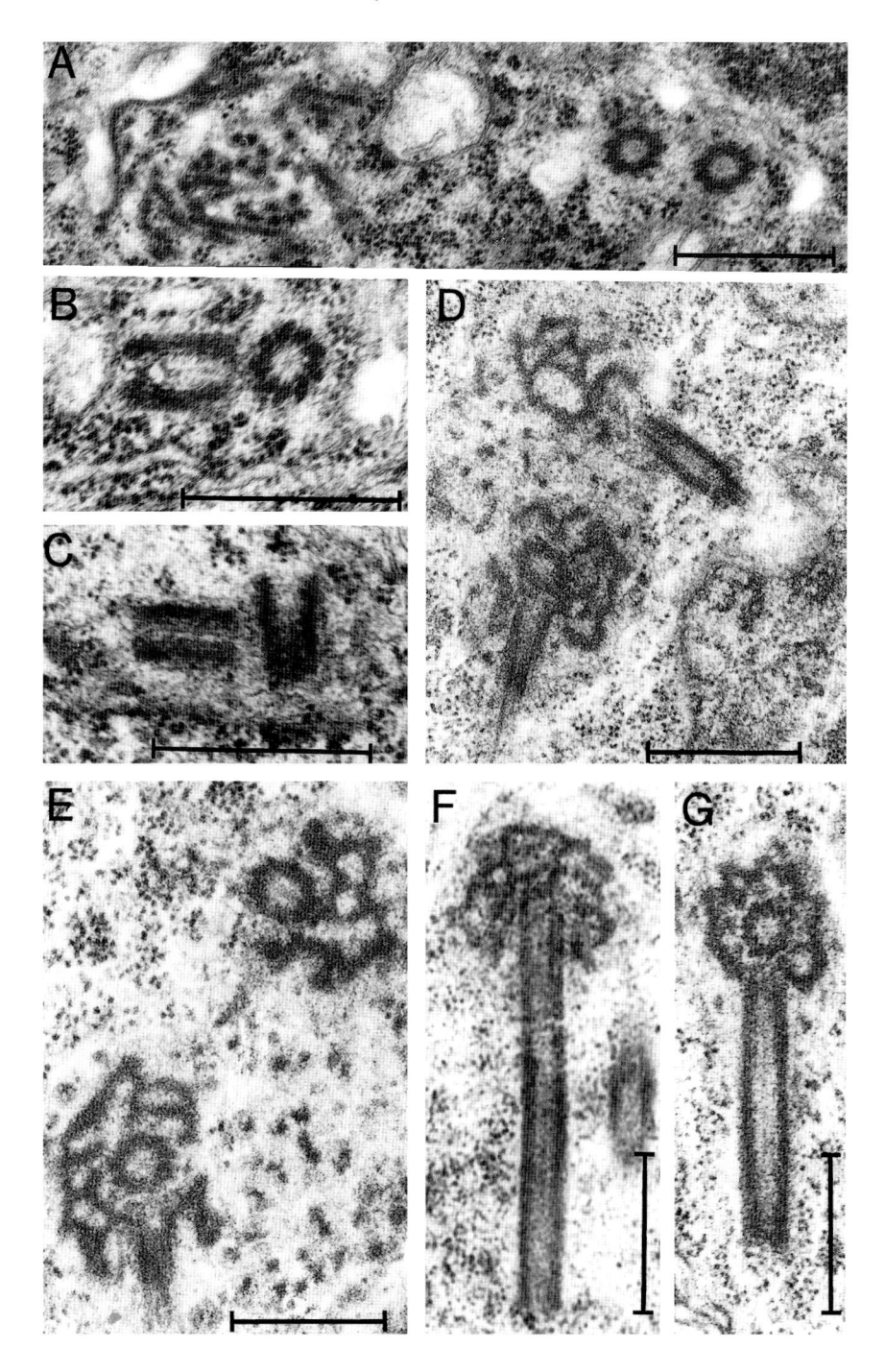

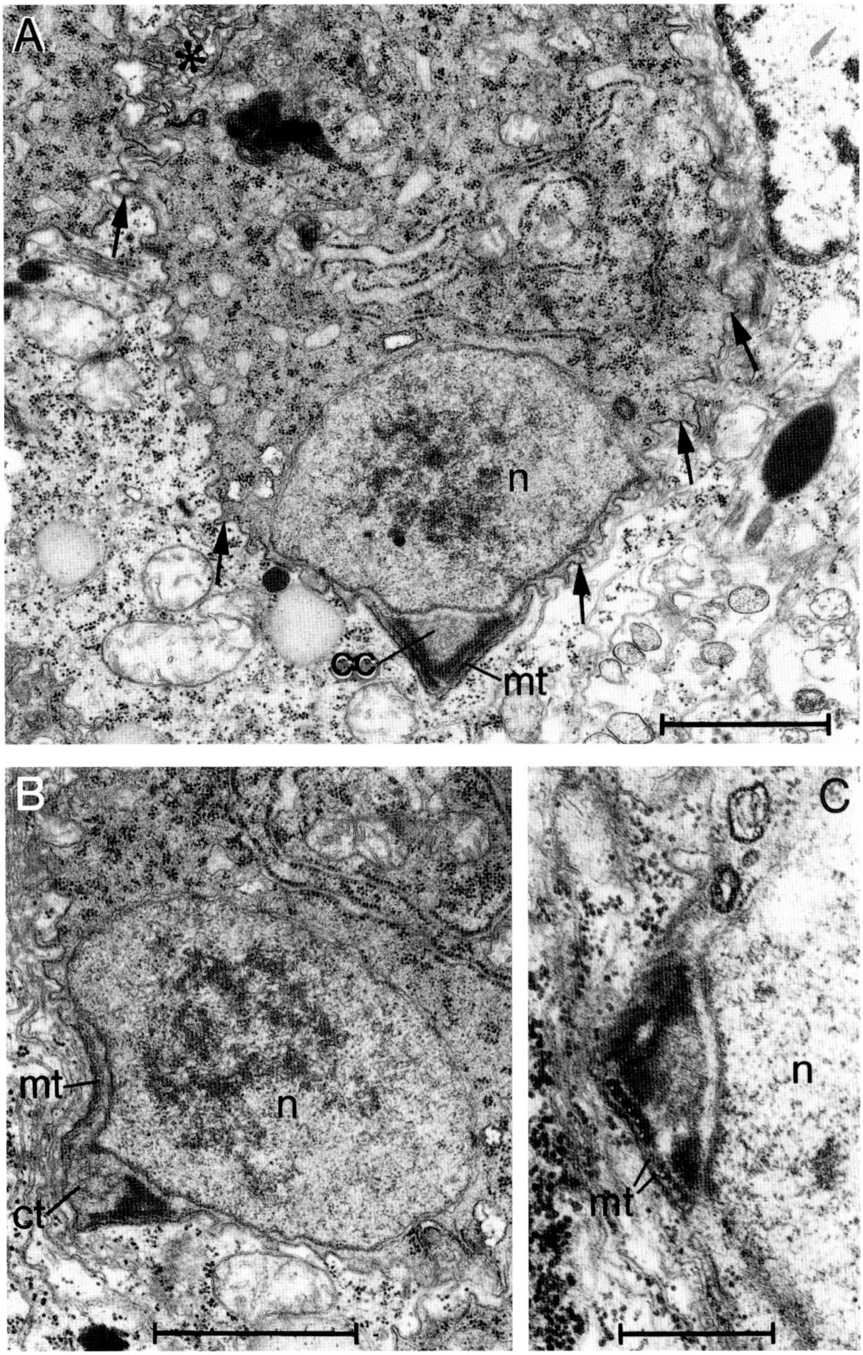
A
*
n
cc
mt
B
mt
n
ct
C
n
mt

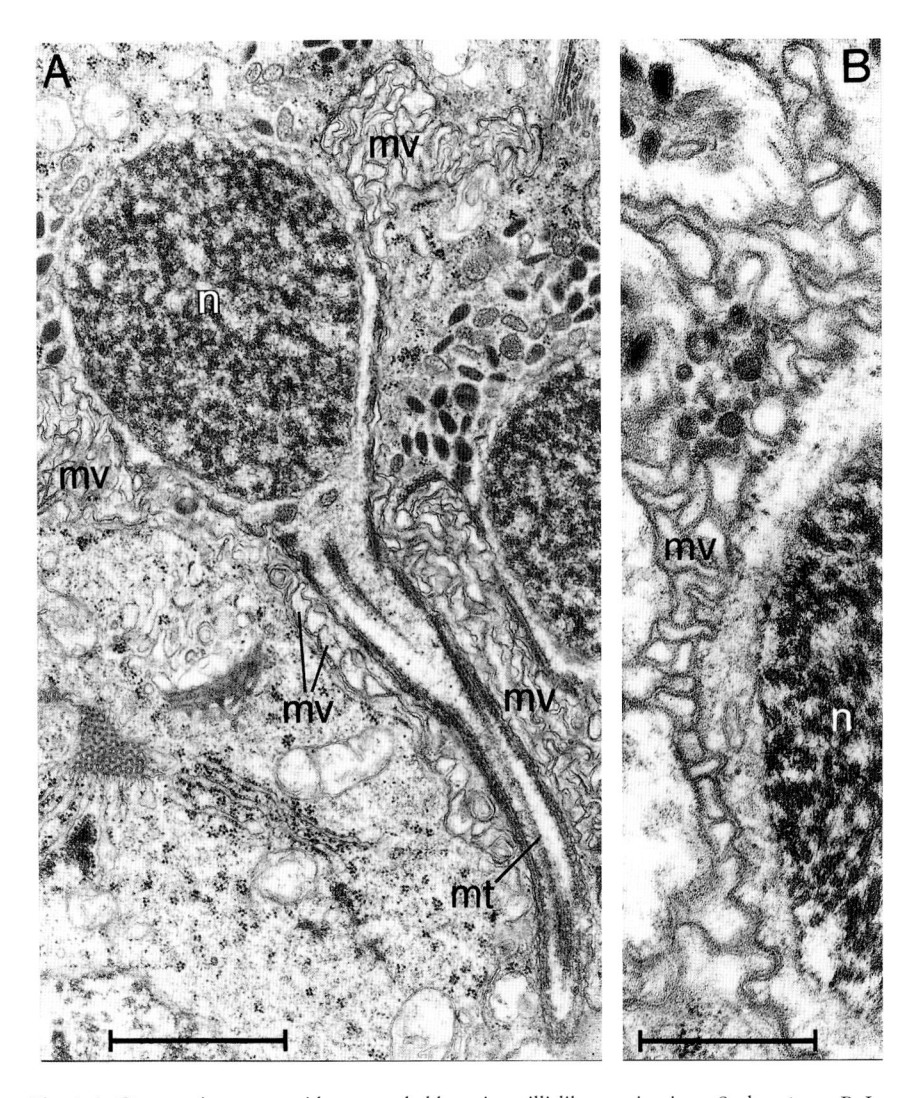

Fig. 4. A. Outgrowing spermatid surrounded by microvilli-like evaginations. Scale = 1 μm. B. Interwoven microvilli-like extensions. Scale = 0.5 μm.

◄ Fig. 3. A. Triangular cytoplasmic cap and first signs of microvilli-like evaginations. Scale = 1 μm. B. Cytoplasmic cap with remnants of a centriole. Scale = 1 μm. C. Cytoplasmic cap with microtubules sheltered by dark material. Scale = 0.5 μm. The arrows and the star in A mark microvilli-like evaginations.

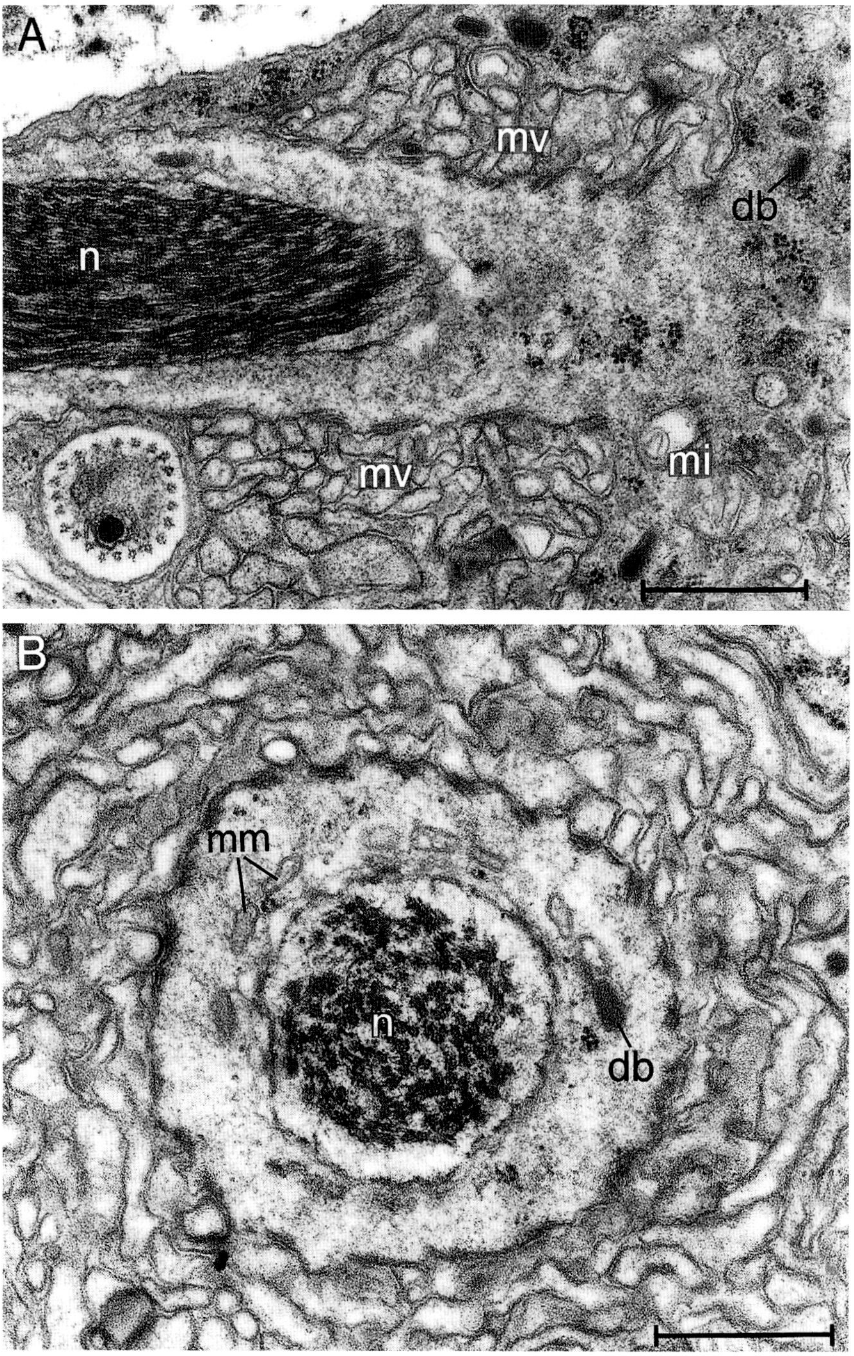
A
mv
db
n
mv
mi
B
mm
n
db

remnants of a centriole could be observed in the cap (Fig. 3B). In this stage of gametogenesis the surface membrane of the cytophore is evaginated to form numerous microvilli-like extensions, which become tightly interwoven with each other (Figs. 3A; 4).

As chromatin condensation proceeds an outgrowth starts to sprout. This outgrowth is lined by microtubules embedded in osmiophilic material and is surrounded by microvilli-like evaginations of the surface membrane (Fig. 4A). During the following steps of maturation the nucleus, dense bodies, mitochindria and "minimitochondria" migrate into the outgrowth (Fig. 5). Chromatin condenses first to skeins, causing a honeycomb-like pattern in transverse sections, then in typical fibrous cords. The microvilli-like elaborations of the surface membrane flatten or are cast off during the process of cell elongation, since they are not present in late spermatids or young spermatozoa, respectively.

Mature male gametes

The spermatozoa of *Ptychopera westbladi* are filiform in shape. The sperm cells are equipped with 25 – 29 long cortical microtubules arranged with comparatively wide spaces in a circle, longitudinal to droplet-shaped dense bodies and "minimitochondria", a single row of roundish mitochondria and a nucleus with twisted strings of fibrous chromatin (Fig. 6A, B). Dense bodies and "minimitochondria" are not arranged in a clear pattern, although 4 – 5 dense bodies could be observed in slightly oblique transverse sections. Glycogen deposits are randomly distributed midway down the length in the cytoplasm. All organelles run in a spiral course.

Since cilia are lacking, it is difficult to define a frontal and a hind end of the cell. In the presumed functional fore-end 18 cortical microtubules form a horseshoe-like semicircle at the periphery closely beneath the cell membrane, while about 10 microtubules are seen in the centre. In a short distance below, a circle of 9 microtubules surrounds an electron-dense core and 19 microtubules, partly accompanied by dark material, run at the periphery (Fig. 7A–B).

The first organelle to appear in transverse sections is a roundish mitochondrion (Figs. 6A; 7C). In a short distance down the length of the spermatozoon a mitochondrion is seen in the centre enclosed by oval "minimitochondria".

◄ Fig. 5. A. Longitudinal section of an outgrowing spermatid with microvilli-like evaginations. B. Transverse section of an outgrowing spermatid surrounded by microvilli-like evaginations. Scales = 0.5 µm.

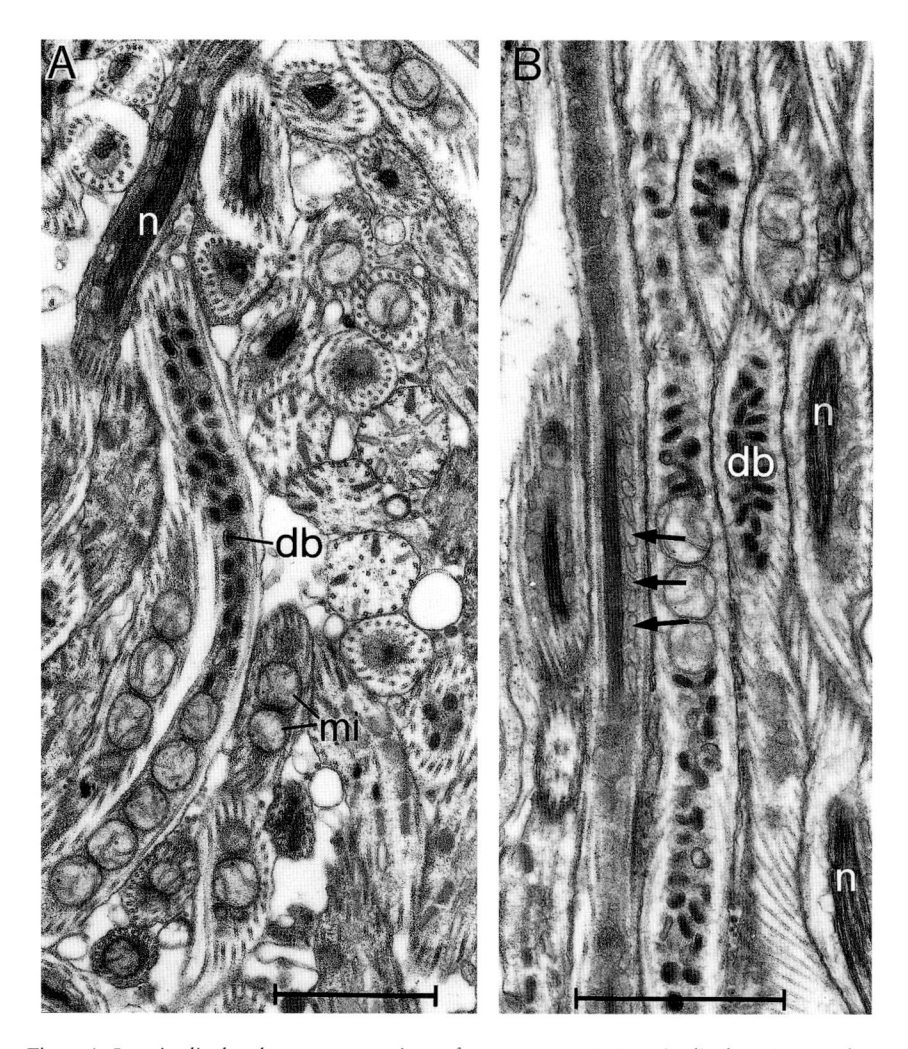

Fig. 6. A. Longitudinal and transverse sections of spermatozoa. B. Longitudinal sections. Scales = 1 µm.

Fig. 7. A, B. Transverse sections through frontal tips of spermatozoa. Scales = 0.5 µm. C. Transverse sections through different regions of spermatozoa. The arrow head marks the tip of a rear end, the small arrow the tip of a fore end. Scale = 0.5 µm.

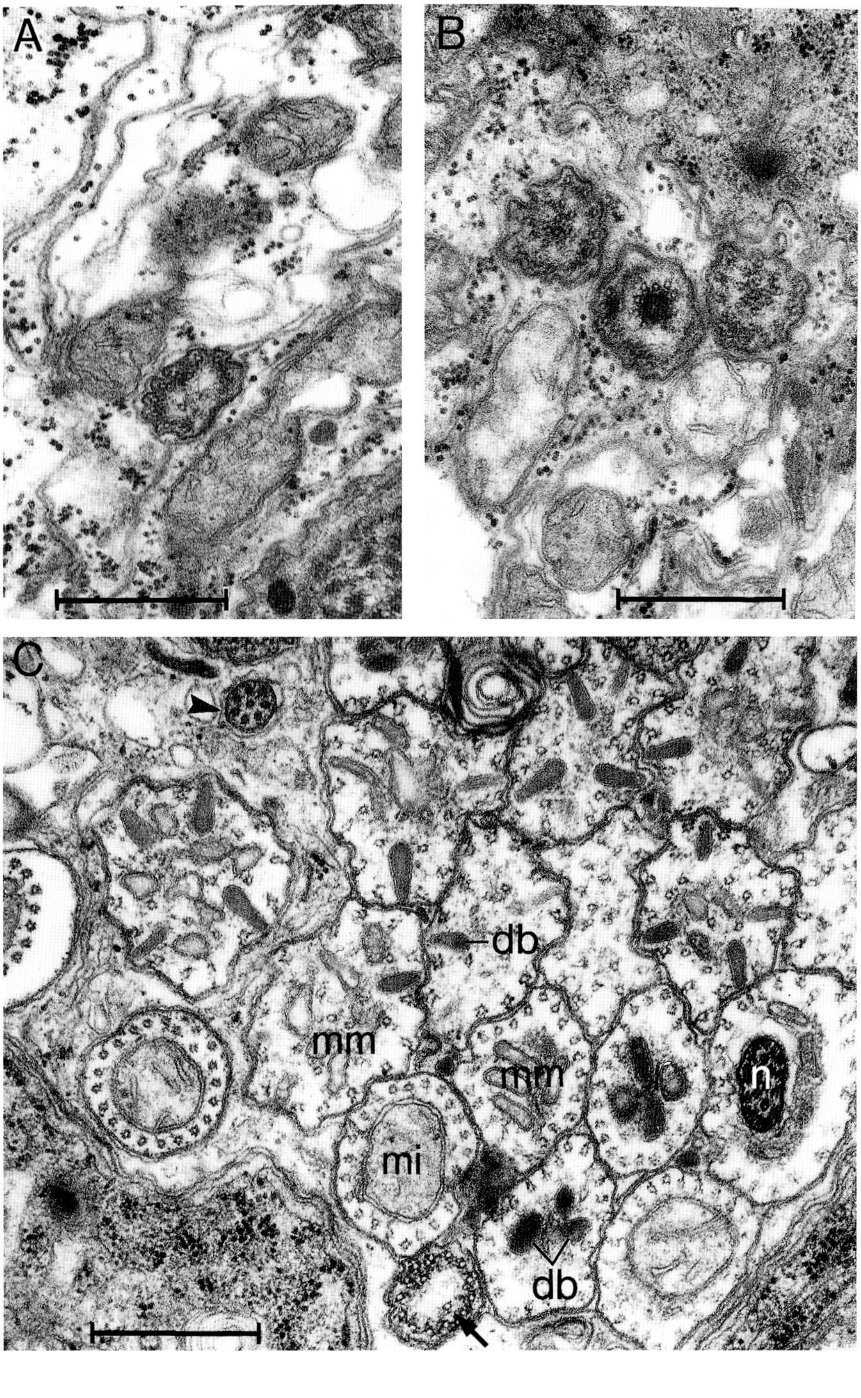
A
B
C
db
mm
mm
mi
n
db

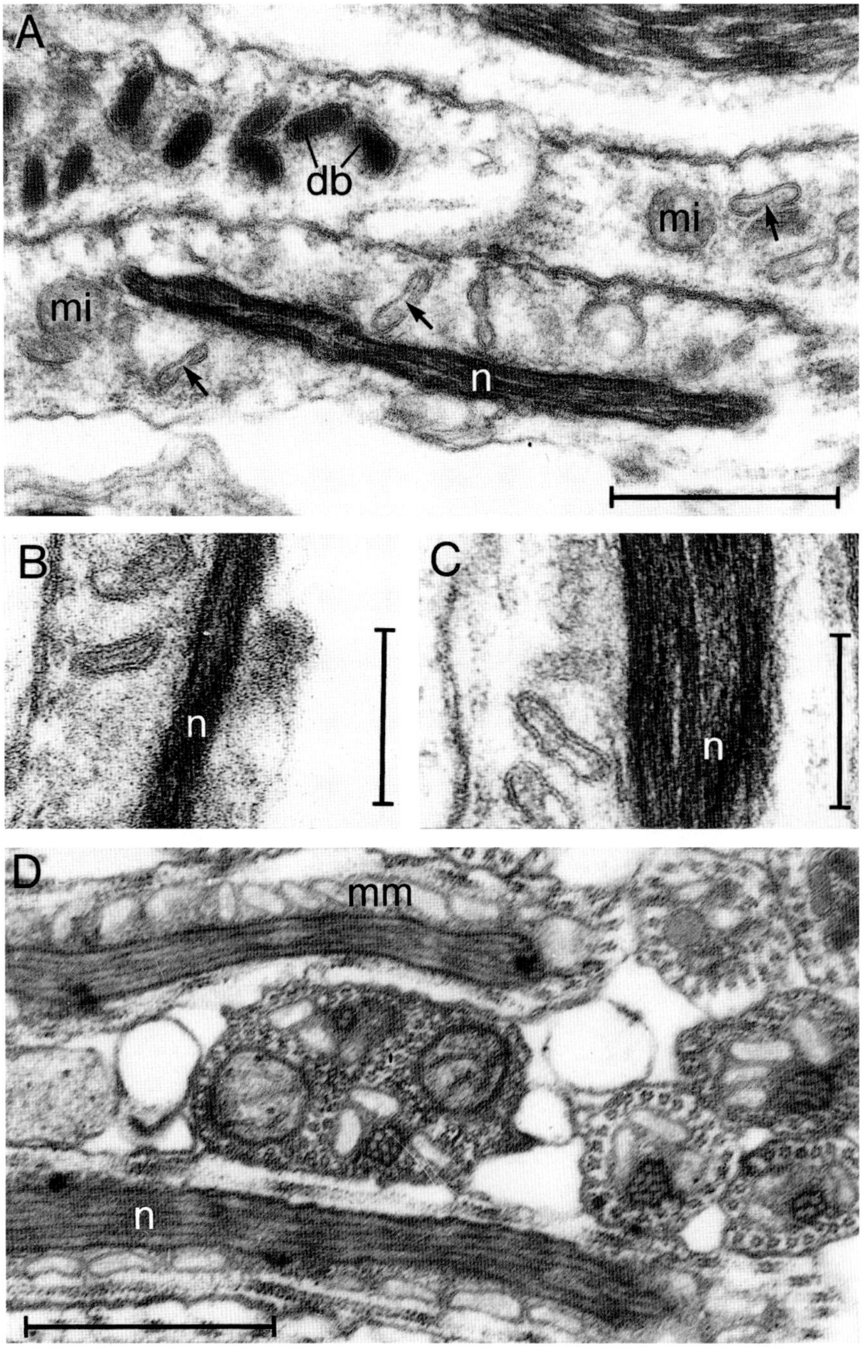
A
db
mi
mi
n
B
n
C
n
D
mm
n

Further down droplet-shaped dense bodies show a peripheral location and are situated in gaps between microtubules. A peripheral ring of 14 cortical microtubules is present and 7 – 8 microtubules are centrally intermingled between dense bodies and "minimitochondria" (Figs. 6A; 7C).

The next differentiation to appear is the nucleus. Dorsally it is encircled by "minimitochondria" and dense bodies, respectively. Ventrally of it the row of roundish mitochondria runs (Fig. 6A, B). A short distance below this middle section the string of mitochondria and the "minimitochondria" cease, and the nucleus is accompanied just by dense bodies. Eventually nucleus and dense bodies disappear almost simultaneously. The presumed functional hind end ceases in a tip with 8 microtubules, one located in the centre, seven at the periphery (Fig. 7C).

In some instances sperm cells equipped with a double or even fourfold set of organelles could be observed (Fig. 8D). Each of these sets was always encircled by a microtubular ring of its own. These configurations are supposed to have their origin in an artificial development.

D. Discussion

Although there exist some features in spermiogenesis and in the organisation of mature gametes as well, which do not correspond to those of other representatives of the Plathelminthes Trepaxonemeta, *Ptychopera westbladi* is, without doubt, a member of this taxon. Early stages of spermatogenesis, spermiogenesis and the equipment with organelles in the sperm shaft, as there are the filiform shape, the existence of dense bodies, isolated mitochondria arranged in a string and a monolayered sheath of cortical microtubules are characteristics corresponding to the basic pattern of the Trepaxonemata, as already mentioned above. Furthermore, the transitory centrioles and especially the unusual elongation of one of these organelles indicate remnants of an ancestral biciliary status. This biciliarity is obviously first initiated, but suppressed during proceeding maturation. The differentiation of microvilli-like evaginations of the surface membrane during cell elongation, the existence of "minimitochondria" and a special arrangement of dense bodies and "minimitochondria"

◄ Fig. 8. A. Longitudinal sections of middle regions with dense bodies, nucleus, "minimitochondria" and mitochondria. The small arrows mark delicate christae of "minimitochondria". Scale = 0.5 µm. B, C. "Minimitochondria" with christae in high magnifications. Scale = 0.25 µm. D. Spermatozoon with a fourfold set of organelles. Scale = 0.5 µm.

intermingled with cortical microtubules in the presumed functional fore end are hypothesized to be autapomorphies of *Ptychopera westbladi* or of a supraspecific taxon (e.g. the taxon *Ptychopera*) including this species.

Based on electronmicroscopical observations, aciliary spermatozoa have been hitherto reported for a few species of the Neodermata (for ref. see JUSTINE 1995), for all representatives of the Prolecithophora, and for some dalyellioid taxa.

Male gametes of Prolecithophora, however, are quite different. There exist some derived features in their organisation indicating, that this aciliarity is an autapomorphy of this taxon (see EHLERS 1981, 1988). The male gametes of dalyellioid species also differ from those in *P. westbladi* (see Cifrian et al. 1988; NOURY-SRAIRI et al. 1989; ROHDE & WATSON 1993; SOPOTT-EHLERS & EHLERS 1995 ; SOPOTT-EHLERS 1995). This indicates, that the aciliary status of male gametes has evolved more than once within the Plathelminthes Trepaxonemata.

Unpublished data of our own on the fine structure of spermatozoa in *Proxenetes deltoides* den Hartog, 1965 show, that this species also has aciliary sperm cells. In respect to highly modified mitochondrial lenses occuring in the eyes of *Proxenetes deltoides* and *Ptychopera westbladi*, as well (see SOPOTT-EHLERS 1996), aciliary spermatozoa are hypothesized to be an autapomorphic characteristic of the Trigonostominae sensu AX (1971) and EHLERS (1974). However, more detailed investigations of species of other taxa like *Beklemischeviella*, *Ceratopera*, *Cryptostiopera*, *Messoplana*, *Petaliella* and *Trigonostomum* are required to substantiate this hypothesis. At present, the known ultrastructural features of male gametes of *P. westbladi* and *P. deltoides* do not allow a new hypothesis for the systematic position of the Trigonostominae within the Rhabdocoela.

Acknowledgements

Financial support was provided by the Akademie der Wissenschaften und der Literatur, Mainz. Mrs. S. Gubert and Mr. B. Baumgart are thanked for technical assistances.

Zusammenfassung

Verschiedene Stadien der Spermiogenese und die ultrastrukturelle Organisation der männlichen Gameten von *Ptychopera westbladi* werden beschrieben. Während der Reifephase der Gameten treten zwei Differenzierungen von besonderem Interesse auf: (1) transitorische Centriole von denen sich eines beträchtlich verlängert und (2) Auswüchse der Zellmembran in Form von Mikrovilli bei auswachsenden Spermatiden. Reife Gameten sind fadenförmig und weisen kortikale Mikrotubuli, eine Reihe von Mitochondrien, "dense bodies", "Minimitochondrien" und einen Kern mit gewundenen Strängen fibrösen Chromatins auf; Cilien fehlen den reifen Gameten. "Minimitochondrien", "dense bodies" und Mikrotubuli vermischen sich im Bereich des mutmaßlichen Zellvorderendes. Die Mikrovilli vergleichbaren Zellauswüchse, die "Minimitochondrien" und ihr spezielles Arrangement mit anderen Zellorganellen lassen sich als Autapomorphien von *P. westbladi* oder eines supraspezifischen Taxons (z.B. *Ptychopera*) hypothetisieren. Aciliäre Spermien sind auch von *Proxenetes deltoides* bekannt; dieser Mangel von Cilien mag eine Autapomorphie des Taxons Trigonostominae darstellen.

Abbreviations

cc	cytoplasmic cap	mv	microvilli-like evaginations
ct	centriole	n	nucleus
db	dense bodies	nl	nucleolus
mi	mitochondrion	spo	spermatogonia
mm	minimitochondria	spy	spermatocytes
mt	microtubules	spt	spermatids

References

Ax, P. (1971): Zur Systematik und Phylogenie der Trigonostominae (Turbellaria, Neorhabdocoela). Mikrofauna Meeresboden **4**, 1-84.

Cifrian, B., S. Martinez-Alos & P. Garcia-Corrales (1988): Ultrastructural study of spermatogenesis and mature spermatozoa of *Paravortex cardii* (Platyhelminthes Dalyellioida). Acta Zool.(Stockh.) **69**, 195-204.

Ehlers, U. (1974): Interstitielle Typhloplanoida (Turbellaria) aus dem Litoral der Nordseeinsel Sylt. Mikrofauna Meeresboden **49**, 1-102.

– (1981): Fine structure of the giant aflagellate spermatozoon in *Pseudostomum quadrioculatum* (Leuckart) (Platyhelminthes, Prolecithophora). Hydrobiologia **84**, 287-300.

– (1985): Das phylogenetische System der Plathelminthes. Fischer, Stuttgart, New York, pp. 1-317.

– (1988): The Prolecithophora - a monophyletic taxon of the Plathelminthes? Fortschr. Zool./ Prog. Zool. **36**, 359-365.

Franzen, A. (1977): Sperm structure with regard to fertilization biology and phylogenetics. Verh. Dt. Zool. Ges. 1977: 123-138.

Justine, J.-L. (1991): Phylogeny of parasitic Platyhelminthes: a critical study of synapomorphies proposed on the basis of the ultrastructure of spermiogenesis and spermatozoa. Can. J. Zool. **69**, 1421-1440.

– (1995): Spermatozoal ultrastructure and phylogeny in the parasitic Platyhelminthes. In: Jamieson, B. G. M., J. Ausio & J.-L. Justine (eds), Advances in spermatozoal phylogeny and taxonomy. Mém. Mus. natn. Hist. nat. **166**, 55-86.

Noury-Srairi., J.-L. Justine & L. Euzet (1989): Ultrastructure comparée de la spermiogenèse et du spermatozoide de trois espèces de *Paravortex* (Rhabdocoela, "Dalyellioida", Graffillidae), turbellariés parasites intestinaux de mollusques. Zool. Scr. **18**, 161-174.

Rohde, K. & N. Watson (1993): Ultrastructure of spermiogenesis and sperm in an undescribed species of Luridae. (Platyhelminthes: Rhabdocoela). Aust. J. Zool. **41**, 13-19.

Sopott-Ehlers, B. (1995): Some data on the ultrastructure of the eyes and spermatozoa in *Provortex psammophilus* (Plathelminthes, Rhabdocoela, "Dalyellioida"). Microfauna Marina **10**, 307-312.

– (1996): First evidence of mitochondrial lensing in two species of the "Typhloplanoida" (Plathelminthes, Rhabdocoela): phylogenetic implications. Zoomorphology **116**, 95-101.

Sopott-Ehlers, B. & U. Ehlers (1995): Modified sperm ultrastructure and some data on spermiogenesis in *Provortex tubiferus* (Plathelminthes, Rhabdocoela): phylogenetic implications for the Dalyellioida. Zoomorphology **115**, 41-49.

Watson, N. & K. Rohde (1995): Sperm and spermiogenesis of the "Turbellaria" and implications for the phylogeny of the phylum Platyhelminthes. In: Jamieson, B.G.M., J. Ausio & J.-L. Justine (eds), Advances in spermatozoal phylogeny and taxonomy. Mém. Mus. natn. Hist. nat. **166**, 37-54.

Dr. Beate Sopott-Ehlers and Prof. Dr. Ulrich Ehlers
II. Zoologisches Institut und Museum der Universität Göttingen
Berliner Straße 28, D-37073 Göttingen

Submicroscopic anatomy of female gonads in *Ciliopharyngiella intermedia* (Plathelminthes, Rhabdocoela, "Typhloplanoida").

Beate Sopott-Ehlers

Abstract

The cytoarchitecture of female gonads in *C. intermedia* is described with special reference to the substructure of eggshell-forming granules in vitellocytes and marginal (cortical) granules in germocytes. The substructure in eggshell-forming granules is not expressed in clear patterns. There exist irregularly running fine lines as well as roundish to dumb-bell-shaped lucent areas. The design in eggshell-forming globules does not correspond to the pattern prevailing in representatives of the Rhabdocoela and Prolecithophora nor to those typical of Proseriata. The substructure of the marginal granules in germocytes resembles the designs found in eggshell-forming granules in vitellocytes in taxa of the Proseriata. From the data described it is concluded that female gonads in *C. intermedia* show rather autapomorphic features than characteristics of the basic pattern of the Proseriata or of the taxon Rhabdocoela + Prolecithophora.

A. Introduction

Ciliopharyngiella intermedia Ax, 1952 was first ascribed to the Proseriata, though with some reservation (see Ax 1952 p. 303 ff.). One of the reasons for this systematic assignment was the general morphology of vitellaria and germaria of this species. Later, however, *C. intermedia* was ascribed to the "Typhloplanoida" and is since then considered as a "link" between Rhabdocoela and Proseriata (see discussion in Ehlers 1972). In the present system of the "Typhloplanoida", the two species *C. intermedia* and *C. constricta* Martens & Schockaert, 1981 constitute the "family" Ciliopharyngiellidae.

During recent years investigations on the ultrastructure of female gonads have provided a lot of data suitable to elucidate phylogenetic relationships between different high-ranked taxa of the Plathelminthes. Among others, SOPOTT-EHLERS (1986, 1992, 1995), EHLERS (1985), GREMIGNI (1988) and GREMIGNI & FALLENI (1991) demonstrated that female gonads, especially the vitellocytes in species of the free-living and symbiotic Rhabdocoela, Neodermata, and Prolecithophora share a common ultrastructure that is quite different from the situations known for species of the Proseriata. The submicroscopic anatomy of vitellaria and germaria in *C. intermedia* was studied to see, whether features exist, which could contribute to come to a more precise hypothesis on the systematic position of the taxon Ciliopharyngiellidae.

B. Material and Methods

The animals derive from beaches of the island of Sylt (North Sea). Fixation, dehydration, embedding, sectioning and staining followed conventional steps. Series of sections were studied using a Zeiss EM 900 electron microscope.

C. Results

Without doubt, *Ciliopharyngiella intermedia* Ax, 1952 is a representative of the Neoophora. Species ascribed to this high-ranked taxon are characterized by female gonads being devided in a part with nutritive cells (vitellocytes) and a section with generative cells (germocytes). In most taxa of the Neoophora, these different segments form separate organs, the vitellaria and the germaria.

1. Vitellaria

The vitellaria in *Ciliopharyngiella intermedia* consist of two lateral strings of serially arranged follicles. These strings run in the praepharyngeal body region. They begin in a short distance behind the brain and cease in front of the pharynx. An epithelium made up by tunica cells enveloping the vitellocytes could not be observed. Just a very delicate layer of extracellular matrix delimits the vitellocytes from surrounding somatic tissues, i. e. adjacent intestinal cells.

Fig. 1. A. Young vitellocyte. B. Yolk platelet. Scales = 1 µm. ▶

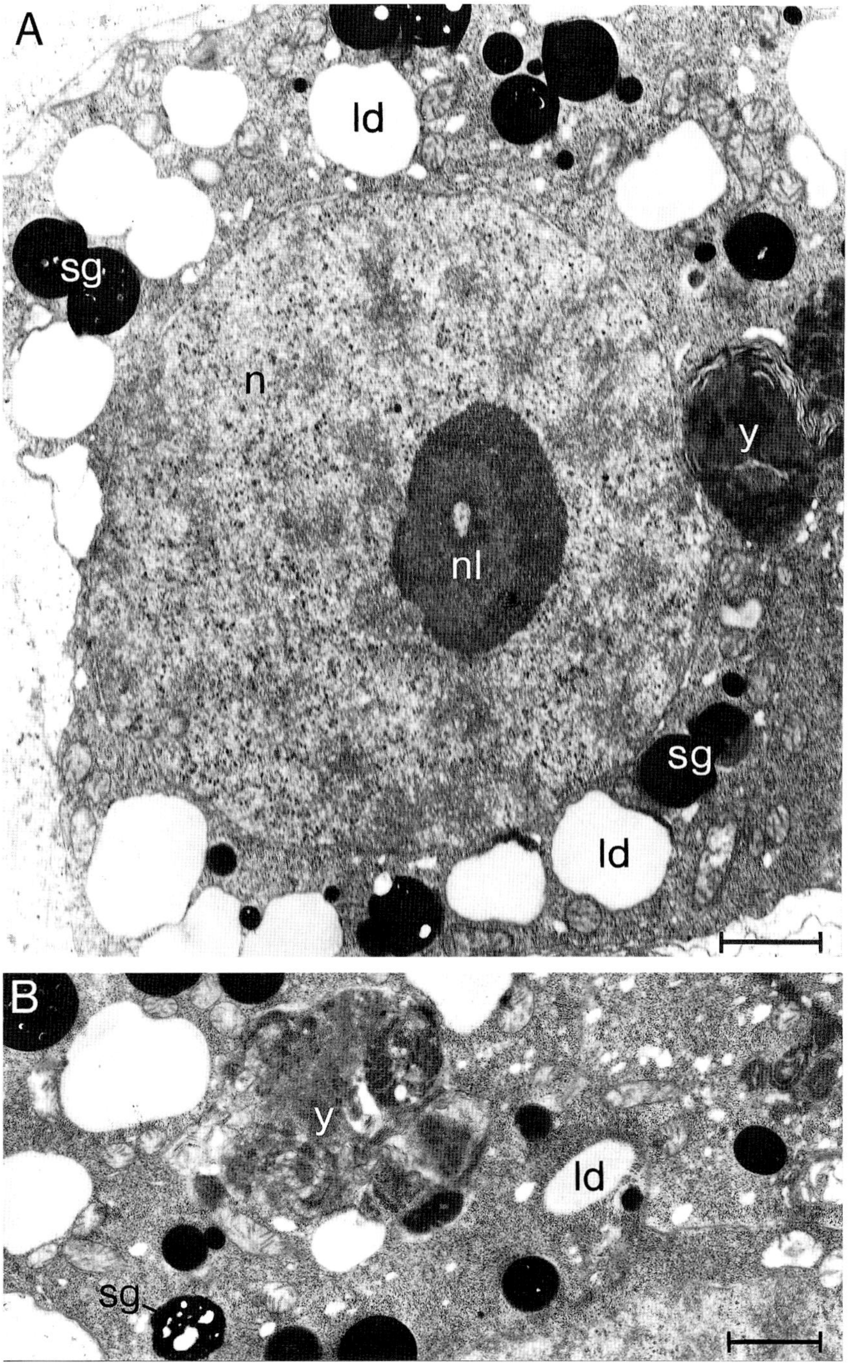
A
ld
sg
n
y
nl
sg
ld
B
y
ld
sg

Young vitellocytes are of a low cytoplasm to nucleus ratio. The nuclei are roundish in shape and have a pronounced nucleolus. The cytoplasm is dense with free ribosomes and glycogen. Plentiful small mitochondria are included.

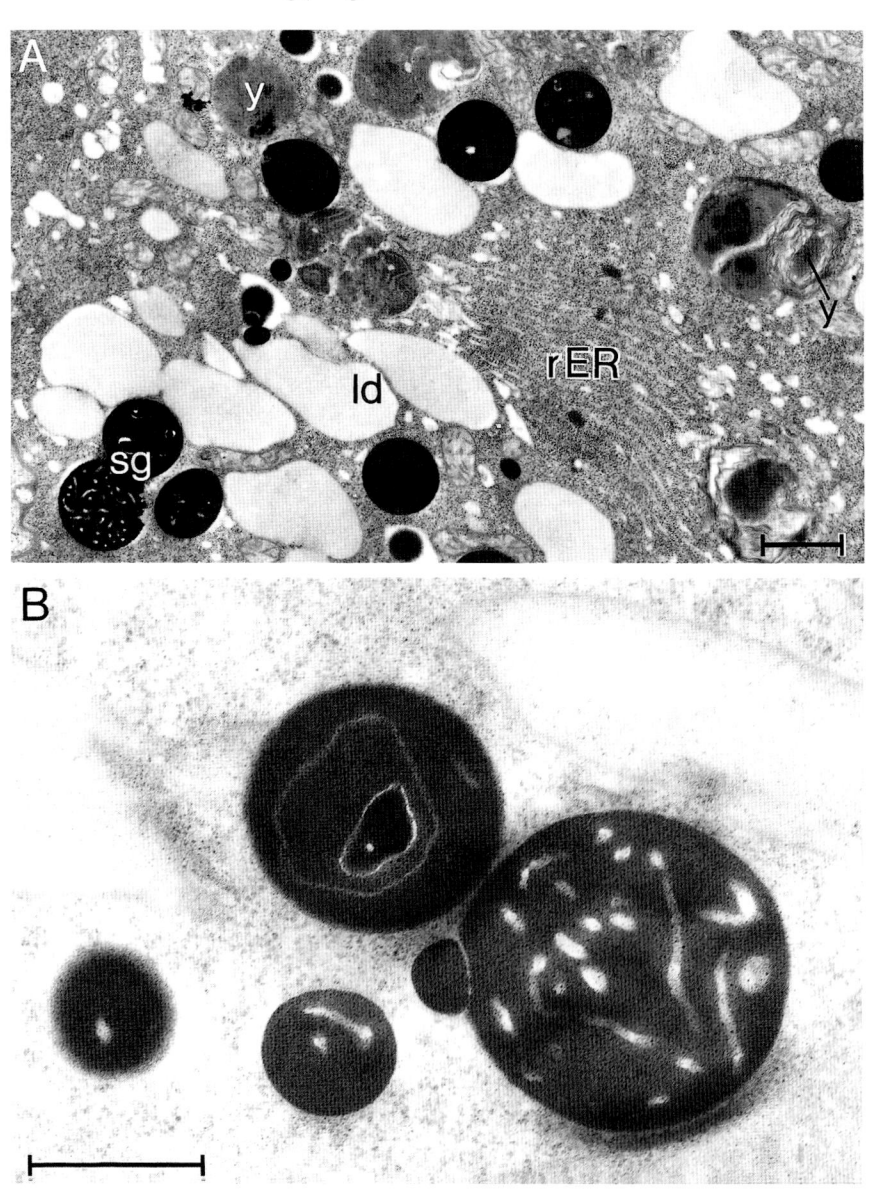

Fig. 2. A. Segment of a vitellocyte with yolk platelets, voluminous rough ER, lipid droplets and eggshell-forming granules. Scale = 1 µm. B. Eggshell-forming granules. Unstained. Scale = 0.5 µm.

In vitellocytes of a more advanced maturation the nucleolus has still a clear demarcation, but besides mitochondria different inclusions are seen in the small cytoplasmic layer surrounding the nucleus (Fig. 1 A).

Mature vitellocytes are characterized in general by a more or less ovoid to lobate nucleus, a high amount of dictyosomes and giant stacks of granular ER densely studded with ribosomes. Beyond this, differentiations such as numerous large non-membrane-bounded lipid droplets, stored nutritive material sequestered in membrane-bordered areas, so-called yolk globules or yolk platelets, and numerous eggshell-forming granules are typical of vitellocytes.

The yolk bodies in *C. intermedia* are irregular in shape, with a diameter of about 1.5 – 2 µm. The development of these yolk compartments is bound to the voluminous profiles of rough ER. In mature yolk globules remnants of stacks of rough ER can be found. Furthermore, glycogen deposits are often included. In some instances transformed mitochondria, lipid droplets, and even

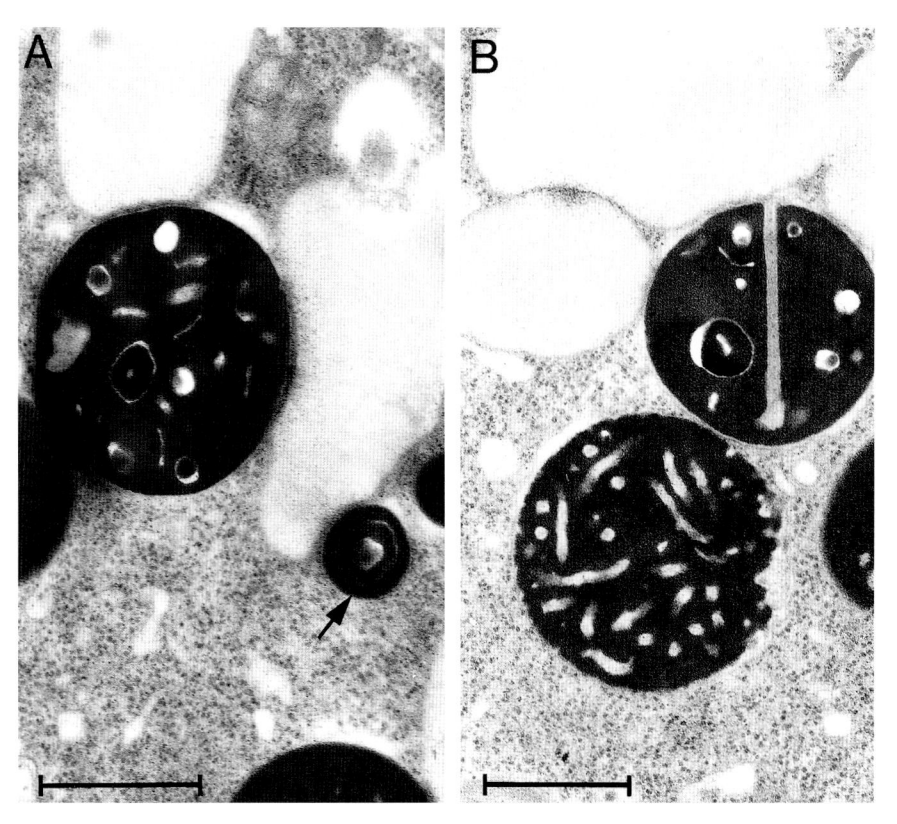

Fig. 3. A,B. Different patterns in eggshell-forming granules. The arrow in A marks a granule with traces of concentric rings. Scales = 0.5 µm.

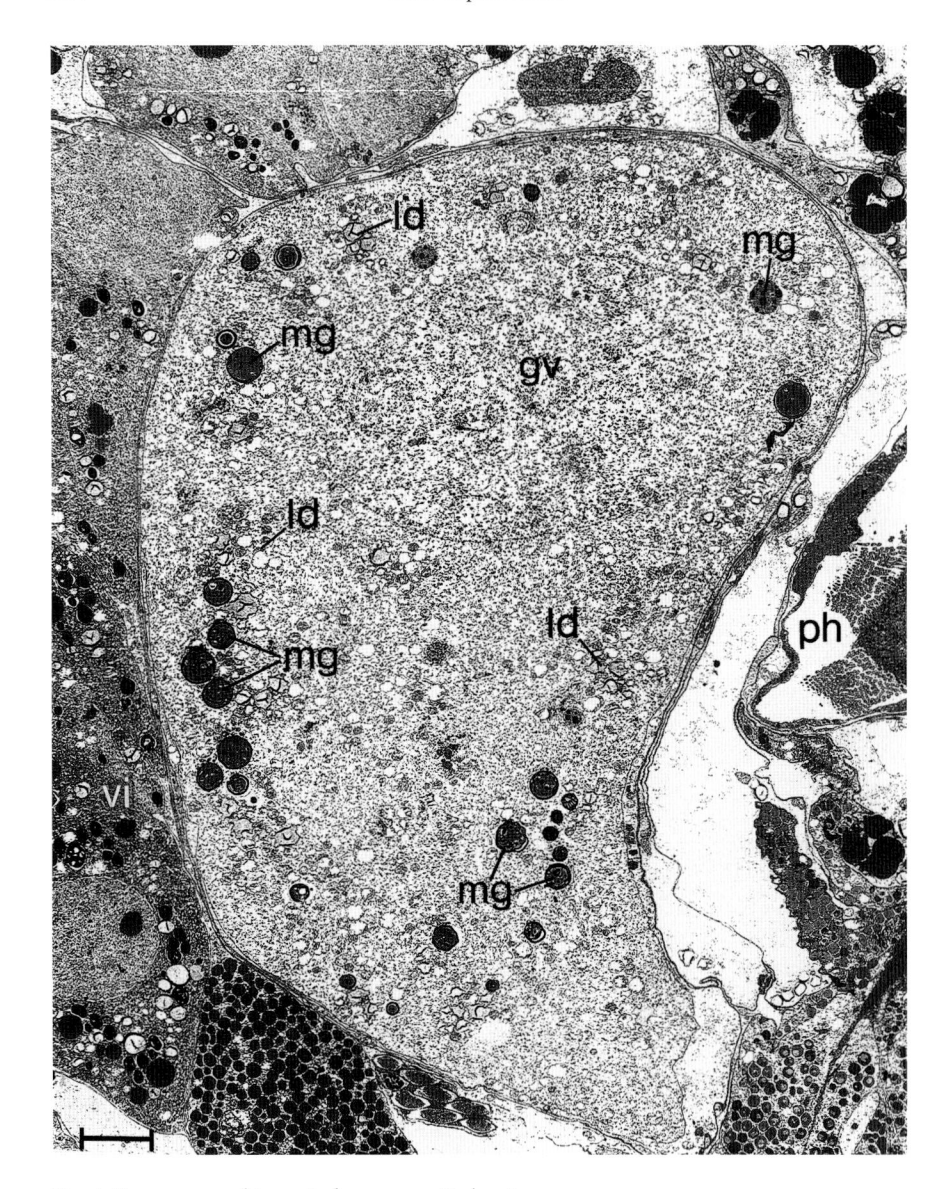

Fig. 4. Germocyte and its typical structures. Scale = 5 µm.

Fig. 5 A. An assemblage of marginal granules closely beneath the oolemm. B. Marginal granules ▶ with different patterns. At the top two granules with more or less centric rings. Uranyl- staining only. Scales = 1 µm.

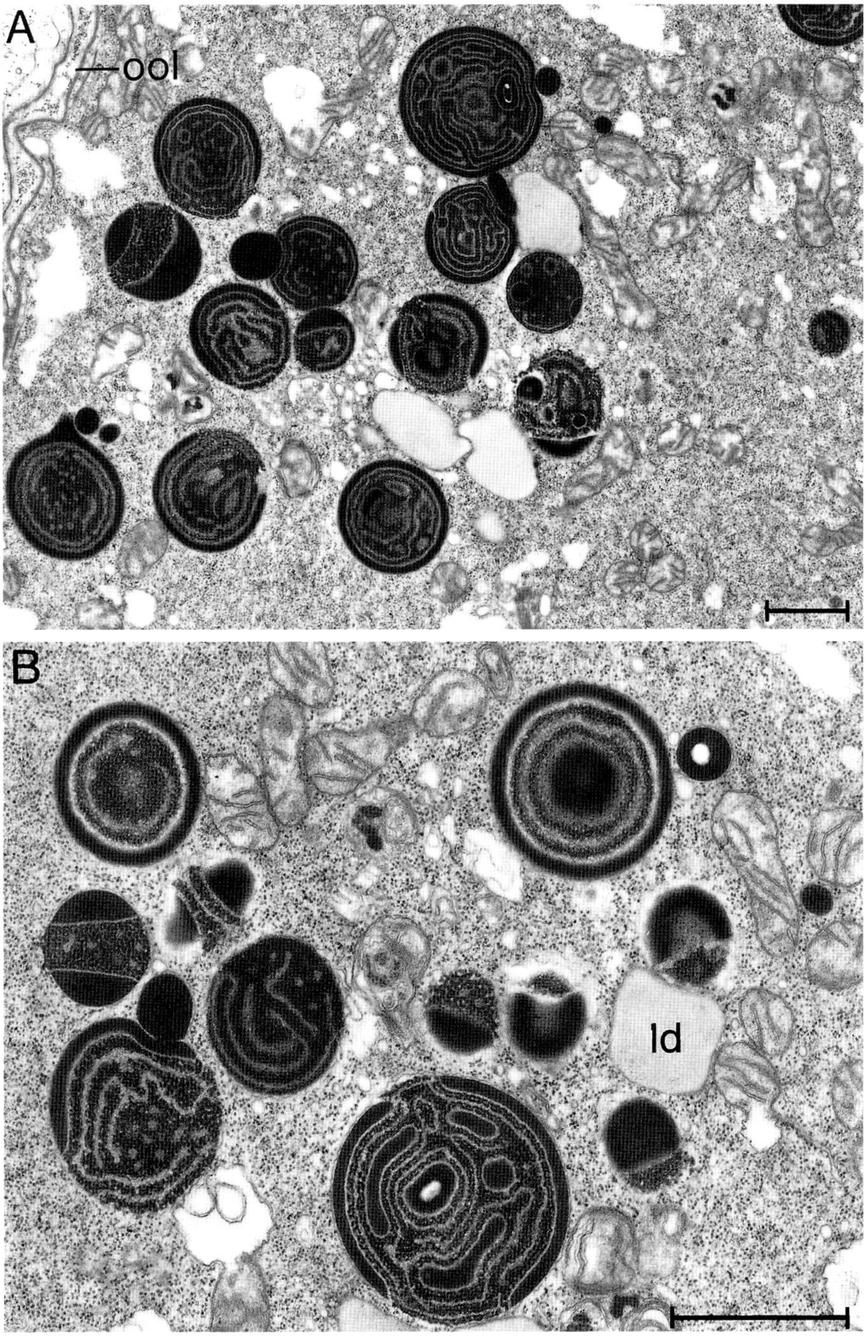
A
—ool
B
ld

a few residual eggshell granules were contained in yolk platelets (Fig. 1 B; Fig. 2 A). Aggregates of a similar appearance are also found in lysosomal inclusions in the cytoplasm of gut cells.

Due to their intense electrondensity caused by osmiophilic, polyphenolic substances, the eggshell-forming granules are the most conspicuous differentiations in vitellocytes. The globules of eggshell substance derive from Golgi complexes and reach a diameter of about 1 µm. They are inclosed by a bordering membrane and predominantly circular in shape. In a few instances, there were elongated granules, presumably derived from a fusion of two granules.

The globules of eggshell-forming material (Fig. 2 B; Fig. 3 A,B) exhibit a special substructure, which, however, is not expressed in clear patterns as known for other neoophoran species. Almost all granules have a special design of their own. There exist very fine electron-lucent lines running in irregular courses. Sometimes, these lines encircle light areas different in shape. Some globules are scattered with roundish to ovoid lucent islets, others show rod- to dumb-bell-shaped areas lacking polyphenolic substances. A few granules seemed to be divided in half by lucent material. Just two small granules containing opaque and elctron-light layers deposited in concentric rings were observed.

2. Germaria

The paired germaria follow the cord of vitellaria. The germaria are enwrapped by extremely flattened tunica cells. A layer of extracellular matrix could not continuously be traced between germaria and adjacent tissues.

Early germocytes display a large roundish nucleus containg a well demarcated nucleolus. The cytoplasm is poor in organelles, but dense with ribosomes. Stacks of annulate lamellae already occur. Mature female germ cells are large in size (Fig. 4). The nucleus or germinal vesicle, respectively, is lobate. Its envelope has numerous clearly expressed nuclear pores. Often large areas of chromatoid bodies are seen in the plasm near the nuclear envelope. Furthermore the ooplasm contains lipid droplets, plentiful mitochondria, small profiles of rough and smooth ER, as well as dictyosomes with densely filled cisternae. Roundish to oval granules, about 1.5 µm in diameter containing material of dark staining quality are especially leaping to the eye. These membrane- bordered differentiations derive from Golgi complexes. In younger germocytes, the granules are seen more centrally, in mature ones, however, peripherally, near the oolemm. Since differentiations of this type surely do not correspond to cortical granules in other taxa, the granules have been termed marginal granules. In these marginal granules layers of electron-light and el-

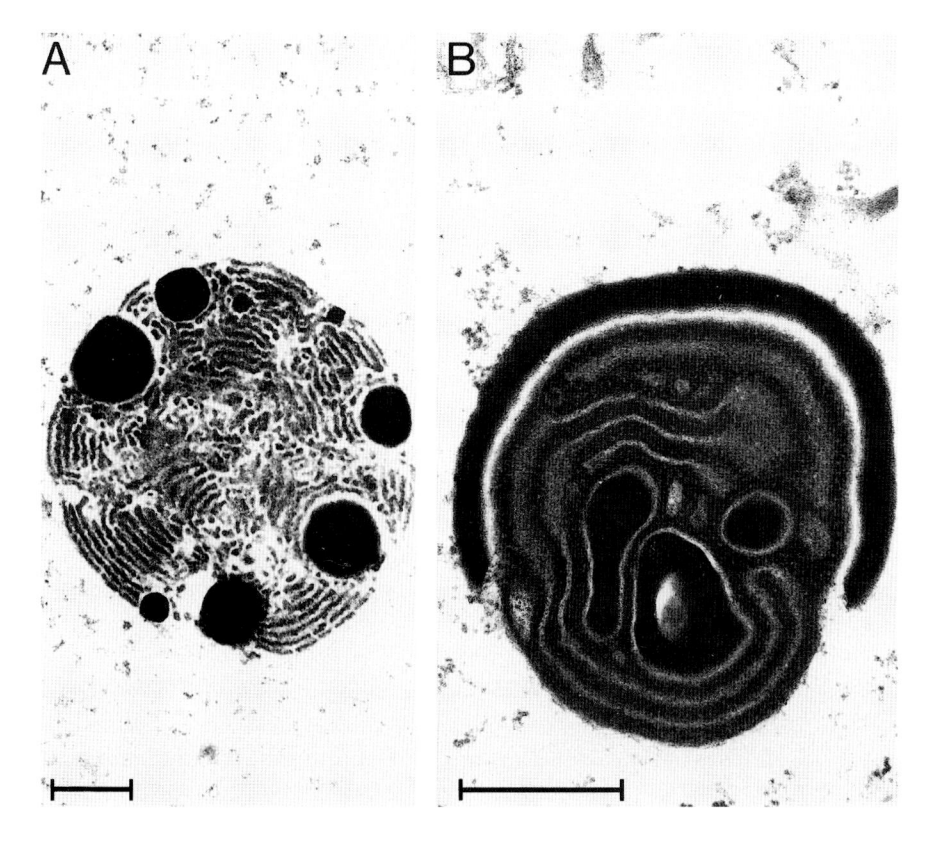

Fig. 6. A,B. Marginal granules in higher magnification. Double staining. Scales = 0.5 µm.

ectron-dense materials are deposited. From this, bizarre patterns in their substructure result (Figs. 5; 6). Almost all granules have a particular design of their own. In sections, some of the lucent layers look like lines of isotherms, others appear like terraces of fields or vineyards. Marginal granules exhibiting segments of homogenous black material and sections of granular, light and dark substances were also observed. In a few instances layers of electron-lucent and opaque materials were deposited in a bowl- shaped manner causing a pattern of more or less irregularly circular rings.

Stored nutritive material sequestered in membrane-bounded areals does not exist in the ooplasm of female germ cells in *Ciliopharyngiella intermedia*.

D. Discussion

Special emphasis has been put on the structure of eggshell-forming granules and yolk platelets in vitellocytes when evaluating the significance of submicroscopic anatomy of female neoophoran gonads for plathelminth phylogeny. Furthermore, the existence or the lack of stored nutritive materials, of eggshell substances and the shape of marginal (cortical) granules in germocytes has been included into such considerations.

Representatives of the Rhabdocoela including the Neodermata studied so far are characterized by vitellocytes containing eggshell-forming granules, which exclusively exhibit a substructure made up by round to polygonal grana (see SOPOTT-EHLERS & EHLERS 1986; GREMIGNI 1988; GREMIGNI & FALLENI 1991). This mosaic design also exists in species ascribed to the Prolecithophora (see EHLERS 1985, 1988) and belongs to the basic pattern of this taxon. In representatives of the Seriata, however, a meandering or convoluted pattern within the granules prevails and a design formed by concentrically arranged rings additionally exists in some supraspecific taxa, *e.g.* Coelogynoporidae (see *i. al.* GREMIGNI 1988; SOPOTT-EHLERS 1986, 1990, 1992).

The pattern found in the eggshell-forming granules of *Ciliopharyngiella intermedia* does not correspond to the one prevailing in species ascribed to the Rhabdocoela and Prolecithophora, nor to the ones typical of Proseriata. It is also different from the situation known for members of the Lecithoepitheliata (FALLENI 1997). Most probably, the way of deposition of eggshell material observed in *C. intermedia* is therefore hypothesized as an autapomorphic feature of the taxon *Ciliopharyngiella* or of *C. intermedia*, respectively.

Sufficiently documented data on the submicroscopic structure of deposited nutritive reserve materials are still quite spotty. From own unpublished observations on differnt taxa of the Rhabdocoela, however, it can be stated, that yolk platelets in vitellocytes of *C. intermedia* do not correspond in their composition to those found in representatives of the Rhabdocoela Kalyptorhynchia, "Dalyellioida" and "Typhloplanoida". Yet, there exists a striking similarity to yolk globules typical of species ascribed to the Proseriata Lithophora (see *i.al.* SOPOTT-EHLERS 1990, 1992).

The substructure of marginal (cortical) granules in the generative cells of *C. intermedia* is totally different from the patterns reported so far for representatives of the Rhabdocoela (cf. GREMIGNI 1988; FALLENI & LUCCHESI 1992; LUCCHESI et. al. 1995). Moreover, the marginal granules in *C. intermedia* do not correspond either in size nor in their substructure to any situation existing in representatives of the Proseriata (see *i. al.* GREMIGNI 1988; GREMIGNI & NIGRO 1983; GREMIGNI et al. 1986; SOPOTT-EHLERS 1990, 1991, 1994, 1995).

This implies that the special substructure in the marginal granules of *Ciliopharyngiella intermedia* might be an autapomorphic feature of this species or of the taxon *Ciliopharyngiella*.

On the other hand, there exists a striking correspondance between the fine architecture of marginal granules in germocytes of *C. intermedia* and the patterns found in **eggshell-forming globules** in vitellocytes of some representatives of the Proseriata Coelogynoporidae (see SOPOTT-EHLERS 1990), species of the Otoplanidae Parotoplaninae (see SOPOTT-EHLERS 1992) and of *Polystyliphora filum*, a representative of the Proseriata Unguiphora (see SOPOTT-EHLERS 1995).

This fact gives support to the idea vividly discussed in recent years (see CHANDLER et al. 1992; SHINN 1993; SOPOTT-EHLERS 1994; LUCCHESI et al. 1995) that marginal granules in neoophoran Plathelminthes may rather play a role in the formation of envelopes enclosing eggs, than have the task to build up a barrier to prevent polyspermy.

Leaving the yet not sufficiently studied compositions of yolk bodies in different taxa out of consideration, it can be concluded from the data described above, that the female gonads in *Ciliopharyngiella intermedia* show rather autapomorphic features than characterists of the basic pattern of the Proseriata or the Rhabdocoela + Prolecithophora, respectively. It seems premature to discuss the exact phylogenetic position of the taxon *Ciliopharyngiella* within the Neoophora. But the present results strengthen the hypothesis that *C. intermedia* is not a member of the monophylum Proseriata nor of a monophylum Rhabdocoela + Prolecithophora.

Acknowledgements

Financial support was provided by the Akademie der Wissenschaften und der Literatur, Mainz. Thanks are due to Mrs. S. Gubert for technical assistance and to Mr. B. Baumgart for help with the figures.

Zusammenfassung

Die Cytologie der weiblichen Gonade, insbesondere die Substruktur der Schalensubstanzvesikel in den Vitellocyten sowie der Marginalgranula (Cortikalgranula) in den Germocyten von *Ciliopharyngiella intermedia* wird beschrieben. Die unregelmäßig geformten Substrukturen in den Schalensubstanzvesikeln erscheinen als irregulär verlaufende Linien (Schichten) wie auch als rundlich bis hantelförmig geformte Areale. Diese Ultrastruktur der Schalensubstanzvesikel unterscheidet sich deutlich von jener der Rhabdocoela und Prolecithophora wie auch jener der Proseriata. Bemerkenswert erscheint, daß die Marginalgranula der Germocyten von *C. intermedia* in der Substruktur jener von Schalensubstanzvesikeln in den Vitellocyten bei Proseriaten ähneln.

Auf der Basis dieser Daten läßt sich der Schluß ziehen, daß die weibliche Gonade in ihrer Ultrastruktur eher autapomorphe Zustände für *C. intermedia* (oder das Taxon *Ciliopharyngiella*) zeigt als Grundmustermerkmale der Proseriata bzw. des Taxons Rhabdocoela + Prolecithophora.

Abbreviations

gv	germinal vesicle	ph	pharynx
ld	lipid droplet	rER	rough endoplasmic reticulum
mg	marginal granule	sg	eggshell- forming granule
n	nucleus	y	yolk platelet
nl	nucleolus	vi	vitellar
ool	oolemm		

References

Ax, P. (1952): *Ciliopharyngiella intermedia* nov. gen. nov. spec. Repräsentant einer neuen Turbellarien-Familie des marinen Mesopsammon. Zool. Jb. Abt. Syst. **81**, 286–312.

CHANDLER, R. M., M. B. THOMAS & J. P. S. SMITH III (1992): The role of shell granules and accessory cells in eggshell formation in *Convoluta pulchra* (Turbellaria, Acoela). Biol. Bull. 182, 54-65.

EHLERS, U. (1972): Systematisch-phylogenetische Untersuchungen an der Familie Solenopharyngidae (Turbellaria, Neorhabdocoela). Mikrofauna Meeresboden **11**, 1–78.

– (1985): Das phylogenetische System der Plathelminthes. G. Fischer, Stuttgart, New York, 317 p.

– (1988): The Prolecithophora – a monophyletic taxon of the Plathelminthes? Fortschr. Zool./Prog. Zool. **36**, 359–365.

FALLENI, A. (1997): Ultrastructural aspects of the germovitellarium of two prorhynchids (Platyhelminthes, Lecithoepitheliata). Invertebr. Reproduct. Developm. **31**, 285–296.

FALLENI, A. & P. LUCCHESI (1992): Ultrastructure and cytochemical aspects of oogenesis in *Castrada viridis* (Platyhelminthes, Rhabdocoela) J. Morphol. **213**, 241–250.

GREMIGNI, V. (1988): A comparative ultrastructural study of homocellular and heterocellular female gonads in free-living Platyhelminthes – Turbellaria. Fortschr. Zool. Prog./Zool. **36**, 245–261.

GREMIGNI, V. & A. FALLENI (1991). Ultrastructural features of cocoon-shell globules in the vitelline cells of neoophoran platyhelminths. Hydrobiologia **227**, 105–111.

GREMIGNI, V. & M. NIGRO (1984): Ultrastructural study of oogenesis in *Monocelis lineata* (Turbellaria, Proseriata). Int. J. Invert. Rprod. Devel. **7**, 105–118.

GREMIGNI, V., M. NIGRO & M. S. SETTEMBRINI (1986): Ultrastructural features of oogenesis in some marine neoophoran turbellarians. Hydrobiologia **132**, 145–150.

LUCCHESI, P., A. FALLENI & V. GREMIGNI (1995): The ultrastructure of the germarium in some Rhabdocoela. Hydrobiologia **305**, 207–212.

SHINN, G.L. (1993): Formation of egg capsules by flatworms (Phylum Platyhelminthes). Trans. Am. Microsc. Soc. **112**, 18–34.
SOPOTT-EHLERS, B. (1986): Fine structural characteristics of female and male germ cells in Proseriata Otoplanidae (Platyhelminthes). Hydrobiologia **132**, 137–144.
– (1990). Feinstrukturelle Untersuchungen an Vitellarien und Germarien von *Coelogynopora gynocotyla* Steinböck, 1924 (Plathelminthes, Proseriata). Microfauna Marina **6**, 121–138.
– (1992): Ultrastructural studies on the vitellocytes of Parotoplaninae (Plathelminthes, Proseriata) with special reference of eggshell-forming granules. Zoomorphology **112**, 125–131.
– (1994): On the fine structure of female germ cells in *Archimonocelis oostendensis* (Plathelminthes, Proseriata). Microfauna Marina **9**, 339–344.
– (1995): Fine structure of vitellaria and germaria in *Polystyliphora filum* (Plathelminthes, Proseriata). Microfauna Marina **10**, 159–171.
SOPOTT-EHLERS, B. & U. EHLERS (1986): Differentiation of male and female germ cells in neoophoran Plathelminthes. In: M. PORCHET, J.-C. ANDRIES & A. DHAINAUT (eds) Advances in Invertebrate Reproduction **4**. Elsevier Science Pub. Amsterdam, pp. 187–194.

Dr. Beate Sopott-Ehlers
II. Zoologisches Institut und Museum der Universität Göttingen
Berliner Straße 28, D-37073 Göttingen

On a small collection of Laophontidae (Copepoda) from Sulawesi, Indonesia

Wolfgang Mielke

Abstract

Five species of Laophontidae were collected from two localities of Sulawesi, Indonesia.

The material from Donggala (west coast of Central Sulawesi) contained, among other species, *Laophonte spinicauda* (Vervoort, 1964), *Laophontina sensillata* Wells & Rao, 1987, and *Langia maculata* Wells & Rao, 1987.

Esola longicauda Edwards, 1891 was found in sublittoral samples just in front of the coast of Bunaken Island (North Sulawesi).

Another laophontid species inhabits the beach of Bunaken Island. It is characterized as a new subspecies of *Quinquelaophonte quinquespinosa* (Sewell, 1924): *Q. q. bunakenensis* nov. subspec.

A. Introduction

It will take tremendous efforts to compile just the outlines of the harpacticoid copepod fauna of Indonesia's enormous assemblage of islets and islands. The present state of information on freshwater and marine species is based on sporadic investigations. In particular Sulawesi, the former Celebes, with its four characteristic peninsulas, offers a multitude of different habitats. Though their treatment is still in its infancy, some contributions already exist (A. Scott 1909, Brehm & Chappuis 1935, Cottarelli et al. 1984, Cottarelli 1985). The following results are no more than a further modest step towards completing the knowledge of the island's copepod stock.

As can be presumed from the laophontid species found, the copepod composition reveals great similarity with the now well-investigated littoral copepod fauna of the Andaman and Nicobar Islands (WELLS & RAO 1987). This statement is underlined by the additional finding of *Parastenhelia oligochaeta* Wells & Rao, 1987 in the beach of Donggala.

On the occasion of a short stay at the beaches of Donggala (west coast of Central Sulawesi) and of Bunaken Island (North Sulawesi, near Manado), a remarkable number of benthic copepod species could be extracted from only a few qualitative samples. Five members of the Laophontidae are presented now.

The material has been deposited in the collections of the Zoological Museum of the University of Göttingen.
The interpretation of body, mouth parts and thoracopods is adopted from LANG (1965). With respect to the mouth parts, the interpretation of the components by HUYS & BOXSHALL (1991) is given in parentheses.

B. Results

Laophontidae T. Scott, 1904
Laophonte Philippi, 1840
Laophonte spinicauda (Vervoort, 1964)

(Figs. 1–3)

Locality and material. Donggala; beach in front of the bungalow property "John Prince Dive Resort" (20 September 1994); 15 ♀♀ (5 of them ovigerous), 19 ♂♂.
3 ♀♀ and 3 ♂♂ were dissected. P.1 – P.5 ♀, furca ♀, endopodite P.3 ♂ and P.5 ♂ are drawn.
Length of dissected ♀♀: 0.37 – 0.39 mm
♂♂: 0.31 – 0.37 mm

Remark: The species was recorded by VERVOORT (1964) as *Paralaophonte spinicauda* from Ifaluk Atoll in the Caroline Islands. The description was based on two ♀♀, i.e., it was unknown whether or not the ♂ possessed a transformed inner seta on distal segment of enp. P.2, typical for *Paralaophonte*. COULL (1971) who thought to have found the corresponding ♂ at Hull Bay, St. Thomas (Virgin Islands) characterized the P.2 "as in female". If so, the systematic assignment of this species as well as the relationship of VERVOORT's and COULL's animals had to be questioned (MIELKE 1981, p. 43). FIERS (1984, 1986) and WELLS & RAO (1987) subsequently allocated the species to the genus *Laophonte* and added some more localities.

Despite some slight differences from the description of VERVOORT, the individuals from Donggala/Sulawesi are considered as also belonging to *L. spinicauda*.

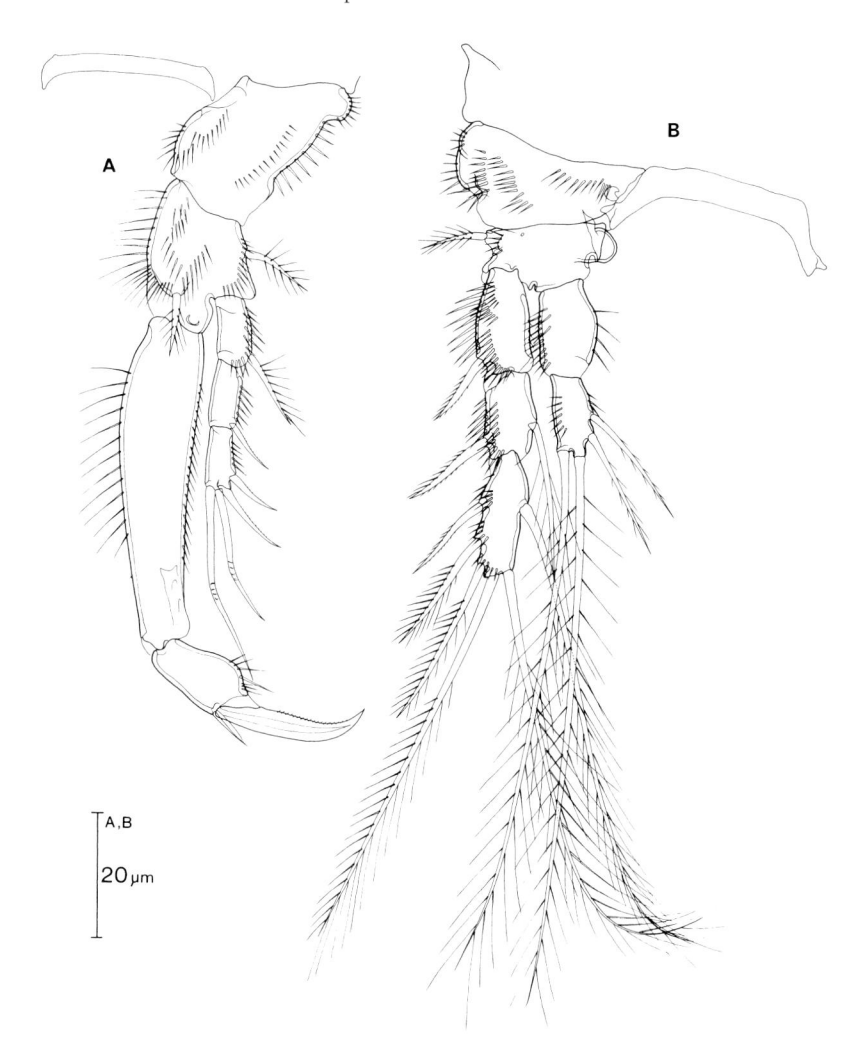

Fig. 1. *Laophonte spinicauda*. ♀. A. P.1. B. P.2.

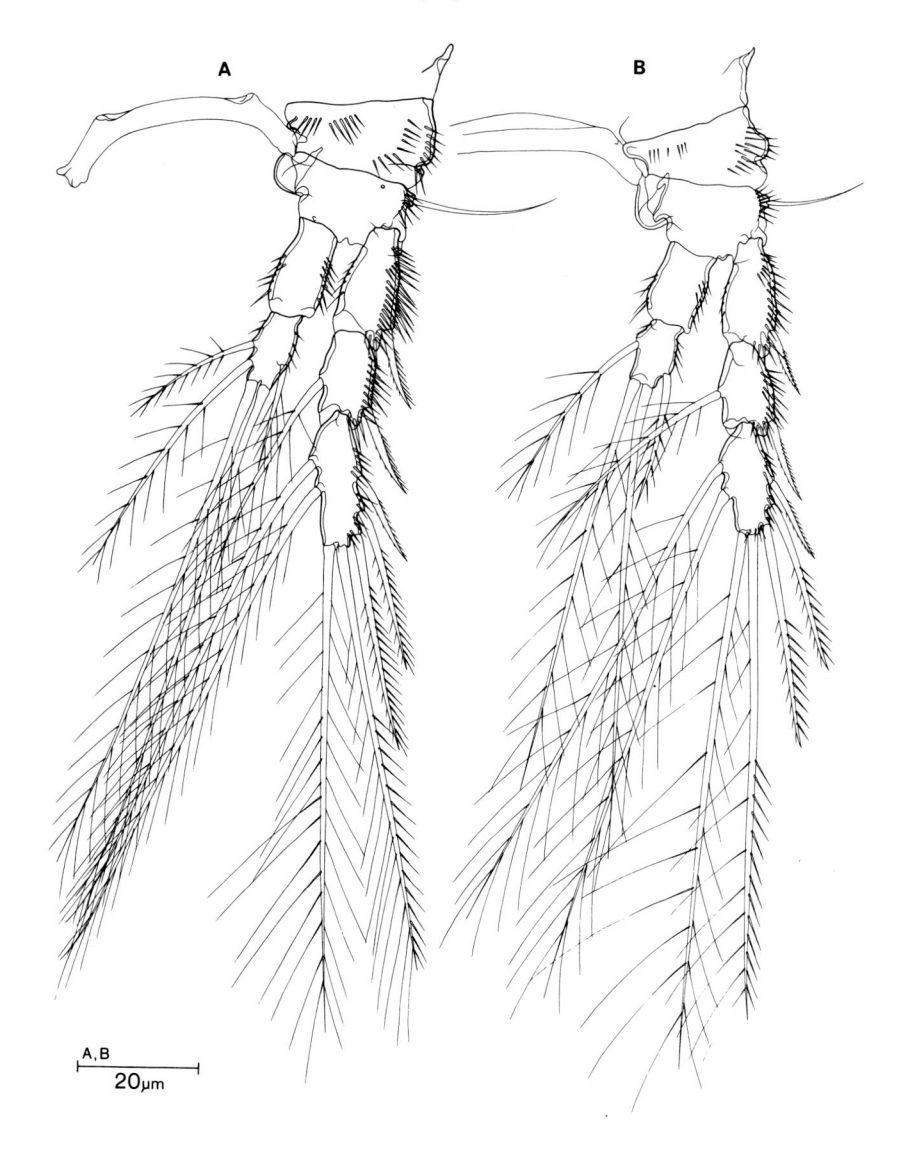

Fig. 2. *Laophonte spinicauda*. ♀. A. P.3. B. P.4.

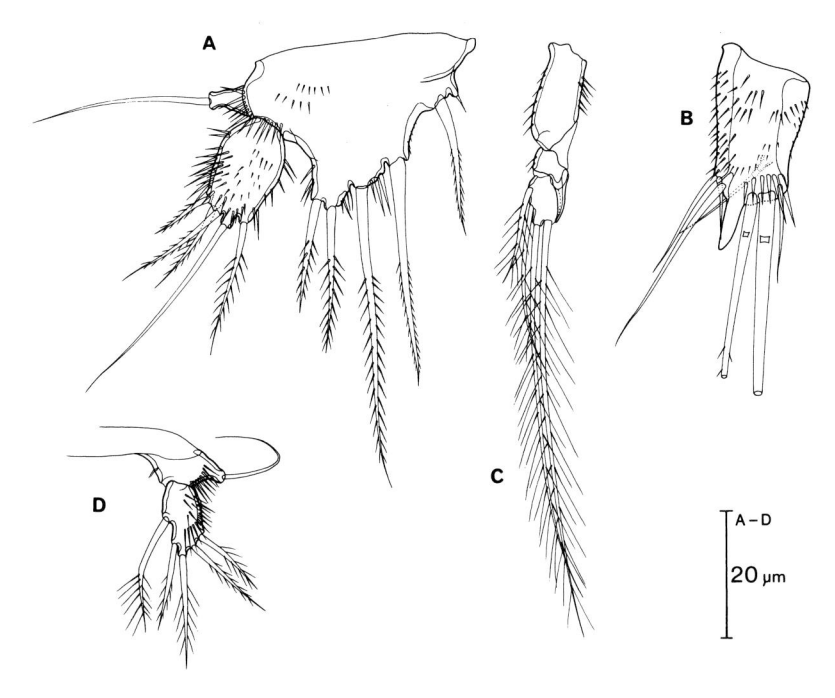

Fig. 3. *Laophonte spinicauda*. A. P.5 ♀. B. Furca ♀, ventral side. C. Endopodite P.3 ♂. D. P.5 ♂.

Laophontina Norman & T. Scott, 1905
Laophontina sensillata Wells & Rao, 1987

(Fig. 4)

Locality and material. Donggala; beach in front of the bungalow property "John Prince Dive Resort" (20 September 1994); 14 ♀♀, 6 ♂♂.
3 ♀♀ and 3 ♂♂ were dissected. P.1 – P.5 ♀, furca ♀ and P.5 ♂ are drawn.
Length of dissected ♀♀: 0.50 – 0.51 mm
♂♂: 0.42 – 0.43 mm

Remark: The animals of Donggala, Sulawesi, largely agree with the ones which WELLS & RAO (1987) have described as *Laophontina sensillata* from several sites of the Andaman and Nicobar Islands (India). In contrast to the description of WELLS & RAO (1) the allobasis of 2nd antenna is furnished with a seta and (2) P.2 bears three instead of only two inner setae. The latter feature is probably variable because one P.2 of a male has only two inner setae; its partner, however, also bears three setae. Moreover, some minor differences as to

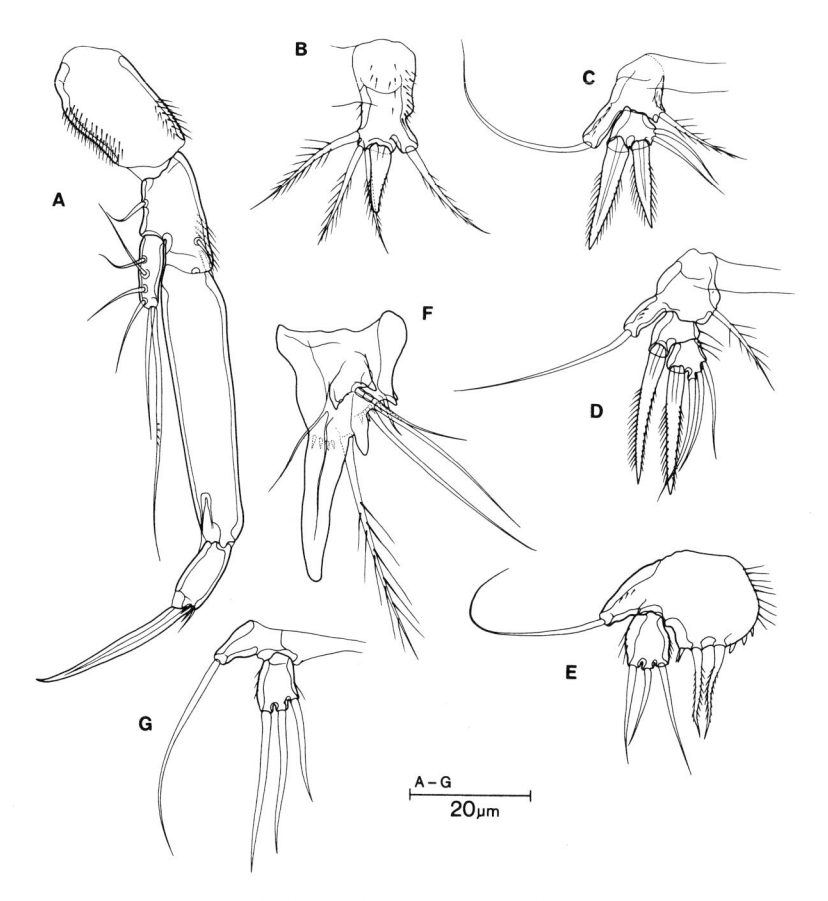

Fig. 4. *Laophontina sensillata*. A. P.1 ♀. B. P.2 ♀. C. P.3 ♀. D. P.4 ♀. E. P.5 ♀. F. Furca ♀, dorsal side. G. P.5 ♂.

the length ratios of setae of the pereiopods or the structure of the mouth parts can be stated. The specimens of Donggala are therefore attributed to *Laophonte sensillata* only with reservation.

Langia Wells & Rao, 1987
Langia maculata Wells & Rao, 1987

(Figs. 5–8)

Locality and material. Donggala; beach in front of the bungalow property "John Prince Dive Resort" (20 September 1994); 6 ♀♀, 6 ♂♂.
3 ♀♀ and 3 ♂♂ were dissected.

Description

Female: Body length of the dissected females from tip of rostrum to end of furcal rami 0.37 – 0.39 mm. Rostrum (Fig. 6A) largely fused with cephalothorax, triangular, somewhat irregularly shaped, with two setules subterminally. Cephalothorax and all following somites with pitted surface dorsally. Caudal margin of cephalothorax smooth and furnished with hairs. Caudal margin of following somites indented dorsally, also furnished with a row of hairs. Furthermore, all dorsal caudal rims – except last two somites – are provided with some short setules. Ventral caudal margins of abdominal somites with short spinules, weakest on genital double-somite. Genital double-somite with a trace of subdivision dorsolaterally, which is also dentated but lacks additional hairs. Genital area see Fig. 5B. Anal operculum semicircular, dentated, proximally – near to the notch – with a slender setule on both sides. Furcal rami about as broad as long, prolonged into a slightly curved spiniform process. Ventral surface set with small spinules; outer margin with 3 setae of different length; subdistal edge also with 3 setae of different length; subdistal edge also with 3 setae grouped together, middle one largest and weakly plumose. Dorsal seta bipartite at base, near ist insertion point an oblique line of denticles extends outwards (Figs. 5A–C).

1st Antenna (Fig. 6A): 5 segments. 1st segment with a short outer tooth. 2nd segment has an outer, caudally bent spiniform process. 4th segment with an indication of a former subdivision. Aesthetascs on 4th and last segments.

2nd Antenna (Fig. 6B): The allobasis has a small seta on anterior margin. Free endopodite segment subapically with some spinules, 1 spine and 1 seta on anterior margin, 1 strong plumose seta on surface, apically with 2 spines of different length, 3 geniculate setae and 1 thin seta, which is fused basally with the most posterior seta. Exopodite 1-segmented, bears 3 setae of about the same length and apparently a short setule on margin.

Mandible (Fig. 6C): Cutting edge with several teeth and a seta nearby. Coxa-basis (basis) with 1 plumose seta. Endopodite seemingly not separated, provided with 3 slender setae. Exopodite represented by 1 seta.

1st Maxilla (Fig. 6D): Arthrite of precoxa with an inner and an outer slender seta and a row of spinules on surface; distal edge with 5 claw-like appen-

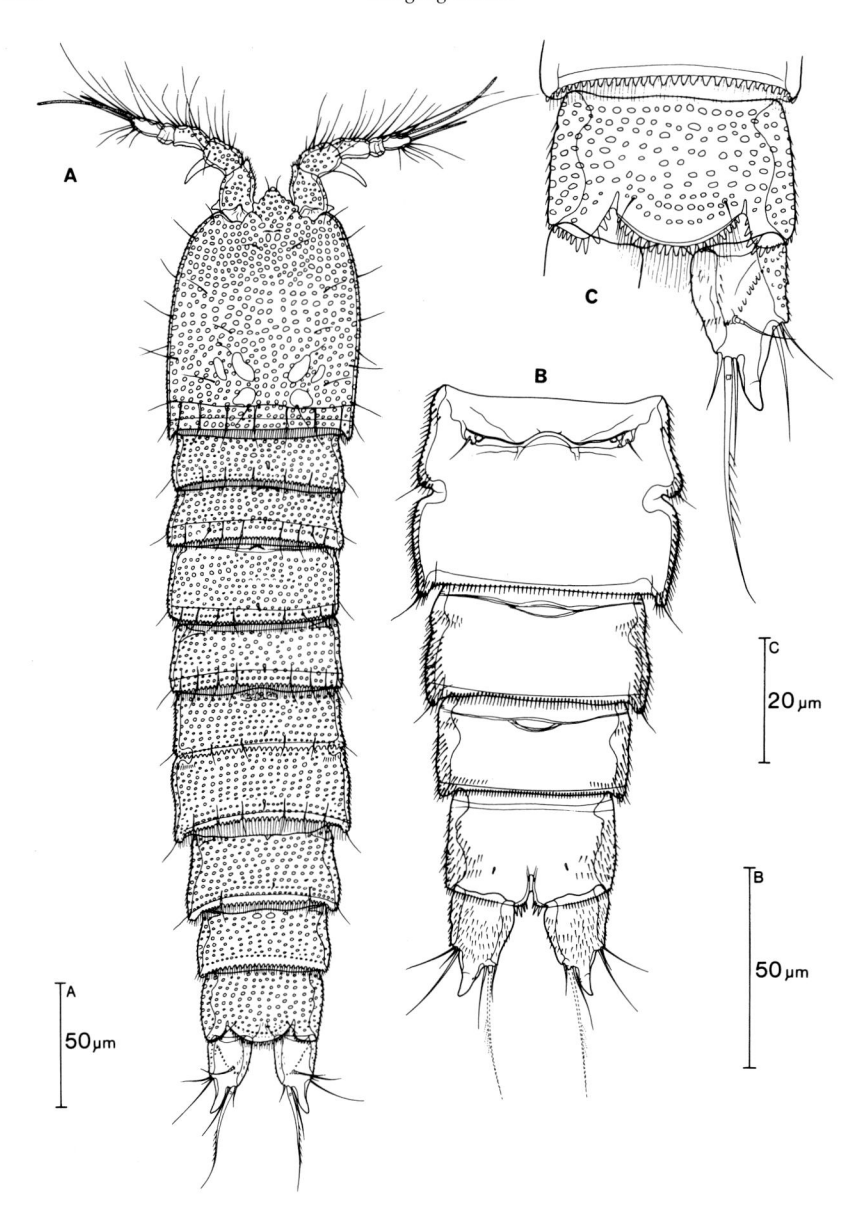

Fig. 5. *Langia maculata*. A. Habitus ♀, dorsal side. B. Abdomen ♀, ventral side. C. Caudal part of Abdomen ♂, dorsal side.

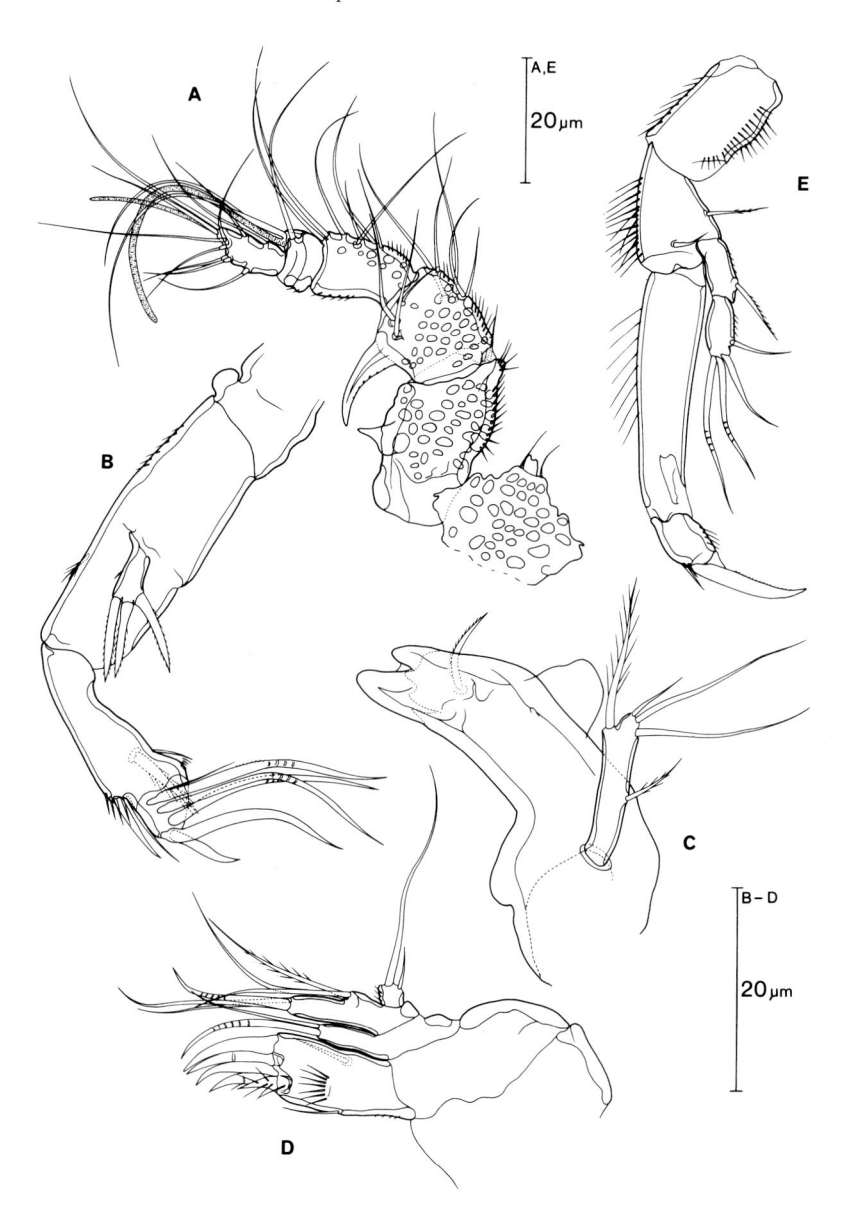

Fig. 6. *Langia maculata*. ♀. A. Rostrum and 1st Antenna. B. 2nd Antenna. C. Mandible. D. 1st Maxilla. E. P.1.

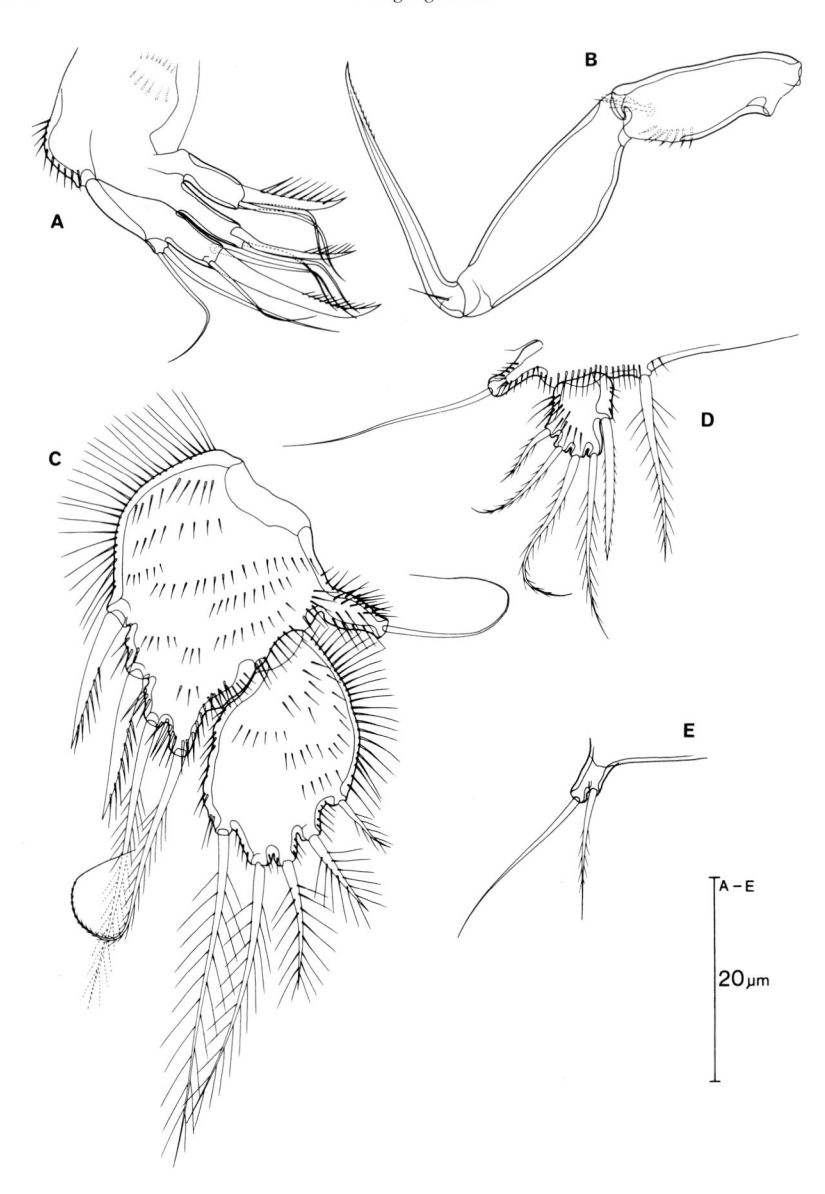

Fig. 7. *Langia maculata*. A. 2nd Maxilla ♀. B. Maxilliped ♀. C. P.5 ♀. D. P.5 ♂. E. P.6 ♂.

dages. Coxa, basis, endopodite and 1-segmented exopodite are furnished with 2, 3, 3, 2 setae, respectively.

2nd Maxilla (Fig. 7A): Syncoxa with 2 endites, each one bearing 1 strong pinnate appendage and 2 slender setae. Basis (allobasis) with 1 strong claw and 2 slender setae. Endopodite 1-segmented, with 2 setae.

Maxilliped (Fig. 7B): Basis (syncoxa) with some spinules and a short seta. Margins of endopodite (basis) apparently smooth. Terminal claw accompanied by a short setule.

P.1 (Fig. 6E): Coxa with slender spinules on inner and outer margin. Basis has slender spinules on inner margin, 1 short seta on surface and 1 seta on outer edge. Exopodite 2-segmented, distal segment with 4 setae. Endopodite 2-segmented. Proximal segment strong, bears a row of hairs on inner edge. Terminal segment short, with some spinules and 1 claw.

P.2 (Fig. 8A): Basis with fine spinules along inner margin. Outer seta plumose. Exopodite 2-segmented. Margins furnished with slender spinules. Proximal segment has 1 outer appendage. Distal segment bears 1 outer subapical seta and 2 apical plumose setae. Endopodite vestigial.

P.3 and P.4 (Figs. 8B, C): Basis carries a long, slender seta inserting on an outer bulge. Exopodite 3-segmented. Proximal and middle segment each with slender spinules on inner margin and on outer area and 1 pinnate outer spine. Distal segment with 2 long pinnate spines on outer margin, 2 apical plumose setae and 1 plumose seta on inner margin. Endopodite 1-segmented, with 2 long plumose setae.

Seta and spine formula:

	Exopodite	Endopodite
P.2	(0.021)	–
P.3	(0.0.122)	(020)
P.4	(0.0.122)	(020)

P.5 (Fig. 7C): Margins and surface spinulose. Baseoendopodite with 1 slender seta on outer lobe; inner part carries 2 stout pinnate setae and 2 slender plumose setae. Exopodite with 5 setae, increasing in length seen from outer to inner seta.

Male: Differs from the female in the following respects:
– Body length of 3 dissected animals 0.33 – 0.36 mm.
– 1st Antenna subchirocer.
– Length ratio of appendages of P.3 and P.4 somewhat different compared to female. Above all both setae of the endopodites are distinctly shorter.

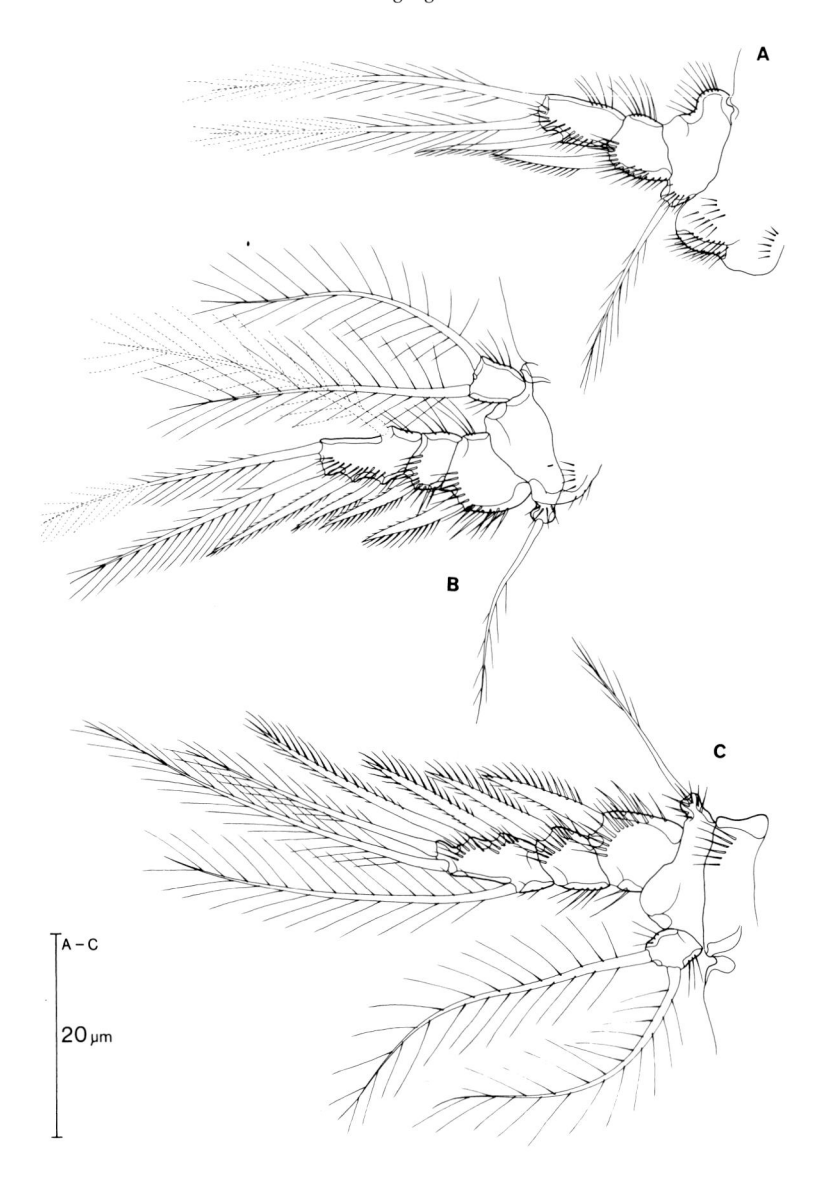

Fig. 8. *Langia maculata*. ♀. A. P.2. B. P.3. C. P.4.

– Baseoendopodite of P.5 with a slender outer seta and an inner plumose seta. Exopodite bears 5 setae (Fig. 7D).
– P.6 with 2 setae, outer one longer than inner one (Fig. 7E).

Discussion. The animals of Donggala/Sulawesi well agree with the ones of the Andaman and Nicobar Islands (WELLS & RAO 1987). Slight differences, e.g. mouthparts, length of innermost seta of exp. P.5 ♀, dentation of anal operculum, length of furcal rami may be merely due to difficulties in the interpretation of the small objects as well as variability of the species.

Esola Edwards, 1891
Esola longicauda Edwards, 1891

(Figs. 9–12)

Locality and material. Bunaken Island, near Manado (North Sulawesi); some sublittoral samples of sand between seagrass and corals (23 September 1994); 4 ♀♀, 5 ♂♂.
2 ♀♀ and 2 ♂♂ were dissected. Rostrum ♀, 1st antenna ♀, exopodite of 2nd antenna ♀, palp of mandible ♀, P.1 – P.5 ♀, furca ♀ and ♂, endopodite of P.3 ♂ and P.5 ♂ are drawn.
Length of ♀♀: 0.49 – 0.53 mm
♂♂: 0.38 – 0.45 mm

Remark: The animals from Bunaken agree with the (partly incorrect) discription of EDWARDS (1891) and LANG's (1948) rectified version of *E. longicauda*. The species seems to be rather variable and wide-spread, but according to the presentations of various authors, one cannot exclude the possibility that it is a conglomeration of a few closely related species. However, this statement cannot be confirmed convincingly solely on morphological level. For example, some related specimens from the Galápagos Islands were previously attributed to *E. longicauda* but considered as a valid subspecies, *E. l. galapagoensis* (see MIELKE 1981), whereas WELLS & RAO (1987) were "not impressed with the criteria . . . used . . . to justifiy . . ." the new subspecies. In any case, the individuals of Bunaken better agree with the "typical" *E. longicauda*.

The material contained another ♀ (0.47 mm) which exhibited some differences compared to the other animals and is therefore designated as *Esola* spec. Such differences concern, e.g., the second segment of 1st antenna, which has a cone on distsal outer margin only in the *Esola* spec. female; the shape and setae length of exopodite of 2nd antenna (especially outer distal seta: marked by an asterisk in Figs. 9C, D); the setation of mandibular palp (5 setae in the "normal" specimens (Fig. 10A), 4 in the deviating ♀); the setae number on the distal segment of exopodite P.1 (5, see Fig. 10B, versus 4); the basal segment of

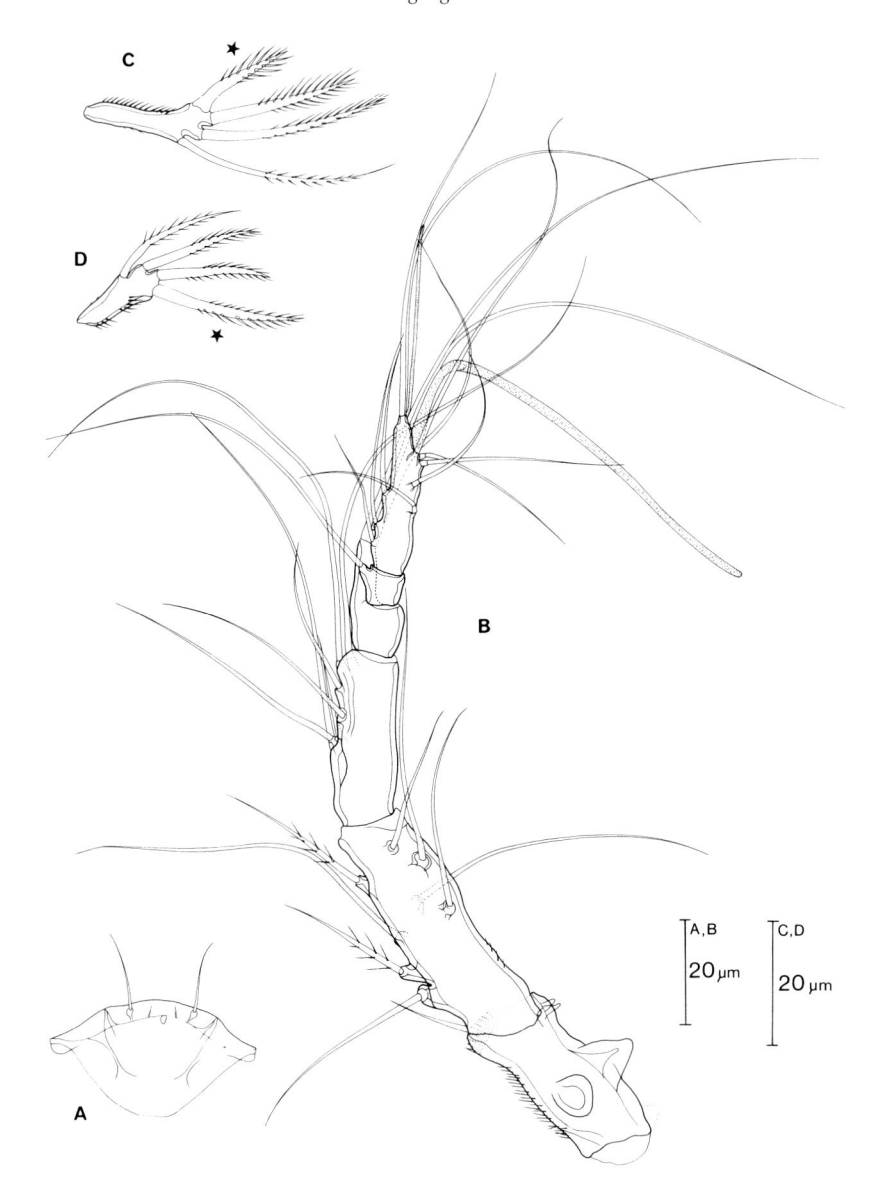

Fig. 9. *Esola longicauda*. ♀. A. Rostrum. B. 1st Antenna. C. Exopodite of 2nd Antenna. D. *Esola* spec.; exopodite of 2nd Antenna (asterisk: outer distal seta).

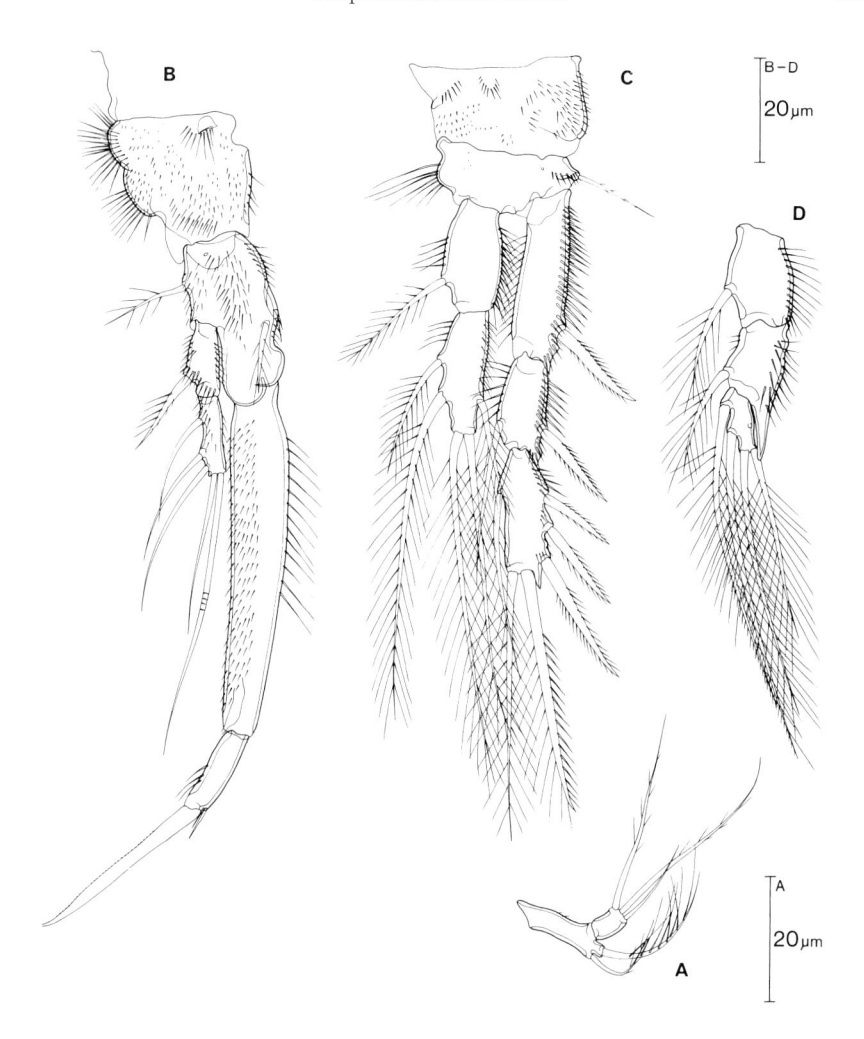

Fig. 10. *Esola longicauda*. A. Palp of Mandible ♀. B. P.1 ♀. C. P.2 ♀. D. Endopodite P.3 ♂.

endopodite P.4 (no seta, see Fig. 11B, versus a rudimentary seta, see Fig. 11C); the form of the furca which is more bulged in the ♀ of *Esola* spec. than in the "normal" ♀♀ (Fig. 12B).

Since the material was collected arbitrarily from several neighbouring sites nothing can be said about distribution and community structure of these two forms, whether they really coexist or occupy different habitats.

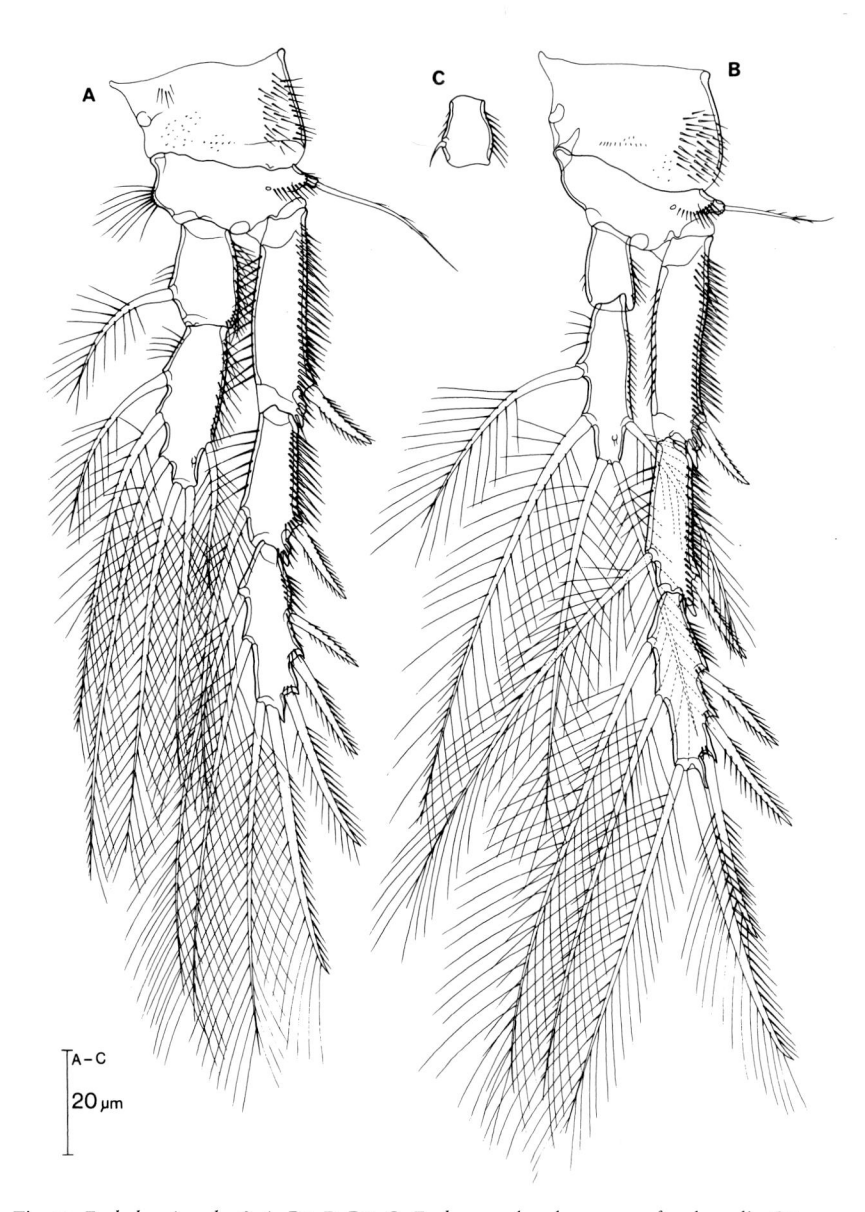

Fig. 11. *Esola longicauda*. ♀. A. P.3. B. P.4. C. *Esola* spec.; basal segment of endopodite P.4.

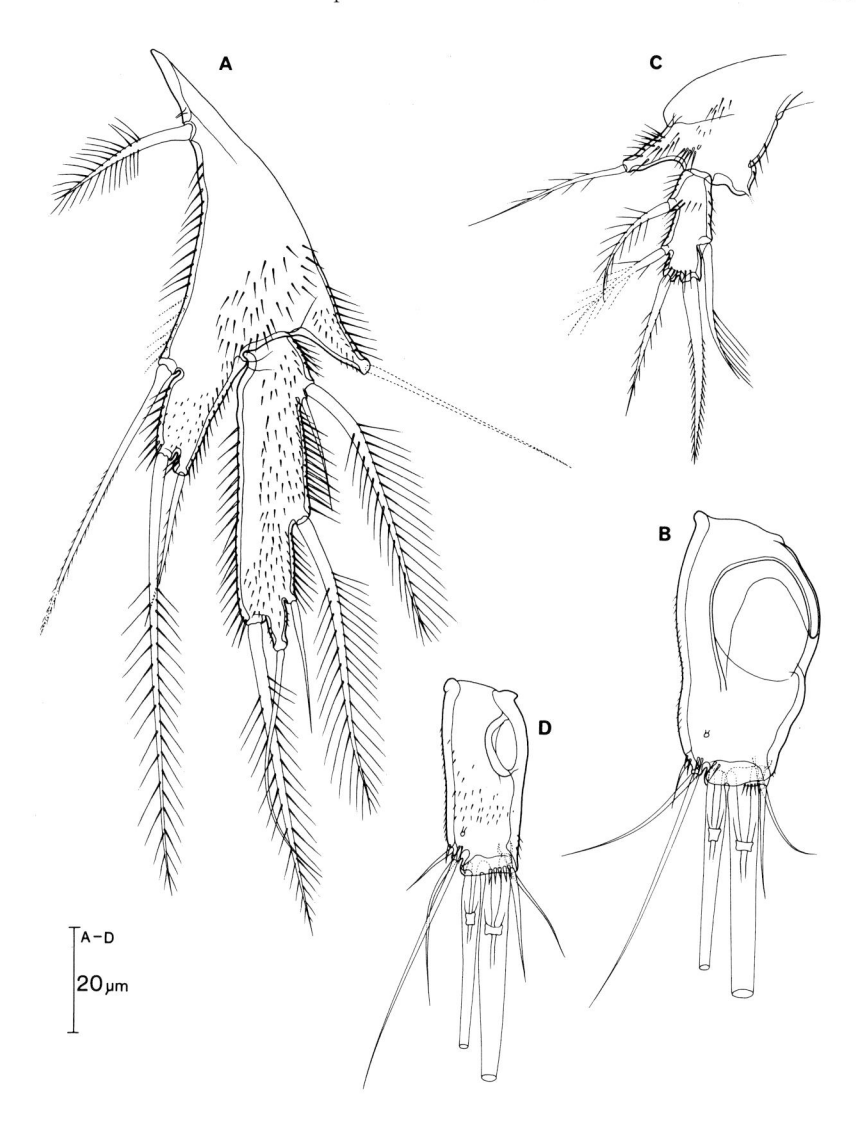

Fig. 12. *Esola longicauda*. A. P.5 ♀. B. Furca ♀, ventral side. C. P.5 ♂. D. Furca ♂, ventral side.

Quinquelaophonte Wells, Hicks & Coull, 1982
Quinquelaophonte quinquespinosa (Sewell, 1924) *bunakenensis* nov. subspec.

(Figs. 13–19)

Locality and material. Beach of Bunaken Island, near Manado (North Sulawesi. Locus typicus; 23 September 1994); 17 ♀♀, 16 ♂♂.
4 ♀♀ and 3 ♂♂ were dissected. Holotype female, reg. no. I SU 1; paratypes are the other dissected animals, reg. no. I SU 2 – 7. With the exception of habitus dorsal, abdomen ventral, rostrum and mandible all drdawings of the female are from the holotype.

Description

Female: Body length from tip of rostrum to end of furcal rami 0.64 – 0.69 mm (dissected ♀♀; holotype 0.65 mm). Rostrum (Fig. 15A) broad, fused with the cephalothorax, with 2 setules subterminally. Dorsal surface of cephalothorax and all following somites with a fur-like trimming. Dorsocaudal margins of all somites seemingly with delicate spinules and – with the exception of the ultimate and penultimate somites – several short setules ("sensillae"). Ventral surface of abdominal somites less pubescent. Ventral caudal margins of abdominal somites with spinules. Genital double-somite subdivided dorsolaterally. Genital area see Fig. 14A; P.6 with 2 seae of different length. Margin of anal operculum set with fine spinules. Furcal rami more than twice as long as broad. Outer margin with 3 setae of different length; distal margin with spinules. The terminal long seta, which is accompanied by an inner short and an outer long seta may have a slight, not very conspicuous swelling. About distal third a dorsal seta, bipartite at base, arises from a socle (Figs. 13A,B, 14A).

1st Antenna (Fig. 15B): 6 segments, surface furry. 4th segment with an aesthetasc. 5th segment narrow.

2nd Antenna (Fig. 15C): Allobasis with a slender seta. Free endopodite segment with spinules on anterior and distal margins, 2 spines (1 is one-sided plumose) and 1 hair-like seta subapically, 2 spines, 3 geniculate setae and 1 short slender seta apically. Exopodite reduced in size, with 3 short setae.

Mandible (Fig. 15D): Chewing edge with several teeth and 1 slender seta. Coxa-basis (basis) furnished with a plumose seta. Exopodite represented by 1, endopodite by 3 setae.

1st Maxilla (Fig. 16A): Arthrite of precoxa with 8 appendages on distal margin and slender spinules on surface. Coxa, basis, endopodite and exopodite with 2, 3, 3, 2 setae, respectively.

2nd Maxilla (Fig. 16B): Syncoxa has 3 endites. Proximal small endite furnished with 1 plumose seta. Middle endite with 1 strong appendage which is fused with the endite and 2 setae. Distal endite with 3 setae. Basis (allobasis) has 1 claw and 3 slender setae. Endopodite carries 2 setae.

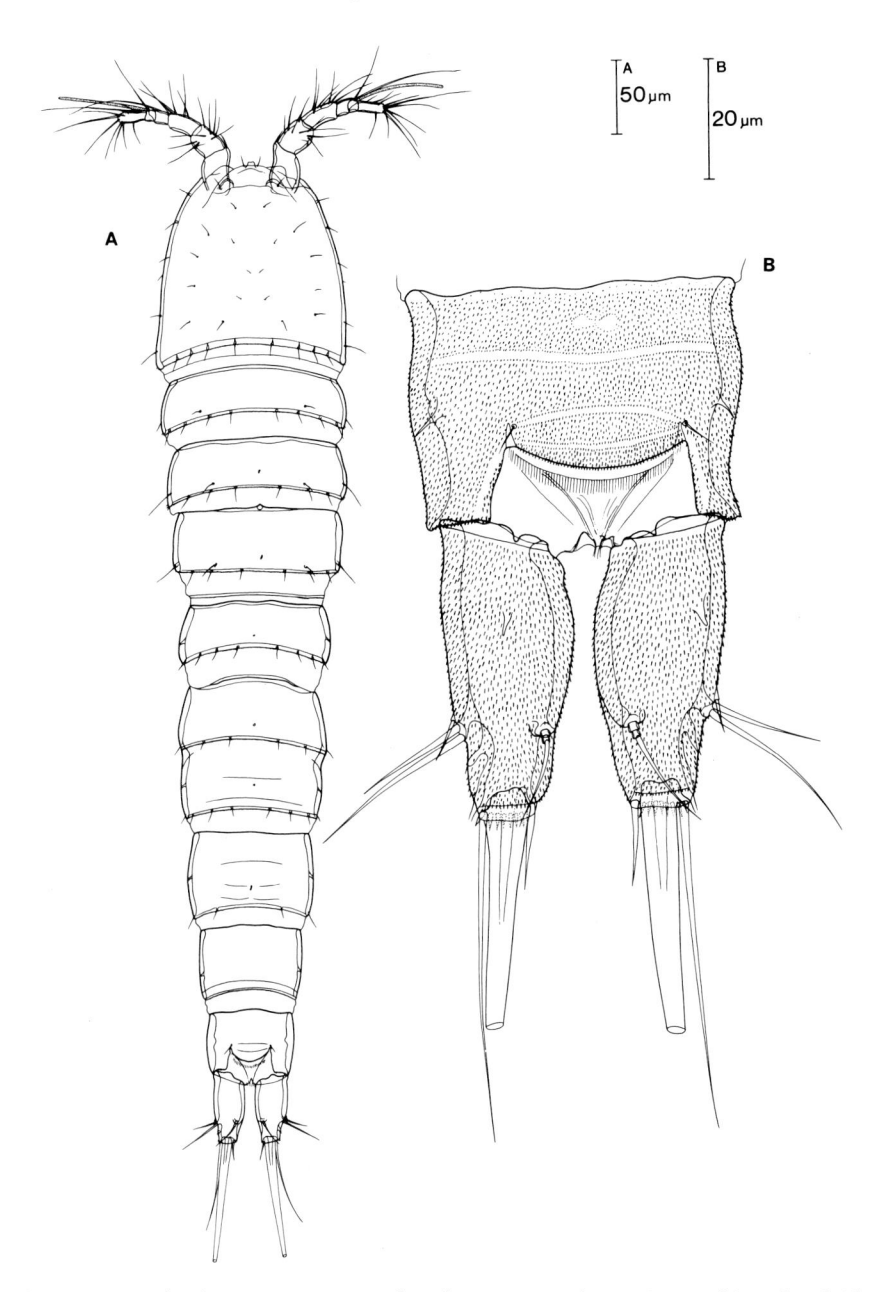

Fig. 13. *Quinquelaophonte quinquespinosa bunakenensis* nov. subspec. ♀. A. Habitus, dorsal side. B. Caudal part of Abdomen, dorsal side.

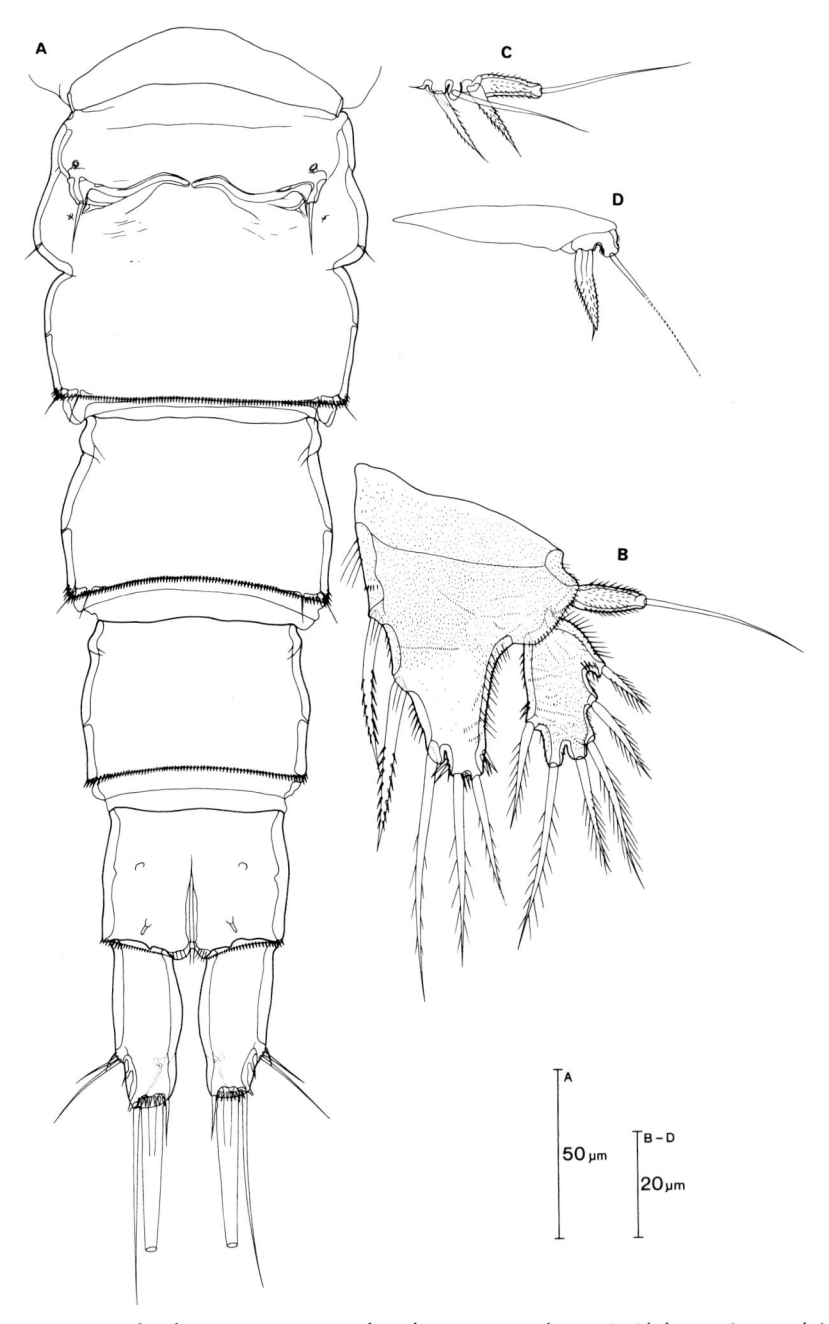

Fig. 14. *Quinquelaophonte quinquespinosa bunakenensis* nov. subspec. A. Abdomen ♀, ventral side. B. P.5 ♀. C. P.5 ♂. D. P.6 ♂.

Maxilliped (Fig. 16C): Basis (syncoxa) with spinules and 1 plumose seta. 1st endopodite segment (basis) hairy on inner margin. 2nd endopodite segment (endopodite) short, furnished with a slender claw which has an accessory small seta.

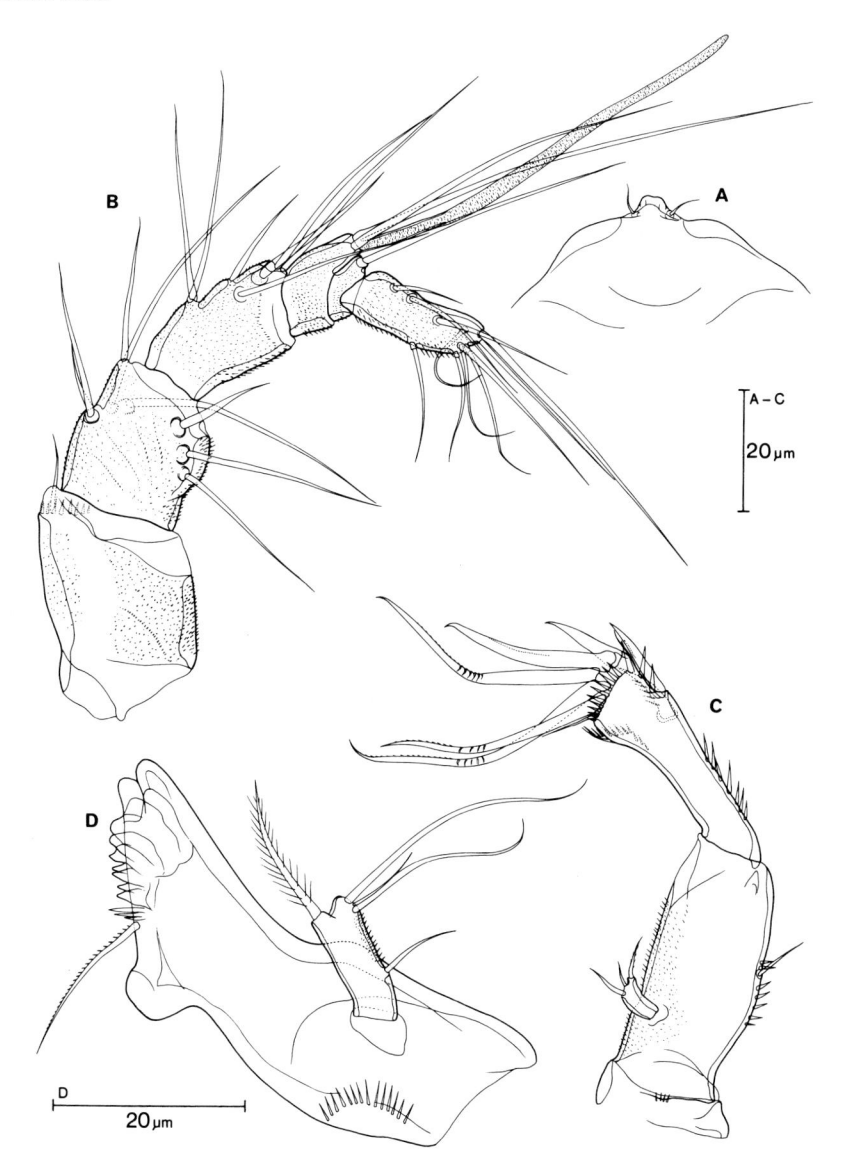

Fig. 15. *Quinquelaophonte quinquespinosa bunakenensis* nov. subspec. ♀. A. Rostrum. B. 1st Antenna. C. 2nd Antenna. D. Mandible.

P.1 (Fig. 16D): Coxa with a row of spinules on outer margin and some spinules proximal on inner surface. Basis with two curved rows of spinules, 1 spine on outer edge and 1 spine on inner surface. Exopodite 2-segmented. Proximal segment with some spinules and 1 spine on outer margin. Distal

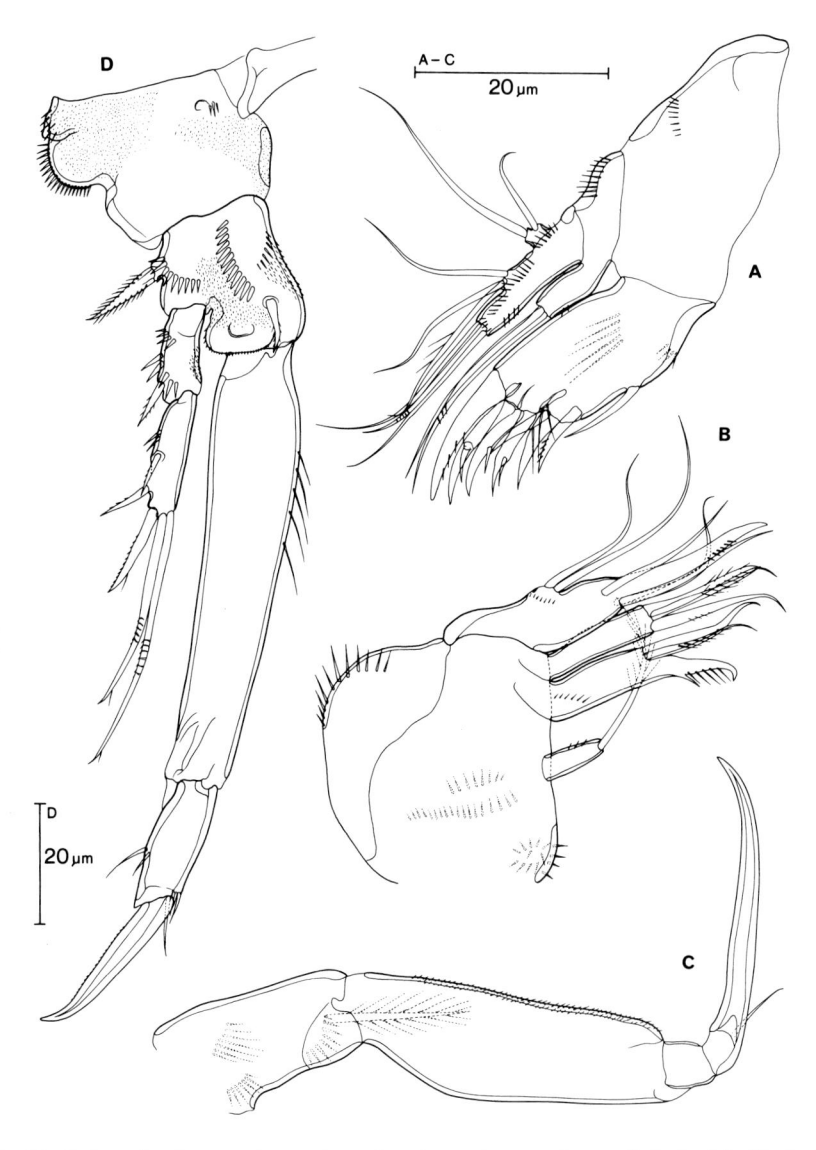

Fig. 16. *Quinquelaophonte quinquespinosa bunakenensis* nov. subspec. ♀. A. 1st Maxilla. B. 2nd Maxilla. C. Maxilliped. D. P.1.

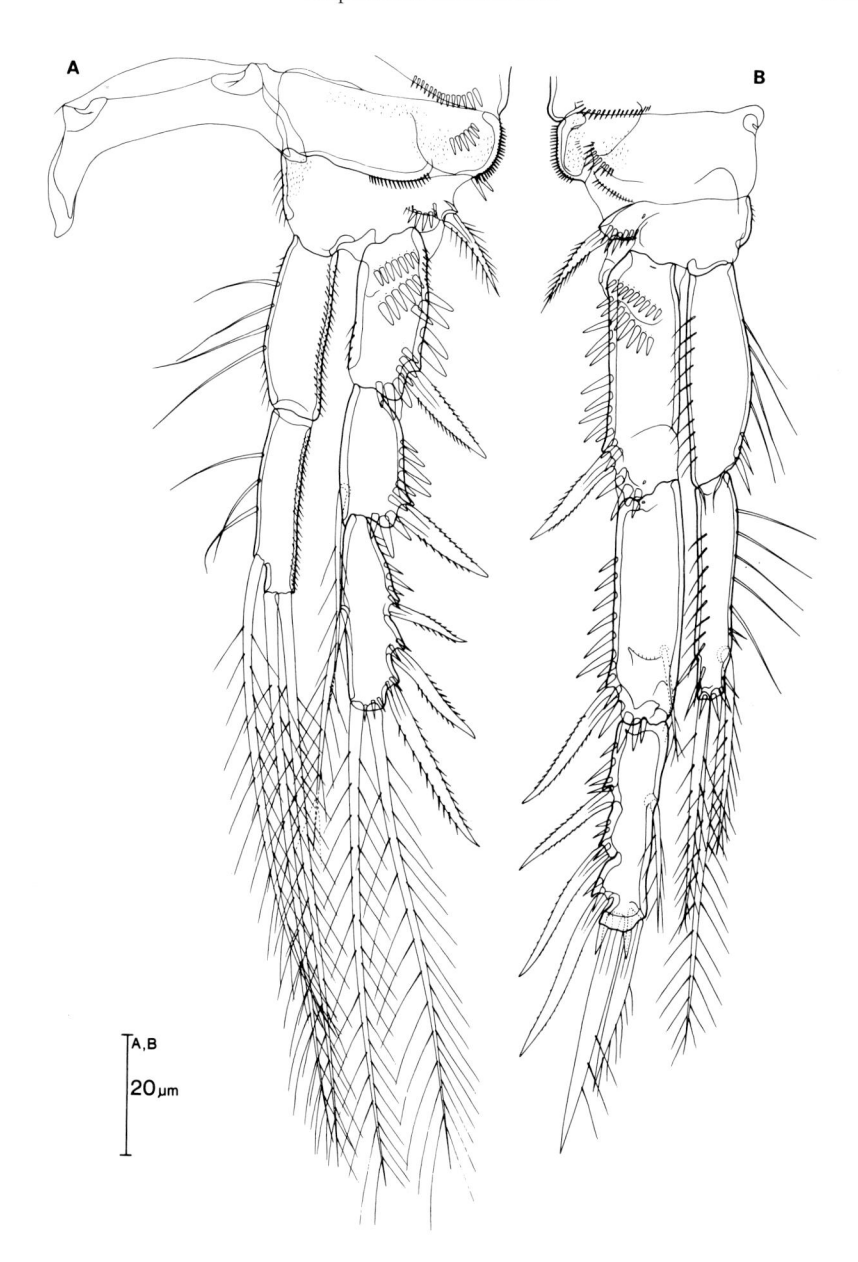

Fig. 17. *Quinquelaophonte quinquespinosa bunakenensis* nov. subspec. A. P.2 ♀. B. P.2 ♂.

segment with 5 appendages. Endopodite 2-segmented. Proximal segment lengthened, with some hair-like setae on inner margin. Distal segment short, furnished with some spinules, 1 apical claw and 1 accompanying setule.

P.2 – P.4 (Figs. 17A, 18A, 19A): Coxa with spinules on outer surface and on

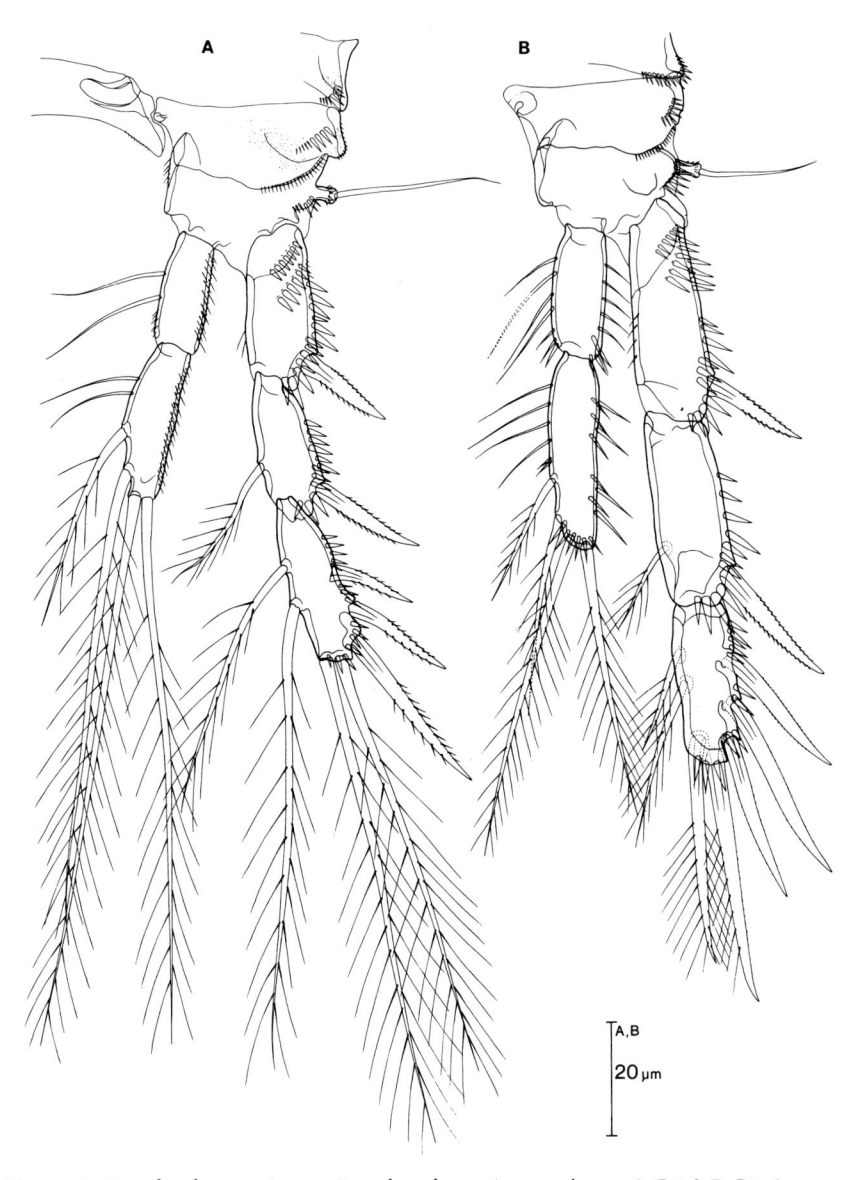

Fig. 18. *Quinquelaophonte quinquespinosa bunakenensis* nov. subspec. A. P.3 ♀. B. P.3 ♂.

medial and outer margin (least in P.4). Basis with spinules on outer surface and an outer appendage which is plumose and stout in P.2 or long and slender in P.3 and P.4. Exopodites 3-segmented. Proximal segment with 1 strong outer spine and rows of spinules on surface and on outer margin. Middle segment with spinules on outer margin and an outer spine. Inner margin has a slender plumose seta. Distal segment long in P.2 and P.3, stout in P.4, armed with 1 inner seta in P.2 and P.4 or 2 setae in P.3. Furthermore, 2 apical plumose setae and 3 outer spines insert on distal segment. Endopodites 2-segmented, reaching

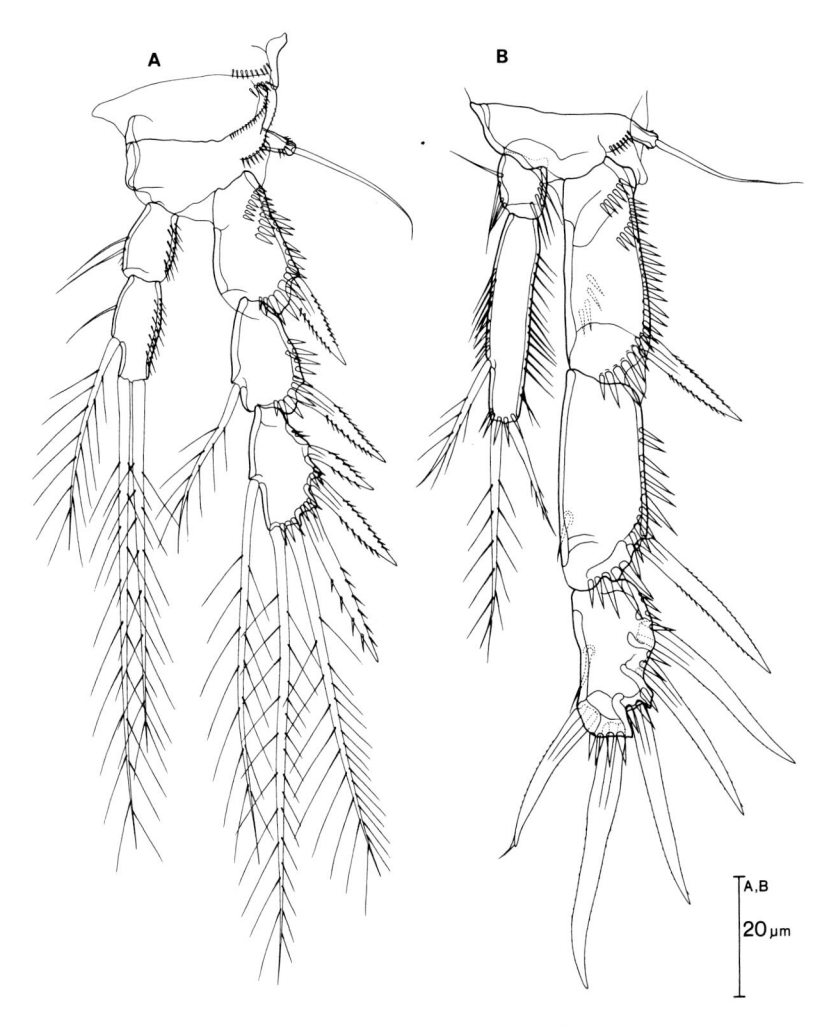

Fig. 19. *Quinquelaophonte quinquespinosa bunakenensis* nov. subspec. A. P.4 ♀. B. P.4 ♂.

beyond middle segment of exopodite in P.2, ending at the same level as middle segment of exp. in P.3 and not reaching to end of middle segment of exp. in P.4. Inner margins of both segments furnished with long slender spinules, outer margins with densely arranged spinules. Distal segment with 2 apical setae and 1 plumose seta subapical on inner margin of P.2 and P.4 or 2 plumose subapical setae in P.3.

Seta and spine formula:

	Exopodite	Endopodite
P.2	(0.1.123)	(0.120)
P.3	(0.1.223)	(0.220)
P.4	(0.1.123)	(0.120)

P.5 (Fig. 14B): Outer lobe of baseoendopodite pubescent, bearing a slender seta. Inner part of benp. with 5 setae, 2 inner ones strongly armed. Exopodite small, furnished with 6 plumose setae.

Male: Differs from the female in the following respects:
– Body length of dissected males 0.57 – 0.65 mm.
– Spinules on ventral caudal margins of abdominal somites more distinct.
– Furcal rami comparatively longer.
– 1st Antenna subchirocer.
– P.2 (Fig. 17B): Exopodite longer than in female; apical appendages of distal segment short, spine-like. Setae of distal segment of endopodite clearly shorter.
– P.3 (Fig. 18B): Inner setae of distal segment of exopodite shorter, apical appendages spine-like. Setae of distal segment of endopodite shorter.
– P.4 (Fig. 19B): Exopodite longer than in female. Inner setae of middle and distal segments distinctly shorter. Both apical appendages of distal segment short and spine-like. Outer spines comparatively longer. Distal segment of endopodite longer, setae shorter than in female.
– P.5 (Fig. 14C): Reduced; outer lobe with a slender seta. Close to that lobe 2 slender and 2 strong setae are arranged.
– P.6 (Fig. 14D): Consisting of an outer slender seta and an inner cone. Both appendages either insert on a bilobed socle which continues in a transverse plate or this socle is fused with the somite.

Variability. The shape of P.5 ♀♀ is slightly variable. The terminal long furcal seta of ♀♀ occasionally has an indistinct swelling. The length of setae or spines varies slightly. Occasionally a seta may be absent; for example, on middle and on distal segment of exopodite P.3 ♂ an inner seta or an outer spine may be lacking. The shape of the distal segment of endopodite P.4 ♂ may be irregular. The transverse plate of P.6 ♂ may be on the right or on the left side.

Discussion. LANG's (1948) "*Quinquespinosa*-Gruppe" of *Heterolaophonte* was raised to the separate genus *Quinquelaophonte* by WELLS et al. (1982). So far, the taxon comprises six species (WELLS et al. 1982, BODIN 1988) or only five, as far as *Q. parasigmoides* is synonymous with *Q. quinquespinosa* (see FIERS 1986). The animals of Bunaken/Sulawesi would best agree with *Q. quinquespinosa* (Sewell, 1924), a broadly distributed species exhibiting some variability according to WELLS & MCKENZIE (1973) or WELLS et al. (1982). However, they differ from all descriptions of specimens attributed to *Q. quinquespinosa* in proportions and length ratios (e.g. distal segment of enp. P.4 ♂) and above all in the form of the distal segment of endopodite P.3 ♀ and ♂. Here, the seta on outer margin has been lost in ♀ and the ♂ lacks an inner "sharp process" (SEWELL 1924) or a "dimorphic pointed spine" (POR 1973). Therefore, the present individuals are introduced as a new subspecies, *Q. quinquespinosa bunakenensis*.

Zusammenfassung

An zwei Fundorten von Sulawesi/Indonesien wurden fünf Laophontiden-Arten gesammelt.

Das Material von Donggala (Westküste von Zentral-Sulawesi) enthielt, zusammen mit anderen Spezies, *Laophonte spinicauda* (Vervoort, 1964), *Laophontina sensillata* Wells & Rao, 1987, und *Langia maculata* Wells & Rao, 1987.

In sublitoralen Proben, die vor der Küste der Insel Bunaken (Nord-Sulawesi) genommen wurden, fand sich *Esola longicauda* Edwards, 1891.

Eine andere Laophontiden-Art siedelt im Strand der Insel Bunaken. Sie wird als neue Unterart von *Quinquelaophonte quinquespinosa* (Sewell, 1924) charakterisiert: *Q. q. bunakenensis* nov. subspec.

References

BODIN, P. (1988): Catalogue des nouveaux Copépodes Harpacticoides marins (Edition 1988). Univ. Bretagne Occ. Brest, 1–288.

BREHM, V. & P. A. CHAPPUIS (1935): Mitteilungen von der Wallacea-Expedition Woltereck. Mitteilung XI. *Diaptomus* s.l. (*Heliodiaptomus*?) *Kieferi* nov. spec. und *Schizopera Tobae* subsp. nov. *Wolterecki*. Zool. Anz. 111, 237–240.

COTTARELLI, V. (1985): Laophontidae di acque interstiziali litorali dell'Indonesia (Crustacea, Copepoda, Harpacticoida). Boll. Mus. civ. St. nat. Verona **12**, 283–297.

COTTARELLI, V., A. C. PUCCETTI & P. E. SAPORITO (1984): Osservazioni sul genere *Psammopsyllus* (Copepoda, Harpacticoida, Cylindropsyllidae) e descrizione di tre nuove specie. Boll. Mus. civ. St. nat. Verona **11**, 1–29.

COULL, B. C. (1971): Meiobenthic Harpacticoida (Crustacea, Copepoda) from St. Thomas, U.S. Virgin Islands. Trans. Amer. Micros. Soc. **90**, 207–218.

EDWARDS, C. L. (1891): Beschreibung einiger neuen Copepoden und eines neuen copepodenähnlichen Krebses, *Leuckartella paradoxa*. Arch. f. Naturgesch. **57**, 75–104.

FIERS, F. (1984): Allocation of *Paralaophonte spinicauda* Vervoort (Copepoda, Harpacticoida) to the genus *Laophonte*. Crustaceana **46**, 317–319.

FIERS, F. (1986): Harpacticoid copepods from the West Indian Islands: Laophontidae (Copepoda, Harpacticoida). Bijdr. Dierk. **56**, 132–164.

HUYS, R. & G. A. BOXSHALL (1991): Copepod Evolution. Ray Soc. London 468 pp.

LANG, K. (1948): Monographie der Harpacticiden, I & II: 1–1682 (Reprint 1975, Otto Koeltz Sci. Publ., Königstein).

LANG, K. (1965): Copepoda Harpacticoidea from the Californian Pacific coast. Kungl. Svenska Vetenskaps. Handl. **10**, 1–566.

MIELKE, W. (1981): Interstitielle Fauna von Galapagos. XXVIII. Laophontinae (Laophontidae), Ancorabolidae (Harpacticoida). Mikrofauna Meeresboden **84**, 1–104.

POR, F. D. (1973): The benthic Copepoda of the Sirbonian Lagoon (Sabkhat el Bardawil). Cah. Biol. Mar. **14**, 89–107.

SCOTT, A. (1909): The Copepoda of the Siboga Expedition. Part I. Free-Swimming, Littoral and Semi-Parasitic Copepoda. Siboga-Expeditie (Siboga Expedition Monographs) **29a**, 1–323.

SEWELL, R. B. S. (1924): Fauna of the Chilka Lake. Crustacea Copepoda. Mem. Ind. Mus. **5**, 771–851 (Plates XLIV–LIX).

VERVOORT, W. (1964): Free-Living Copepoda from Ifaluk Atoll in the Caroline Islands with Notes on Related Species. Smiths. Inst. U.S. Nat. Mus. Bull. **236**, 1–431.

WELLS, J. B. J. & J. G. MCKENZIE (1973): Report on a small collection of benthic copepods from marine and brackish waters of Aldabra, Indian Ocean. Crustaceana **25**, 133–146.

WELLS, J. B. J., G. R. F. HICKS & B. C. COULL (1982): Common harpacticoid copepods from New Zealand harbours and estuaries. N. Z. Journ. Zool. **9**, 151–184.

WELLS, J. B. J. & G. C. RAO (1987): Littoral Harpacticoida (Crustacea: Copepoda) from Andaman and Nicobar Islands. Mem. Zool. Surv. India **16**, 1–385.

Dr. Wolfgang Mielke
II. Zoologisches Institut und Museum der Universität Göttingen
Berliner Straße 28, D-37073 Göttingen

Fine-structural features of male and female gonads in *Jensenia angulata* (Plathelminthes, Rhabdocoela, "Dalyellioida")

Beate Sopott-Ehlers

Abstract

The cytoarchitecture of male and female gonads and gametes is described. Spermatozoa are biciliated, display cortical microtubules, dot-like dense granules arranged in semicircles, dense bodies, the nucleus and strings of mitochondria. During spermiogenesis, an intercentriolar body-like structure and an icb associated with microtubules and dense disc-like plates are differentiated. Vitellocytes are characterized by yolk bodies and globules of eggshell-forming material that is arranged in a mosaic-like pattern. Marginal (cortical) granules are present in the germocytes. The ultrastructure of the female gametes substantiates the hypothesis that the Prolecithophora + Rhabdocoela (with the Neodermata) form a monophylum to which the name Eulecithophora can be assigned. Two differentiations of male gametes, dot-like granules in a regular arrangement and dense disc-like plates, are widespread in many but not all species ascribed to the "Typhloplanoida" and the "Dalyellioida" (including the Dalyelliidae with the Temnocephalida) and not known from species of the Kalyptorhynchia and the Neodermata.

A. Introduction

In many conventional systems, the free-living and symbiotic Rhabdocoela are splitted into the Typhloplanoida, Kalyptorhynchia, Dalyellioida and Temnocephalida. But the monophyly of these groups, the Kalyptorhynchia excepted, could not been substantiated until now, neither at the light microscopical nor at the ultrastructural level. Indeed, there are arguments for the hypothesis of a paraphyletic status of the "Typhloplanoida" and "Dalyellioida" (e.g. EHLERS 1985; SOPOTT-EHLERS 1996).

The characteristic "lack of a duo-gland adhesive system" was first hypothesized as an autapomorphy of a taxon Doliopharyngiophora, including "dalyellioid" species and the parasitic Neodermata (EHLERS 1985). But such a hypothesis is invalid, as was demonstrated by TYLER (1988), who mentioned an adhesive system for a freshwater "dalyellioid" species, and later also by EHLERS & SOPOTT-EHLERS (1993) with the description of such a system for *Jensenia angulata* (Jensen, 1878). This species and the one studied by TYLER (l.c.) belong to the monophyletic subtaxon Dalyelliidae as characterized by EHLERS & SOPOTT-EHLERS (l.c.). Up to now, this supraspecific taxon is the only "dalyellioid" one with adhesive organs that are, in comparison to other "dalyellioids", plesiomorphous features of the Dalyelliidae. *J. angulata* shows rhabdomeric photoreceptors, another plesiomorphous feature of the Dalyelliidae, but the photoreceptors of *J. angulata* lack a pigmentation (SOPOTT-EHLERS 1997a). Therefore, a study of more characteristics of the Dalyelliidae and especially of its few marine representatives like *Jensenia angulata* are desirable for a better understanding of the phylogenetic relationships within the "Dalyellioida" and of kinships with other taxa of the Rhabdocoela. This paper deals with the cytoarchitecture of male and female gonads and gametes of *J. angulata*.

B. Materials and Methods

The specimens derive from sand samples taken on the small island Hallig Langeneß (North Sea). Fixation, dehydration, embedding, sectioning and staining followed conventional steps. Series of sections were studied using an electron microscope Zeiss EM 900.

C. Results

The male gonads of *Jensenia angulata* consist of two sack-shaped testes situated in the posterior part of the body. The testes are enwrapped by a tissue made up by tunica cells and a thin layer of ECM. Thus the male gonads are delimited from the neighbouring somatic cells.

Fig. 1. A. B. Frontal tips of spermatozoa in the vesicula seminalis. The arrow head in B marks microtubules running in two s-shaped lines. Scales in A, B = 0.5µm. C. D. Raquet-shaped ciliary bases. Scale in C = 0.5µm, in D = 0.25µm. E. F. Tips of axonemata with splitting microtubules. Scales in E, F = 0.5µm.

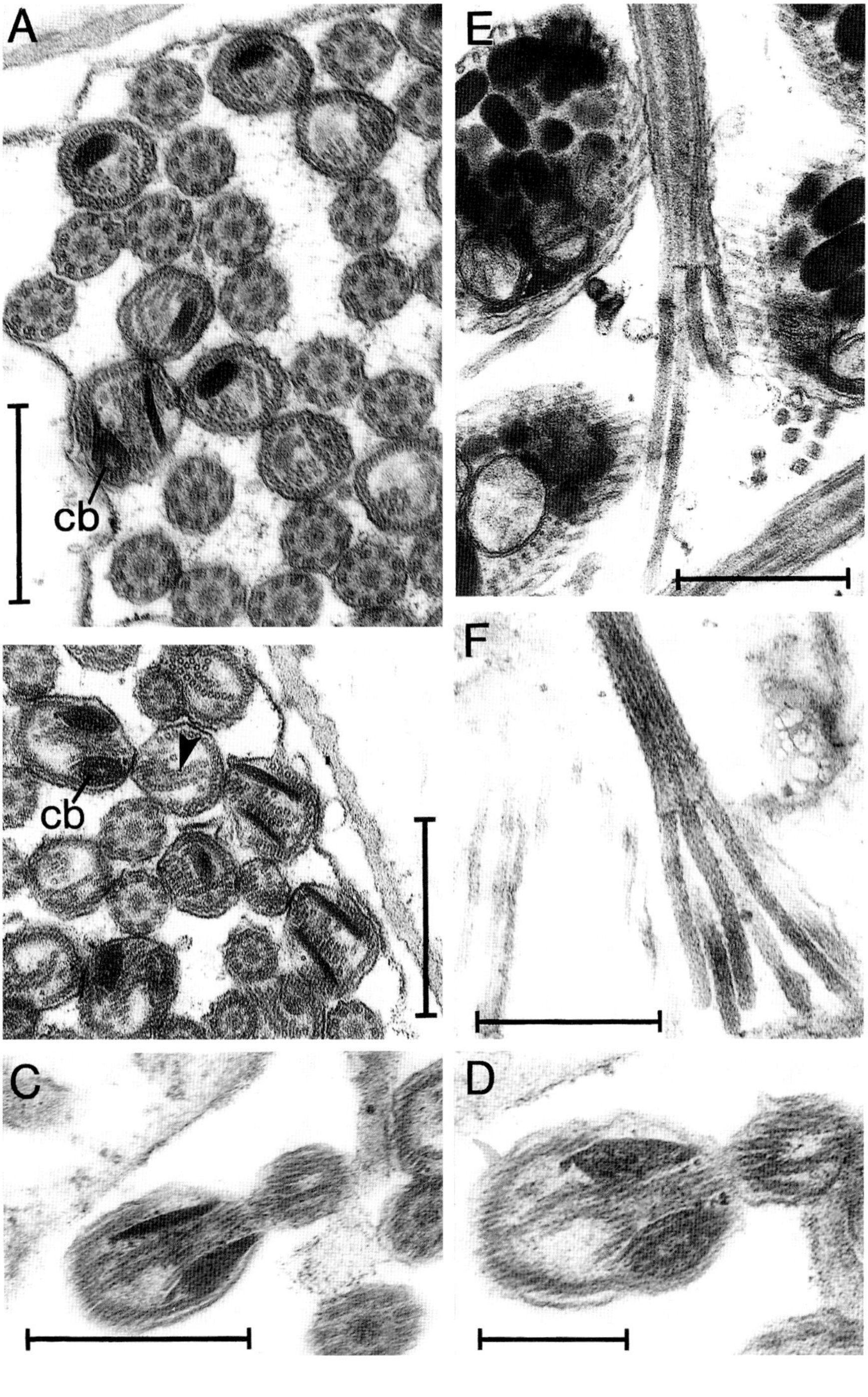
A
cb
E
F
cb
C
D

Mature male gametes

The filiform spermatozoa bear two free cilia or flagella being anchored a short distance below the tip of the functional fore end. The axonemata show the 9 X 2 +"1" microtubular pattern typical of Trepaxonemata. A skid- or keel-shaped dilatation of electron-dense material is attached to the ciliary bases (Fig. 1 A-D). From this a somehow racquet-like picture results in transverse sections. The tips of the cilia split into single cords containing a single microtubule each. Thus the tips of the axonemata have the appearance of a tuft (Fig. 1 E, F).

The sperm body is totally enclosed by about 40 longitudinal cortical microtubules running in a slightly spiral course beneath the surface membrane from one tapering tip to the other. In the region of the ciliary insertion the microtubules lie in a double row or run in two s-shaped lines, respectively (Fig. 1 A-D). When the nucleus commences to taper, the microtubules start to form a spiral (Fig. 2 D). This is to say, they overlap. In the fore end and in the rear end as well, the microtubules touch each other, whereas they are rather widely spaced in the middle region.

Closely beneath the sheath of cortical microtubules a layer of dot-like dense granules is differentiated (Fig. 2 A-D). These granules form a semicircle in transverse sections and extend arranged in regular rows from a short distance below the tip of the functional fore end up to the last third of the male gamete.

Directly beneath the cilia insertion the first dense body surrounded by a semicircle of dense granules is contained in the cytoplasm of spermatozoa (Fig. 2 B). A short distance down the length dense granules, dense bodies and a roundish mitochondrion appear, followed soon by the frontal tip of the nucleus. Profiles of two mitochondria, the nucleus, dense bodies and dense granules are seen further down (Fig. 2 C, D).

There exist two strings of oval mitochondria. These cords flank the nucleus on its ventral side at the right and left in the middle region (Fig. 2 D above). Further down the sperm shaft the cytoplasm contains dense bodies, dense granules and the nucleus. A few glycogen deposites are randomly scattered

---

Fig. 2. A. Longitudinal section through the middle region of a sperm cell in a vas deferens. The small ▶ arrows point to rows of dense granules. Scale = 1μm. B. Transverse sections through functional fore ends with dense bodies and dense granules (small arrows). Scale = 0.25μm. C. Transverse section through the middle region with nucleus, two mitochondria, dense bodies and dense granules (small arrows). Scale = 0.5μm. D. Transverse section through the middle region (above) and slightly oblique transverse section through the hind end (below) with nucleus, overlapping cortical microtubules and remnants of dense bodies. (small arrows = dense granules. Scale = 0.5μm.

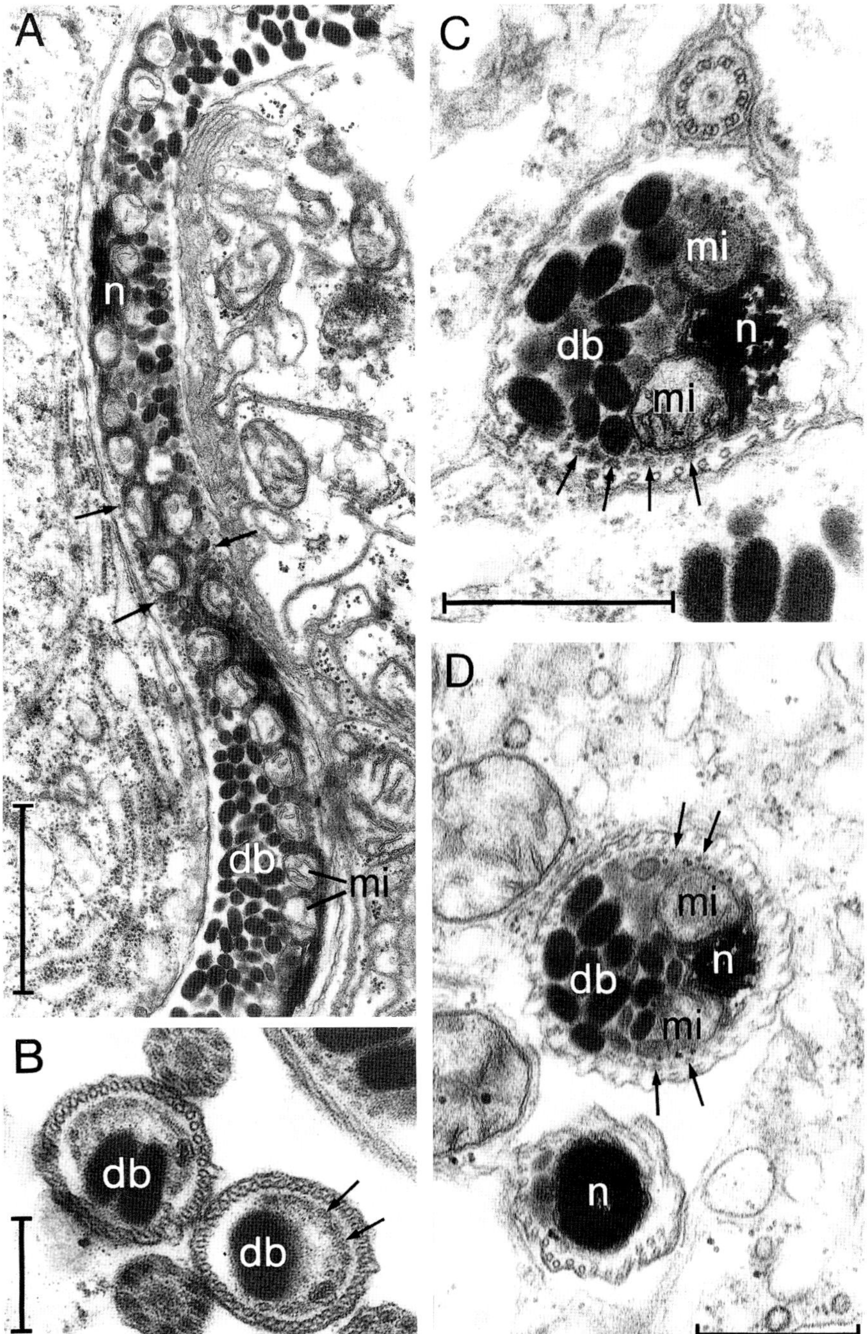
A
n
db
mi
B
db
db
C
mi
n
db
mi
D
mi
n
db
mi
n

throughout the cytoplasm. Soon the dense bodies cease, and just the nucleus and the rows of dense granules are left. Eventually just the nucleus is present. Here the cortical microtubules run in a spiral. The hind end tappers in a tip including anyother differentiations but microtubules.

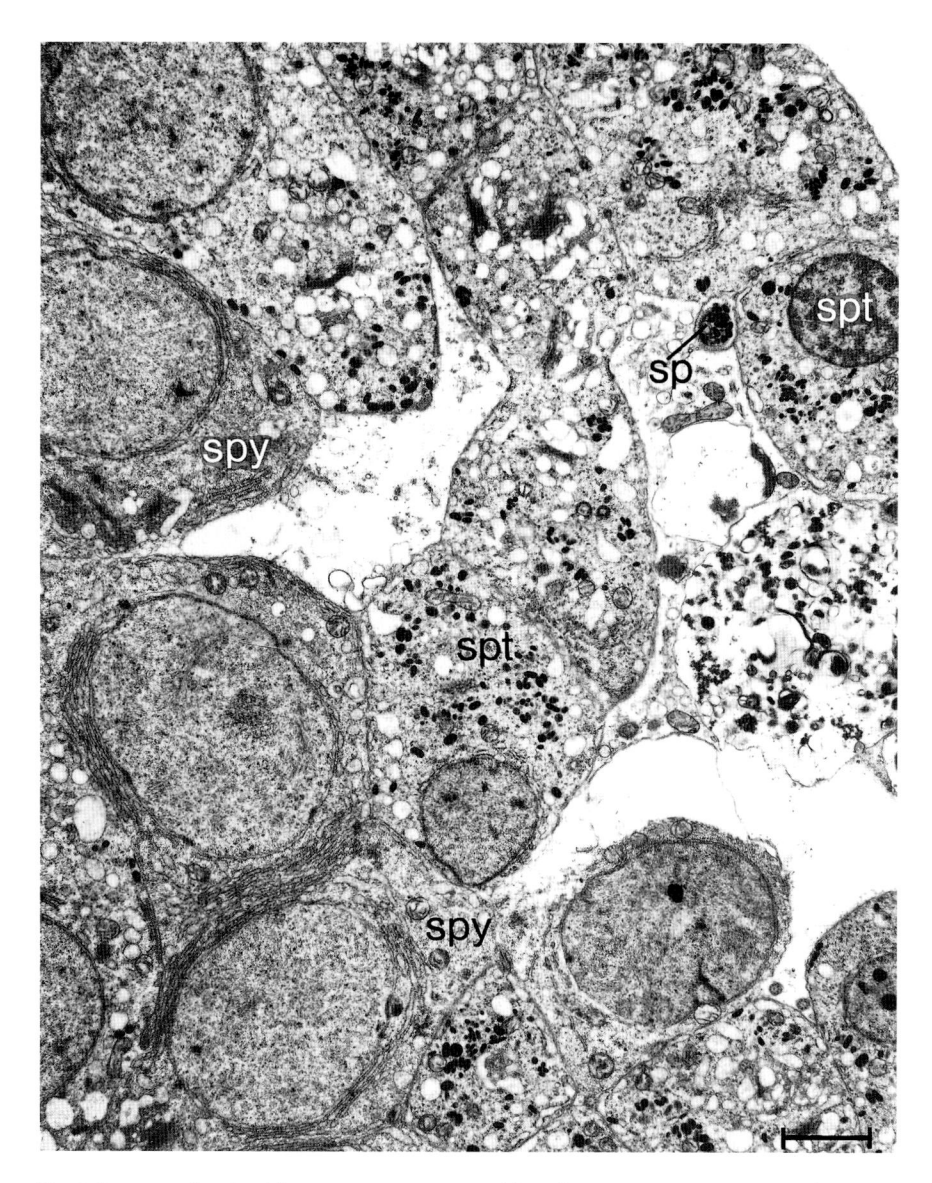

Fig. 3. Segment of testis with spermatocytes, spermatids and a young spermatozoon. Scale = 2µm.

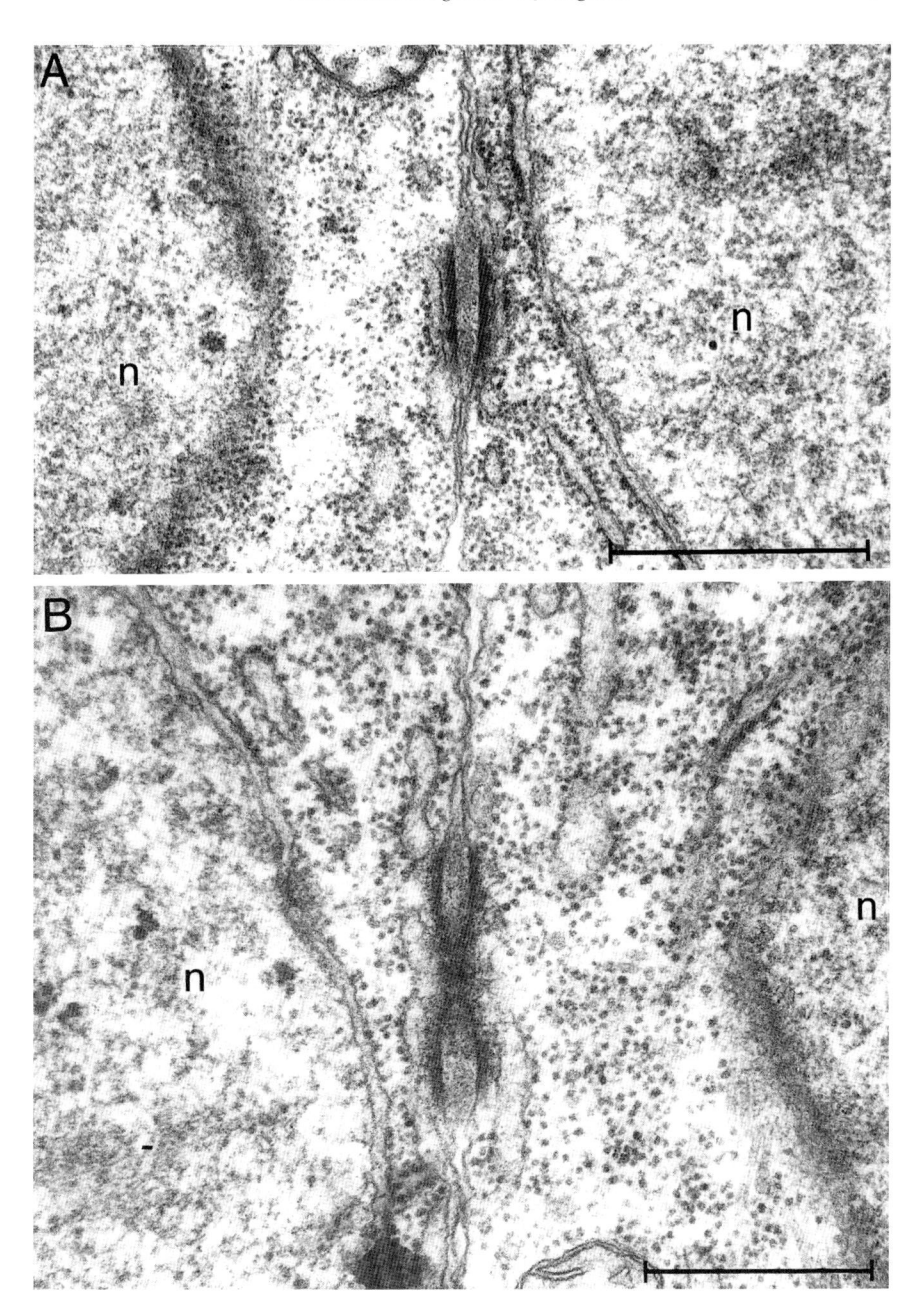

Fig. 4. A. B. Cell junctions between spermatocytes. Scale in A = 1μm, in B = 0.5μm.

Spermiogenesis

Just a few data on maturation of male gametes in *J. angulata* could be obtained from the specimens studied.

There were a few spermatocytes left, some connected by cell junctions or intercellular bridges (Fig. 3).

The cytoplasm of spermatids was scattered with ribosomes and glycogen islets. Well expressed profiles of rough endoplasmic reticulum and Golgi complexes were seen. Numerous small mitochondria and oval dense bodies frequently occured (Fig. 4). The nuclei of spermatids had peripheral positions in rosettes.

Intercentriolar bodies consisting of a central dense double ribbon, two dark bands at each side of this central unit and three dense ribbons at each side near the basal bodies lie in a groove of the nucleus of spermatids. Two cilia spring from the basal bodies. Obviously there exists another differentiation, similar to the intercentriolar body instead of rootlets. This structure runs at a right angle to the intercentriolar body (Fig. 5 A, B). A bundle of microtubules inserts at the peripheral dense plates of the intercentriolar body (Fig. 5 C).

Microtubules accompanied by a peripheral layer of dark material lie opposite to the intercentriolar body, closely beneath the surface membrane (Fig. 5 D).

The onset of cell elongation is characterized by the appearance of two disc-shaped dense differentiations forming a socket between basal bodies and intercentriolar body. Simultaneously this organelle starts to split (Fig. 5 D).

It could not be observed, whether spermatids grow out with the anchor apparatus of the cilia ahead, or whether these structures stay near the cytophore during cell elongation.

Female gonads

A pair of tube-like vitellaria extends throughout the body from the pharynx to the rear end. The vitellaria are delimited from the adjacent tissues by a sheath of flattened cells and a delicate layer of ECM.

The cytoplasm of the vitellocytes (Fig. 6) is dark with free ribosomes and

Fig. 5. A. Intercentriolar body and structure similar to an icb. B. Intercentriolar body and basal bodies. The arrow heads mark peripheral ribbons of the icb-like structure. D. First signs of splitting of the intercentriolar body. Electron-dense discs between icb and basal bodies and microtubules with dark material (small arrows). Scales in A – D = 0.5µm.

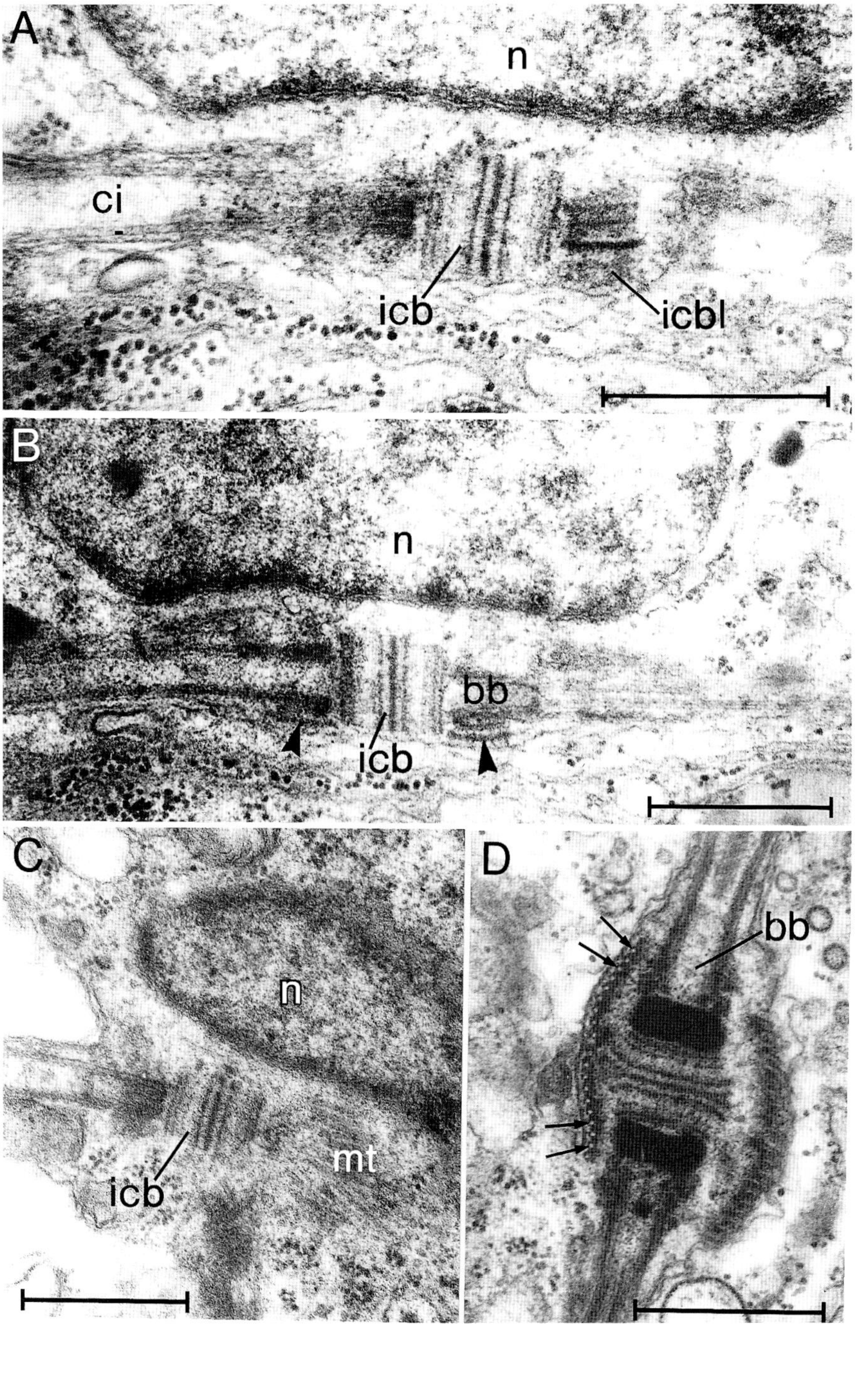
A
n
ci
icb
icbl
B
n
bb
icb
C
n
icb
mt
D
bb

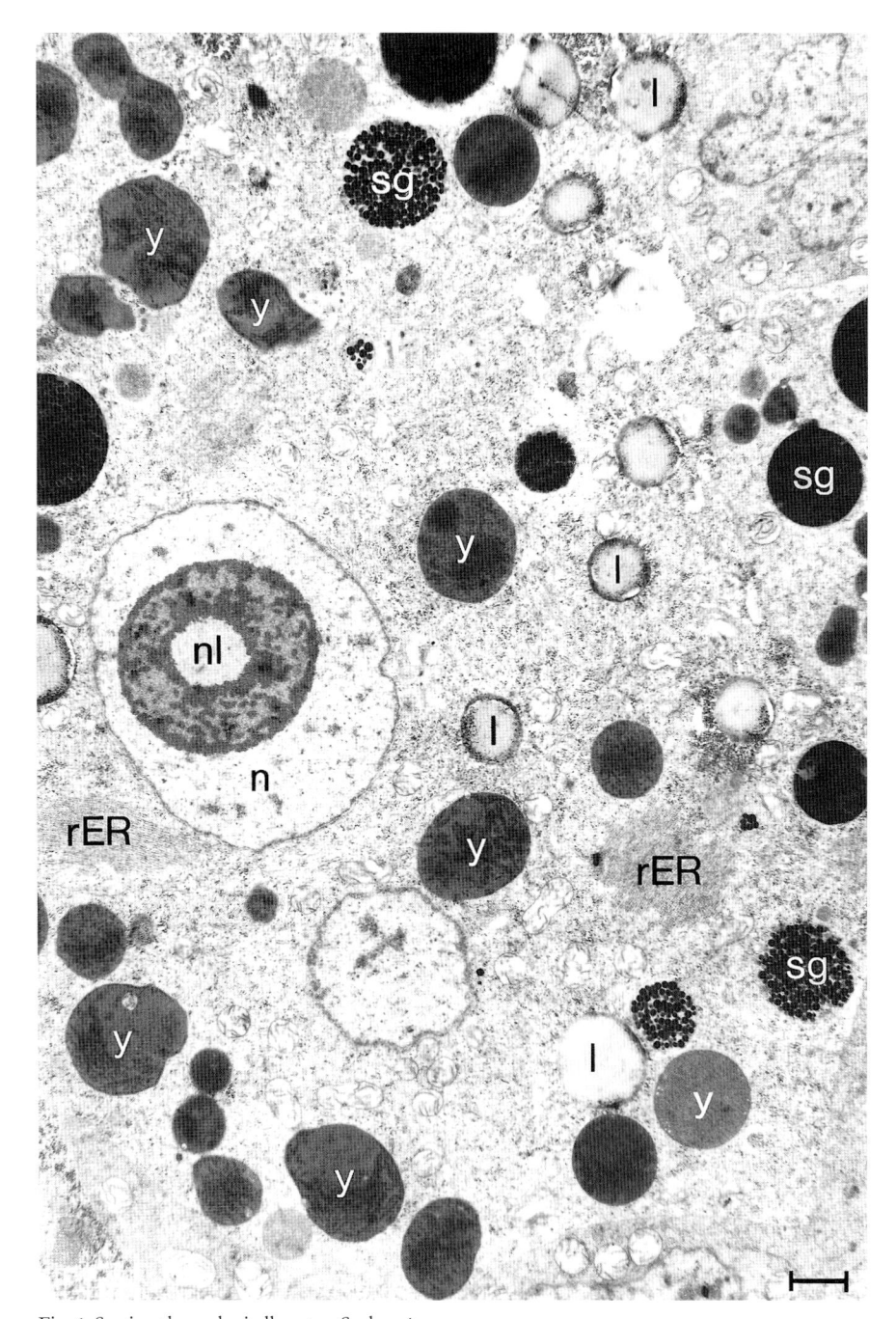

Fig. 6. Section through vitellocytes. Scale = 1µm.

glycogen. Giant stacks of rough endoplasmic reticulum densely trimmed with ribosomes are especially conspicuous (Fig. 7 A). Apart from these giant stacks small profiles of granular ER also exist. Golgi complexes frequently occur. Small mitochondria and lipid droplets are numerous.

Large membrane-bordered areals of stored nutritive material occupy a great portion of the whole vitellocyte volume. The so-called yolk bodies or yolk platelets are more or less roundish in shape. The yolk bodies consist of a grey ground substance, proteinaceous in nature and have a diameter of about 1 – 1.5 µm. More electron-dense strings are embedded in this ground substance (Fig. 7 B). These strings are presumably residual profiles of granular endoplasmic reticulum. Other inclusions, such as relictant mitochondria or residual eggshell materials were never observed.

The globules of eggshell- forming material are up to 2.5 – 3 µm in diameter and almost circular. Their contents consist of polygonal electron-dark units (Fig. 7 C). Thus a mosaic-like pattern is caused. In nascent eggshell globules or in not that densely filled ones, the units appear roundish, not polygonal (Fig. 7 D).

The generative section of the female gonads in *Jensenia angulata* consists of an unpaired germarium situated in the hind end. The germarium (Fig. 8) is enveloped by a tunica of flattened cells and a layer of ECM. In the following, just the cytoarchitecture of mature germ cells will be considered.

Mature germocytes (oocytes) are comparatively large and occupy a great deal of the body diameter. Their nucleus or germinal vesicle is more or less lobated. The nuclear membrane exhibits clearly expressed nuclear pores and apparent invaginations. With exception of a few small dense clumps the chromatin is mostly dispersed as euchromatin (Fig. 9 A).

Free ribosomes and glycogen islets are abundant and can be found in all parts of the ooplasm. Chromatoid bodies are seen near the germinal vesicle. Annulate lamellae are distributed throughout the cytoplasm. Small, spherical mitochondria with few christae and moderately dense to lucent matrices are particularly numerous. The same is true for short profiles of granular endoplasmic reticulum. Golgi complexes are also frequently found. Lipid spheres occur randomly scattered in the ooplasm.

Marginal (cortical) granules about 1 µm in diameter and spherical in shape are concentrated in the periphery of the cell where they appear in a regular arrangement immediately beneath the oolemm (Fig. 9 A). Their content is granular and electron-dense. In some instances a peripheral dark ring or sections with small stacks of dense rods were found (Fig. 9 B, C).

Stored nutritive material sequestered in membrane-bordered areals does not exist in the female generative cells of *Jensenia angulata*.

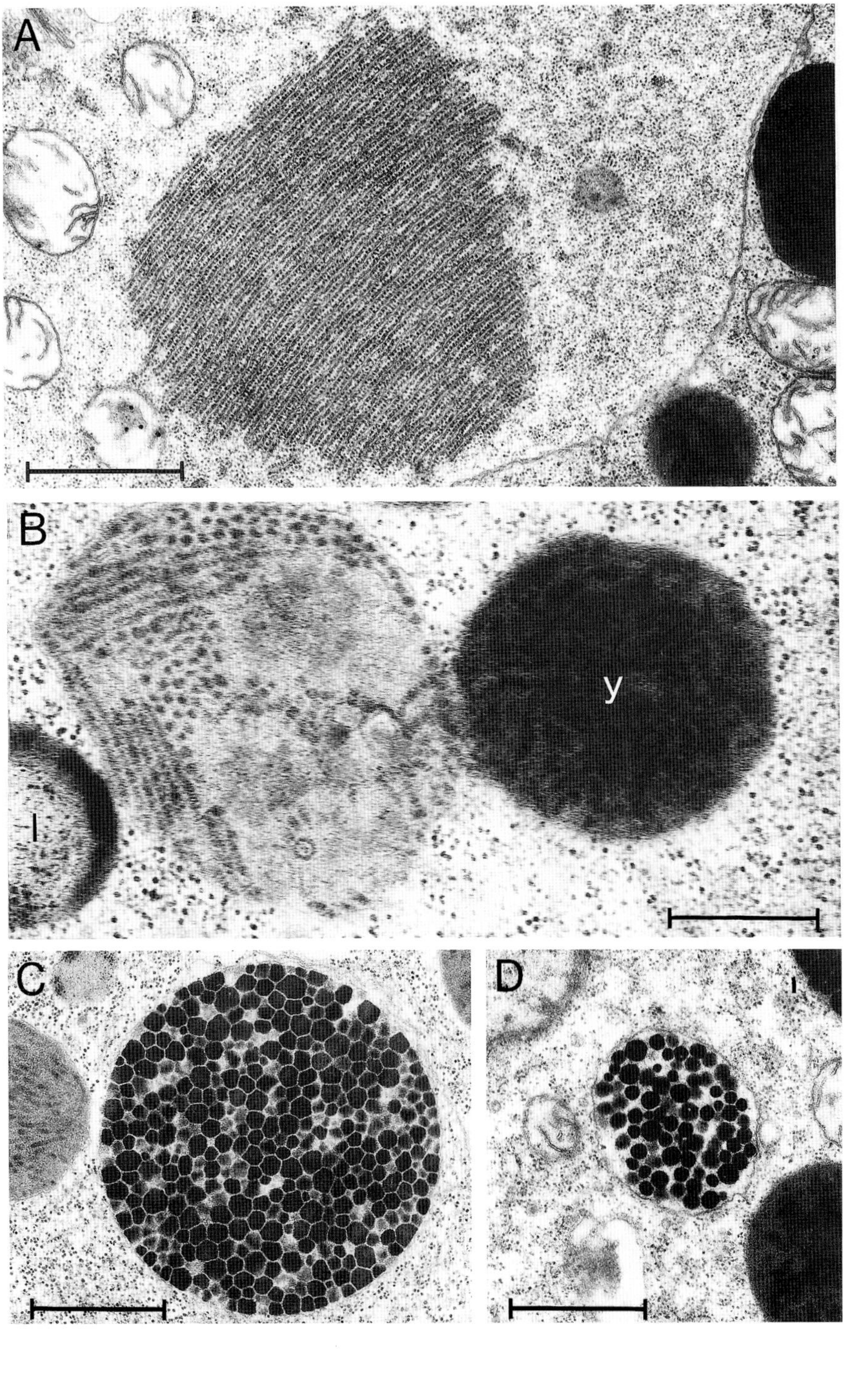
A
B
y
l
C
D

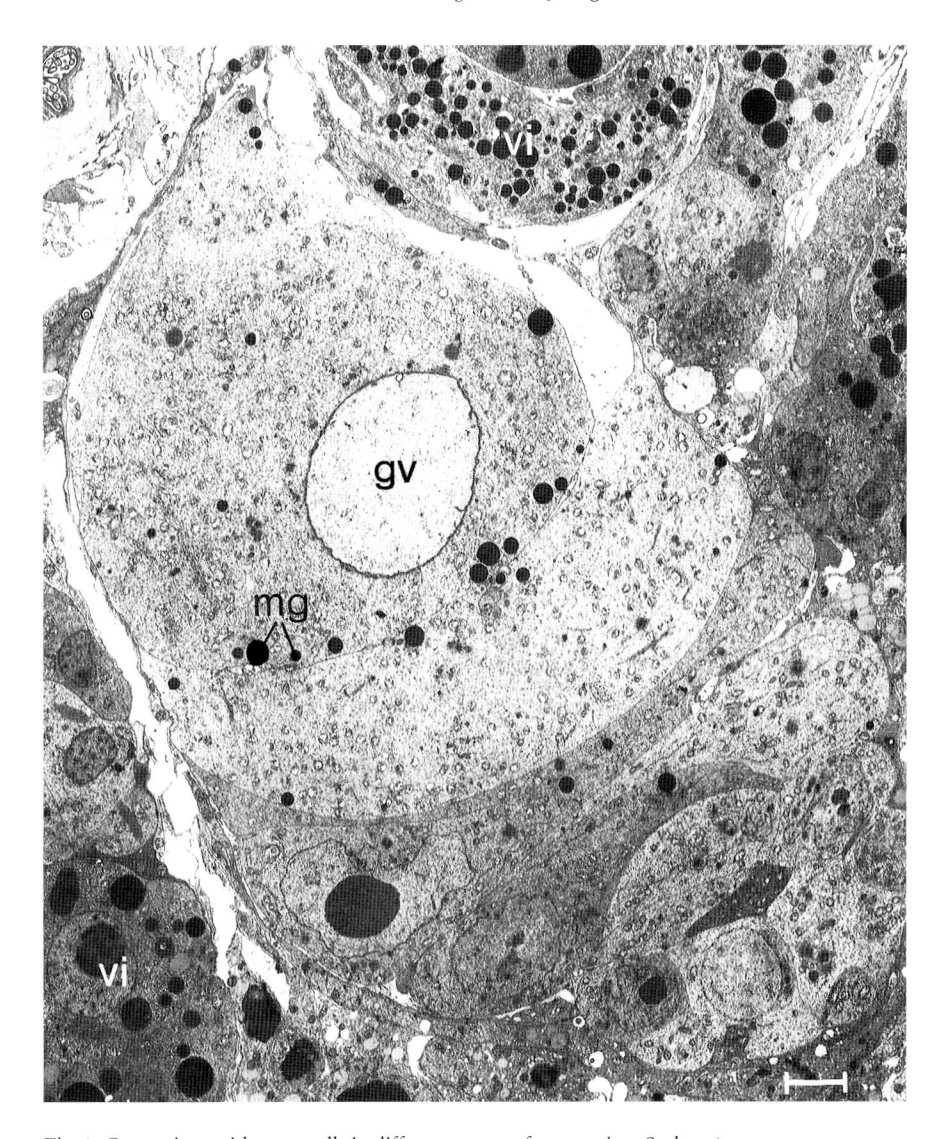

Fig. 8. Germarium with germ cells in different stages of maturation. Scale = 4µm.

◀ Fig. 7. A. Voluminous stack of granular ER. Scale = 1µm. B. Yolk platelets. Scale = 0.5µm. C. Eggshell-forming granule with polygonal grana. Scale = 1µm. D. Nascent eggshell granule with roundish grana. Scale = 1µm.

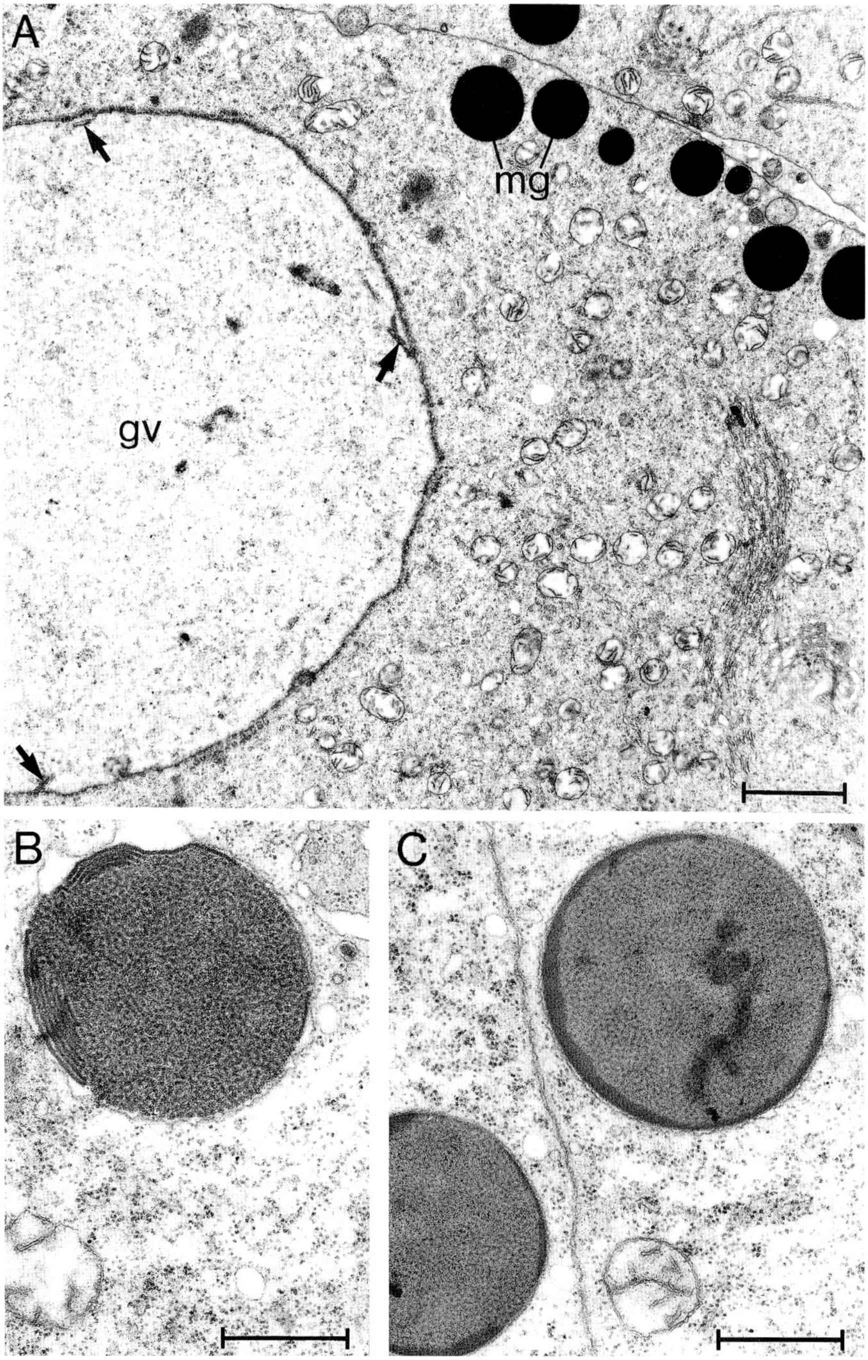
A
mg
gv
B
C

D. Discussion

Male gametes

The male gametes in *Jensenia angulata* correspond in respect to their filiform shape, the existence of two free flagella of the 9 x 2 +"1" axonemal pattern, a sheath of cortical microtubules, dense bodies, numerous mitochondria running in rows and a thread-like nucleus to the basic pattern typical of spermatozoa in Plathelminthes Trepaxonamata. Features of more special interest are the existence of dot-like dense granules arranged in regular longitudinal rows, keel-shaped appositions of electron-dense material to the ciliary bases, and the splitting of the ciliary tips.

Data obtained during last years have shown, that dot-like dense granules in a regular arrangement are widespread in species ascribed to the Rhabdocoela "Dalyellioida", "Typhloplanoida" and Temnocephalida as well (see i. al. CIFRIAN et al. 1988; NOURY-SRAIRI et al. 1989; JOMINI et al. 1994; SOPOTT-EHLERS 1994; SOPOTT-EHLERS & EHLERS 1995b; WATSON & JONDELIUS 1995; WATSON & ROHDE 1995a; see also WATSON & ROHDE 1995b, tab. 1, p. 40 f.; WILLIAMS 1994), but have never ever been observed in representatives of the Kalyptorhynchia. Against the background of these data available up to now the hypothesis that dot-like dense granules are an autapomorphic feature of a monophylum comprising taxa of the "Typhloplanoida", "Dalyellioida" and Temnocephalida (see also EHLERS 1985) appears more probable than the idea that these structures may have evolved more than once (see SOPOTT-EHLERS 1994, p. 38). The same may be true for the skein-like appositions of dense material at the basal bodies of the cilia in mature sperm cells.

A splitting of ciliary tips is known for representatives of triclads (see ISHIDA & TESHIROGI 1988; ISHIDA et al.1991; LI et al. 1992; ROHDE & WATSON 1995) and *Syndisyrinx punicea*, a species of the "Dalyellioida" Umagellidae (see ROHDE & WATSON 1988). From the distribution of this characteristic it can be concluded that split tips have evolved convergently.

As far as features appearing during spermiogenesis could be observed, the following ones will be considered: the existence of an intercentriolar body-like structure instead of rootlets, the differentiation of disc-like dense plates (dense heels in the terminology of WATSON & ROHDE) in an early phase of cell elongation and the bundle of microtubules at the intercentriolar body.

◄ Fig. 9. A. Segment of a mature germocyte. The small arrows mark invaginations of the nuclear envelope. Scale = 2µm. B. Marginal (cortical) granule with stacks of rod-shaped dense material. Scale = 0.5µm. C. Marginal granules with a peripheral circular layer of dense substance. Scale = 0.5µm.

The differentiation of an icb-like structure instead of rootlets has hitherto not been reported for other species and is therefore hypothesized to be an autapomorphy of *J. angulata* or of the supraspecific taxon *Jensenia* with *J. angulata* and *J. parangulata* Ax & Armonies, 1990. Since dense plates show a similar distribution as dense granules (see i. al. IOMINI et al 1994; WATSON & JONDELIUS 1994; WATSON et al. 1995; own unpublished data), this feature could be also an autapomorphy of a taxon comprising taxa of the "Dalyellioida", "Typhloplanoida" and Temnocephalida.

Bundles of microtubules inserting at the intercentriolar body have been reported for several taxa (see i.al. WATSON & ROHDE 1993; ROHDE & WATSON 1995) and may have evolved more than once within the Rhabdocoela.

Female gametes

Structures in female gametes being suitable to elucidate phylogenetic relationships are eggshell-forming granules in vitellocytes and marginal (cortical) granules in germocytes.

A mosaic-like pattern in eggshell-forming granules in vitellocytes as existing in *J. angulata* is widespread within neoophoran Plathelminthes. It exists in representatives of the Neodermata and prevails in species ascribed to the "Dalyellioida", "Typhloplanoida", Kalyptorhynchia and Prolecithophora (for ref. see GREMIGNI 1988; GREMIGNI & FALLENI 1991). This design has to be considered as autapomorphy of a taxon Rhabdocoela + Prolecithophora (see SOPOTT-EHLERS 1995, 1997b).

Although there exist some small variations in different species, marginal granules in germocytes exhibiting a granular content are typical of representatives of the Rhabdocoela "Dalyellioida", "Typhloplanoida", Kalyptorhynchia and for some species of the Prolecithophora as well, (see GREMIGNI 1988; LUCCHESI et al. 1995). From this it can be hypothesized, that a granular substructure is as an apomorphic feature of a taxon Rhabdocoela + Prolecithophora as eggshell-forming granules with a mosaic-like design are.

Concluding remarks

The analysis of the fine structure of male and female gametes of *J. angulata* does not reveal any characteristic that can be hypothesized as an autapomorphy of the Dalyelliidae as characterized by EHLERS & SOPOTT-EHLERS (1993). The known autapomorphies of this taxon are the special type of the spiny copulatory organ (cirrus-like stylet) and only one germarium instead of paired germaria.

Nevertheless, the present study substantiates these hypotheses of relationships:

(1) The "Typhloplanoida", Kalyptorhynchia, "Dalyellioida", Temnocephalida, Neodermata and Prolecithophora constitute a monophylum, based on distinct structures of the female gametes (eggshell-forming granules in vitellocytes and marginal granules in germocytes). The name Eulecithophora de Beauchamp, 1961 can be assigned to this monophylum Prolecithophora + Rhabdocoela (including the Temnocephalida and Neodermata).

(2) Species of the "Typhloplanoida", "Dalyellioida" and Temnocephalida share two unique features of male gametes, the dot-like granules and dense plates, that are not known from any species of the Kalyptorhynchia and the Neodermata and several other taxa as discussed by SOPOTT-EHLERS (1994). At present, these alternative hypotheses are reasonable: (a) Both features are basic characteristics of the Rhabdocoela and have been lost in the lineages of the Kalyptorhynchia and Neodermata (the species of both taxa have spermatozoa with incorporated ciliary axonemes) and also of "typhloplanoid" and "dalyellioid" taxa (see SOPOTT-EHLERS & EHLERS 1995a, 1997) with modified spermatozoa or (b) both features are an autapomorphy of a monophylum within the Rhabdocoela, encompassing distinct taxa of the "Typhloplanoida" and of the "Dalyellioida" (for example the Dalyelliidae with the Temnocephalida), but not the Kalyptorhynchia, the Neodermata and several groups of the "Typhloplanoida" and "Dalyellioida".

Acknowledgements

Financial support was provided by the Akademie der Wissenschaften und der Literatur, Mainz. Mrs. K. Lotz is thanked for gathering the sand samples. The technical assistance of Mrs. K. Lotz, Mrs. S. Gubert and Mr. B. Baumgart is greatly acknowledged.

Zusammenfassung

Die Ultrastruktur der männlichen wie weiblichen Gonaden und Gameten wird beschrieben. Die Spermatozoen sind biciliär, besitzen corticale Mikrotubuli, spezifische, zu Halbzirkeln arrangierte elektronendichte Granula, „dense bodies", den Nucleus und reihige Mitochondrien. Während der Spermiogenese tritt neben dem Intercentriolarkörper eine einem icb vergleichbare Differenzierung auf, zudem elektronendichte scheibenförmige Platten. Dottersubstanzen und Vesikel mit mosaikartig angeordnetem Eischalensekret charakterisieren die Vitellocyten. Marginalgranula kennzeichnen die Germocyten. Die ultrastrukturellen Merkmale der weiblichen Gameten stützen die Hypothese zur Monophylie eines Taxons Prolecithophora + Rhabdocoela (einschließlich der Neodermata); diesem Monophylum wird der Name Eulecithophora zuerkannt. Zwei Differenzierungen der männlichen Gameten, die elektronendichten Granula und die scheibenförmigen Platten, sind von mehreren, aber nicht allen Arten der „Typhloplanoida" und der „Dalyellioida" (einschließlich der Temnocephalida) und auch nicht von Vertretern der Kalyptorhynchia und der Neodermata bekannt, so daß stammesgeschichtliche Bewertungen hier noch Schwierigkeiten bereiten.

Abbreviations

bb	basal body	mt	microtubule
cb	ciliary bases	n	nucleus
ci	cilia	nl	nucleolus
db	dense bodies	rER	rough endoplasmic reticulum
gv	germinal vesicle	sg	eggshell-forming granule
icb	intercentriolar body	sp	spermatozoon
icbl	intercentriolar body-like structure	spt	spermatids
l	lipid droplet	spy	spermatocytes
mg	marginal granule	vi	vitellarium
mi	mitochondrion	y	yolk platelet

References

CIFRIAN, B., P. GARCIA-CORRALES & S. MARTINEZ-ALOS (1988): Ultrastructural study of spermatogenesis and mature spermatozoa of *Paravortex cardii* (Plathelminthes, Dalyellioida). Acta Zool. (Stockholm) **69**, 195–204.

EHLERS, U. (1985): Das phylogenetische System der Plathelminthes. Fischer, Stuttgart New York, pp. 1–317.

EHLERS, U. & B. SOPOTT-EHLERS (1993): The caudal duo-gland adhesive system of *Jensenia angulata* (Plathelminthes, Dalyelliidae): ultrastructure and phylogenetic significance. (With comments on the phylogenetic position of the Temnocephalida and the polyphyly of the Cercomeria). Microfauna Marina **8**, 65–76.

GREMIGNI, V. (1988): A comparative ultrastructural study of homocellular and heterocellular female gonads in free-living Plathelminthes-Turbellaria. Fortschr. Zool./Prog. Zool. **36**, 245–261.

GREMIGNI, V. & A. FALLENI (1991): Ultrastructural features of cocoon-shell globules in the vitelline cells of neoophoran platyhelminths. Hydrobiologia **227**, 105–111.

Iomini, C., M. Ferraguti, G. Melone & J.-L. Justine (1994): Spermiogenesis in a Scutariellid (Platyhelminthes). Acta Zoologica (Stockholm) **75**, 287–295.

Ishida, S. & W. Teshirogi (1988): Comparison of spermatozoa among freshwater planarian species. Fortschr. Zool./Prog. Zool. **36**, 297–302.

Ishida, S., Y. Yamashita & W. Teshirogi (1991): Analytical studies of the ultrastructure and movement of the spermatozoa of freshwater triclads. Hydrobiologia **227**, 95–104.

Li, M.M., N. A. Watson & K. Rohde (1992): Ultrastructure of sperm and spermatogenesis of *Artiopostia* sp. (Platyhelminthes: Tricladida: Terricola). Aust. J. Zool. **40**, 667–674.

Lucchesi, P., A. Falleni & V. Gremigni (1995): The ultrastructure of the germarium in some Rhabdocoela. Hydrobiologia **305**, 207–212.

Noury-Srairi, N. J.-L. Justine & L. Euzet (1989): Ultrastructure comparée de la spermiogenèse et du spermatozoïde de troi espèces de *Paravortex* (Rhabdocoela, "Dalyellioida", Graffillidae), Turbellariés parasites intestinaux de Mollusques. Zool. Scri. **18**, 161–185.

Rohde, K. & N. Watson (1988): Ultrastructure of epidermis and sperm of the turbellarian *Syndisyrinx punicea* (Hickman, 1956) (Rhabdcoela, Umagillidae). Aust. J. Zool. **36**, 131–139.

– (1995): Ultrastructure of sperm and spermiogenesis of two species of the Tricladida (Platyhelminthes): *Romankenkius libidinosus* (Paludicola) and an unidentified species of the Maricola. Invert. Reprod. Develop. **27**, 181–196.

Sopott-Ehlers, B. (1994): Fine structure of spermatozoa in *Anthopharynx sacculipenis* (Plathelminthes, Solenopharyngidae). Zoomorphology **114**, 33–38.

– (1995): Fine structure of vitellaria and germaria in *Polystyliphora filum* (Plathelminthes, Proseriata). Microfauna Marina **10**, 159–171.

– (1996): First evidence of mitochondrial lensing in two species of the "Typhloplanoida" (Plathelminthes, Rhabdocoela): phylogenetic implications. Zoomorphology **116**, 95–101.

– (1997a): First report on the fine structure of unpigmented rhabdomeric photoreceptors in a free-living species of the "Dalyellioida" (Plathelminthes, Rhabdocoela). Microfauna Marina **11**, 27–34.

– (1997b): Submicroscopic anatomy of female gonads in *Ciliopharyngiella intermedia* (Plathelminthes, Rhabdocoela, "Typhloplanoida"). Microfauna Marina **11**, 209–221.

Sopott-Ehlers, B. & U. Ehlers (1995a): Modified sperm ultrastructure and some data on spermiogenesis in *Provortex tubiferus* (Plathelminthes, Rhabdocoela): phylogenetic implications for the Dalyellioida. Zoomorphology **115**, 41–49.

– (1995b): Ultrastructural features of *Bresslauilla relicta* (Plathelminthes, Rhabdocoela). The spermatozoa. Microfauna Marina **10**, 235–247.

– (1997): Electronmicroscopical investigations of male gametes in *Ptychopera westbladi* (Plathelminthes, Rhabdocoela, "Typhloplanoida"). Microfauna Marina **11**, 193–208.

Tyler, S. (1988): The role of function in determination of homology and convergence – examples from invertebrate adhesive organs. Fortschr. Zool./Prog. Zool. **36**, 331–347.

Watson, N. A. & J. Jondelius (1995): Comparative ultrastructure of spermiogenesis and sperm in *Maehrenthalia* sp. and *Bresslauilla relicta* (Platyhelminthes, Rhabdocoela). Invertebr. Reprod. Develop. **28**, 103–112.

Watson, N. A. & K. Rohde (1993): Ultrastructure of sperm and spermiogenesis of *Kronborgia isopodicola* (Platyhelminthes, Fecampiidae). Int. J. Parasitol. **23**, 737–744.

– (1995a): Ultrastructure of spermiogenesis and spermatozoa in the platyhelminths *Actinodactylella blanchardi* (Temnocephalida, Actinodactylellidae), *Didymorchis* sp. (Temnocephalida, Didymorchidae) and *Gieysztoria* sp. (Dalyelliida, Dalyelliidae), with implications for the phylogeny of the Rhabdocoela. Invertebr. Reprod. Develop. **27**, 145–158.

– (1995b): Sperm and spermiogenesis of the "Turbellaria" and implications for the phylogeny of the phylum Platyhelminthes. In: Jamieson, B.G.M., J. Ausio & J.-L. Justine (eds), Advances in spermatozoal phylogeny and taxonomy. Mém. Mus. natn. Hist. nat. **166**, 37–54.

Watson, N. A., K. Rohde, & K. Sewell (1995): Ultrastructure of spermiogenesis and spermatozoa of *Decadidymus gulosus*, *Temnocephala dendyi*, *T. minor*, *Craspedella* sp., *C. spenceri* and *Diceratocephala boschmai* (Platyhelminthes, Temnocephalida, Temnocephalidae), with emphasis on the intercentriolar body and zone of differentiation. Invertebr. Reprod. Develop. **27**, 131–143.

WILLIAMS, J.B. (1994): Comparative study of sperm morphology and reproductive biology of *Kronborgia* and *Temnocephala* (Platyhelminthes, Neoophora): implications for platyhelminth phylogeny. New Zeal. J. Zool. **21**, 179–194.

Dr. Beate Sopott-Ehlers
II. Zoologisches Institut und Museum der Universität Göttingen
Berliner Straße 28, D-37073 Göttingen

New findings of interstitial Copepoda from Punta Morales, Pacific Coast of Costa Rica

Wolfgang Mielke

Abstract

Four species of interstitial Copepoda have been collected in the sandy beach of Punta Morales, Pacific coast of Costa Rica. Though the specimens reveal slight morphological differences in comparison with the original descriptions they are identified as species which are already known from other geographical regions, viz. the coasts of California/USA, Panamá, the Galápagos Archipelago, and Campeche/Mexico. *Zausodes septimus* Lang, 1965 belongs to the Harpacticidae. The other species are representatives of the Laophontidae: *Klieonychocamptoides itoi* Mielke, 1981; *Afrolaophonte schmidti* Mielke, 1981, and *Mexicolaophonte arganoi* Cottarelli, 1977.

A. Introduction

Each systematist of any given taxon has been frequently confronted with the "classical dilemma" (REED 1991): how to interpret slight morphological differences of newly found specimens in comparison with the description of a probably identical species? Even if it is possible to make direct comparisons between the holotype or other material from the locus typicus and the new individuals, different interpretations of concerned specialists cannot be excluded. Modern methods like biochemical analyses or genetic studies may lead to more precise and reliable results (but see DAHMS & SCHMINKE 1995). However, the systematist working solely on morphological features soon encounters the point when objective, unequivocal evidences are replaced by subjective, questionable arguments. Is it a "separate species or a single variable one?" (e.g. GOTTO 1990, p. 308). It is indisputable that many non-uniform "species" in reality represent species groups. On the other hand, many widely distributed species exist whose subpopulations, of course, show all phenomena of geographic variation, e.g. body length, colour, proportions etc. Hence, they come under "the bounds of variability between widely separated population(s)" (GEE & FLEEGER 1986, p. 164). Generally, material from only one locality is under consideration. That leads to the erroneous conception that individuals of one and the same species are morphologically constant. However, evolution is a dynamic process and variability is the necessary precondition for the adaptation to changing environmental conditions. Phenotypes are indicative for responses to alterations. Local populations (local races, subspecies) that actually represent one species may develop into distinct species in the future due to the interruption of gene exchange. Every systematist has to take into consideration that he is working on individuals that only exist during a momentary time section of continuous processes of speciation.

Animals of one species living in a limited area, e.g. a single marine sandy beach, usually do not – or only to a small degree – exhibit intraspecific variability. On the other hand, conspecific individuals occurring in distant localities show less morphological constancy. Furthermore, morphological and geographical distance must not implicitly correspond. Two examples:

a) In his contribution to the marine copepods of the Island of Sylt (Germany) NOODT (1952) reported on the variability of "*Paraleptastacus spinicauda* (T. & A. Scott, 1895)", the most frequent species of all beaches. When investigating the distribution patterns, abundance dynamics and life cycles of interstitial copepods of the east coast beach of List/Sylt (MIELKE 1975, 1976) it could be demonstrated that "*P. spinicauda*" in reality represented three species (taxonomical revision by WHYBREW 1986). Subsequent experience showed that the variability of a copepod species to such a degree is rather a hint on co-occurring, closely related species.

b) Morphological divergences of widely distributed species have to be weighted differently – if endemism is not assumed as a principle for benthic Copepoda. In the Galápagos Islands a new taxon, *Galapalaophonte*, could be established (*G. pacifica*; MIELKE 1981). Despite some variable features concerning body length, ornamentation, relative length of setae/spines, etc. I interpreted the animals from various beaches of the archipelago as belonging to the same species. This species I believe to have found again in Panamá (MIELKE 1982; as *Laophontina triarticulata* Coull & Zo, 1980). However, the finding of some animals in two sampling sites on the Chilean coast was rather problematical. Although several morphological differences between these animals and those from the Galápagos Islands could be documented (MIELKE 1985), I objected to the establishment of the Chilean individuals as a distinct species. This was later done by FIERS (1991; *G. chilensis*). Morphological divergences are surely obvious, but what he considered to be "clearly distinct" species may also be a state at the level of subspecies or geographical races. Up to now, the beaches situated between Galápagos/Panamá and central Chile have not been investigated for *Galapalaophonte*. I think dozens of problematical findings are to be expected.

The procedure of declaring that each morphological deviation represents a distinct species is useful for pragmatic reasons such as making keys, species lists, etc. However, it is too static and not appropriate for the dynamic processes of speciation. It is important to point out all morphological differences, yet, their interpretation should not be given apodictically.

In the last few years the tendency has increased to establish new species solely based on minor structural differences rather than to interpret them as expressing intraspecific variability. The purpose of this study is to present four interstitial copepod species from Costa Rica, exemplarily demonstrating the systematical problems broached above. The material was collected on 28 and 29 August 1990, at locality 4 in MIELKE 1992.

B. Results

Harpacticidae Sars, 1904
Zausodes Wilson, 1932
Zausodes septimus Lang, 1965

Material. Frequent species. Dissected were 3 ♀♀ and 3 ♂♂.
Length of dissected ♀♀: 0.37 – 0.38 mm; ♂♂: 0.28 – 0.31 mm.

Remark. Apart from minor differences in proportions the animals from Punta Morales largely agree with the ones described by LANG (1965) from the Monterey Bay, Pacific coast of the U.S.A. and the specimens studied previously from Panamá (see MIELKE 1990).

Laophontidae T. Scott, 1904
Klieonychocamptoides Noodt, 1958
Klieonychocamptoides itoi Mielke, 1981

Material. 3 ♀♀ and 2 ♂♂. All specimens were dissected.
Length of ♀♀: 0.33 – 0.37 mm; ♂♂: 0.25 – 0.29 mm.

Remark. The present animals correspond with those studied previously from Galápagos (MIELKE 1981) and from Panamá (MIELKE 1982). The most conspicuous difference is the exopodite P.4. Here, the middle of the three apical spines is developed rather weakly. Furthermore, the inner edge of basis P.3 and P.4 has a somewhat variable shape. It can be elongated thread-like, i.e. with a fused, rudimentary setule. This, however, also occurs occasionally with the animals of Galápagos and Panamá.

Afrolaophonte Chappuis, 1960
Afrolaophonte schmidti Mielke, 1981

Material. Several adult ♀♀ and ♂♂. Dissected were 8 ♀♀ and 4 ♂♂. Length of dissected ♀♀: 0.38 – 0.46 mm; ♂♂: 0.35 – 0.40 mm.

Remark. The species was first recorded from the Galápagos Archipelago (MIELKE 1981). Furthermore, 1 ♀ and 1 ♂ were collected at the Island of Taboga, Pacific side of Panamá (MIELKE 1982) which – with reservations – were also assigned to this species. The differences between the Panamanian animals and the ones of Galápagos are likewise true for the specimens from Punta Morales:
– 1st Antenna ♀ with 6 segments, i.e. 4th and 5th segment are clearly separated.
– Endopodite P.3 ♀ weakly developed and apparently fused with the basis.
– Slight differences in proportions and ornamentation.

Mexicolaophonte Cottarelli, 1977
Mexicolaophonte arganoi Cottarelli, 1977

(Figs. 1–4)

Material. 3 ♀♀. All specimens were dissected.
Length of ♀♀: 0.49 – 0.65 mm!
Drawings of all relevant characteristics are presented.

Remark. At present four species belong to the genus *Mexicolaophonte*: *M. arganoi* from "near the Champoton village (Campeche state, Mexico)" (see COTTARELLI 1977); *M. osellai* Cottarelli, 1985 from Celebes Island (now: Sulawesi), Indonesia (only the ♀ is known, see COTTARELLI 1985); *M. creola* Cottarelli & Forniz, 1990 from an "intertidal habitat on Sussex beach", Sierra Leone (only the ♂ is known, see COTTARELLI & FORNIZ 1990); *M. mielkei* Fiers, 1991. This species (only the ♂ is known) was found in the beach of Naos Island on the Pacific side of Panamá, briefly characterized as ?*Laophontina*

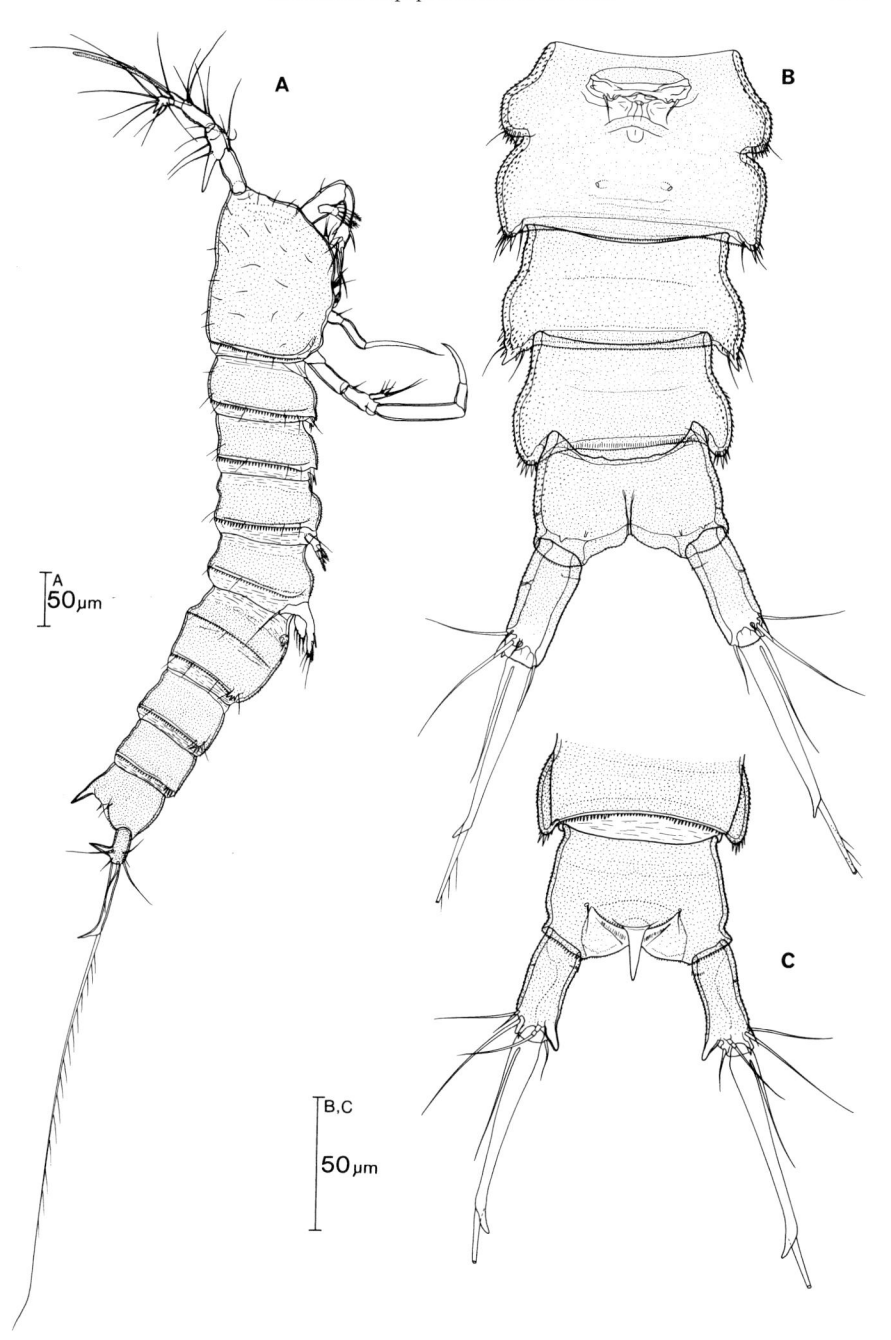

Fig. 1. *Mexicolaophonte arganoi*. ♀. A. Habitus, lateral side. B. Abdomen, ventral side. C. Caudal part, dorsal side..

276 Wolfgang Mielke

spec. (MIELKE 1982) and subsequently placed in the genus *Mexicolaophonte* by FIERS (1986, 1991) and WELLS & RAO (1987).

According to the species of e.g. *Laophontina, Pseudolaophonte, Afrolaophonte, Klieonychocamptoides* also in *Mexicolaophonte* reductions, fusions and sexual dimorphisms are remarkable features (compare FIERS 1986). Be-

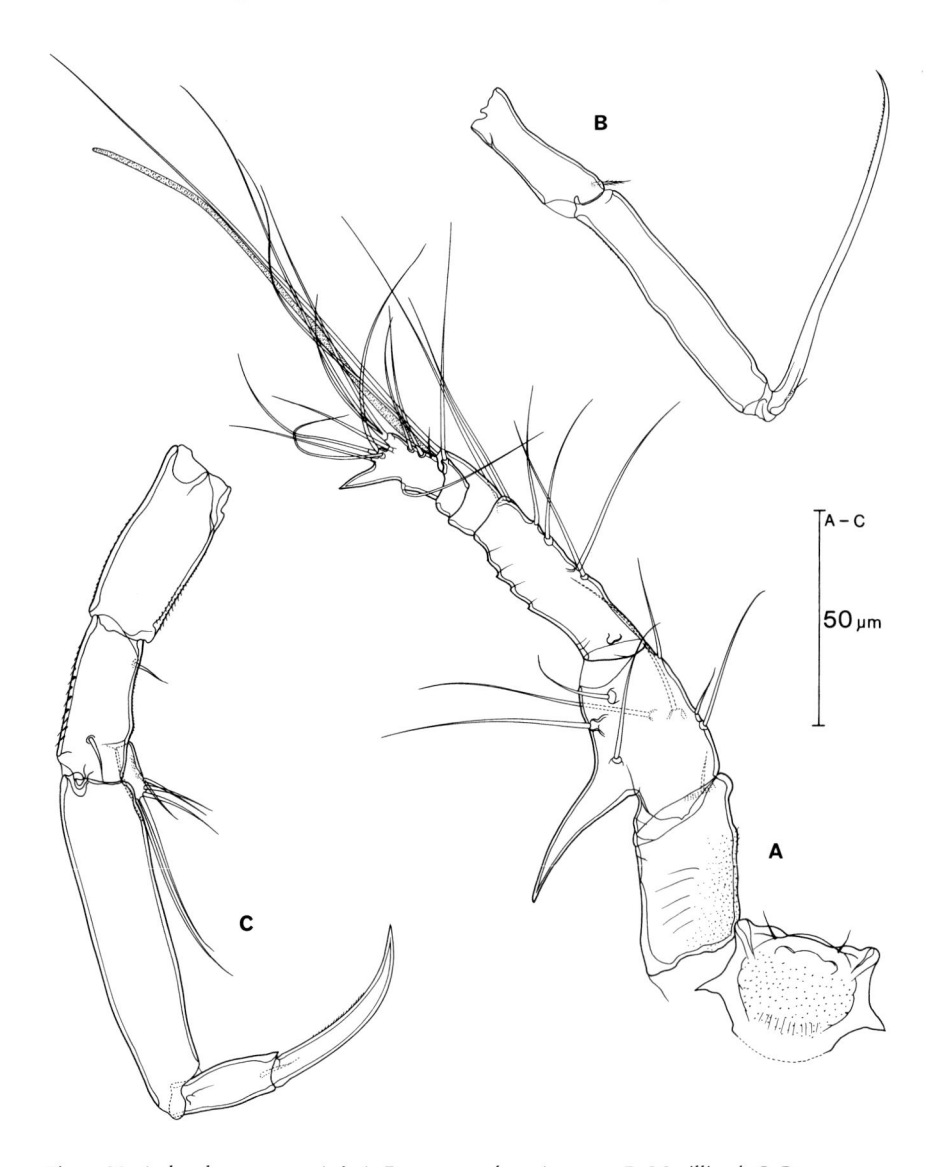

Fig. 2. *Mexicolaophonte arganoi.* ♀. A. Rostrum and 1st Antenna. B. Maxilliped. C. P.1.

cause of the minute size of the appendages and a sometimes only partial fusion of individual pereiopodal segments with each other (for example the narrow distal parts of the exopodites) or fusion with the basis, respectively, it is difficult to obtain unequivocal results. From this point of view both species with known females, i.e. *M. arganoi* and *M. osellai*, do not exhibit fundamental differences.

Only three ♀♀ were found in the beach at Punta Morales in two sampling

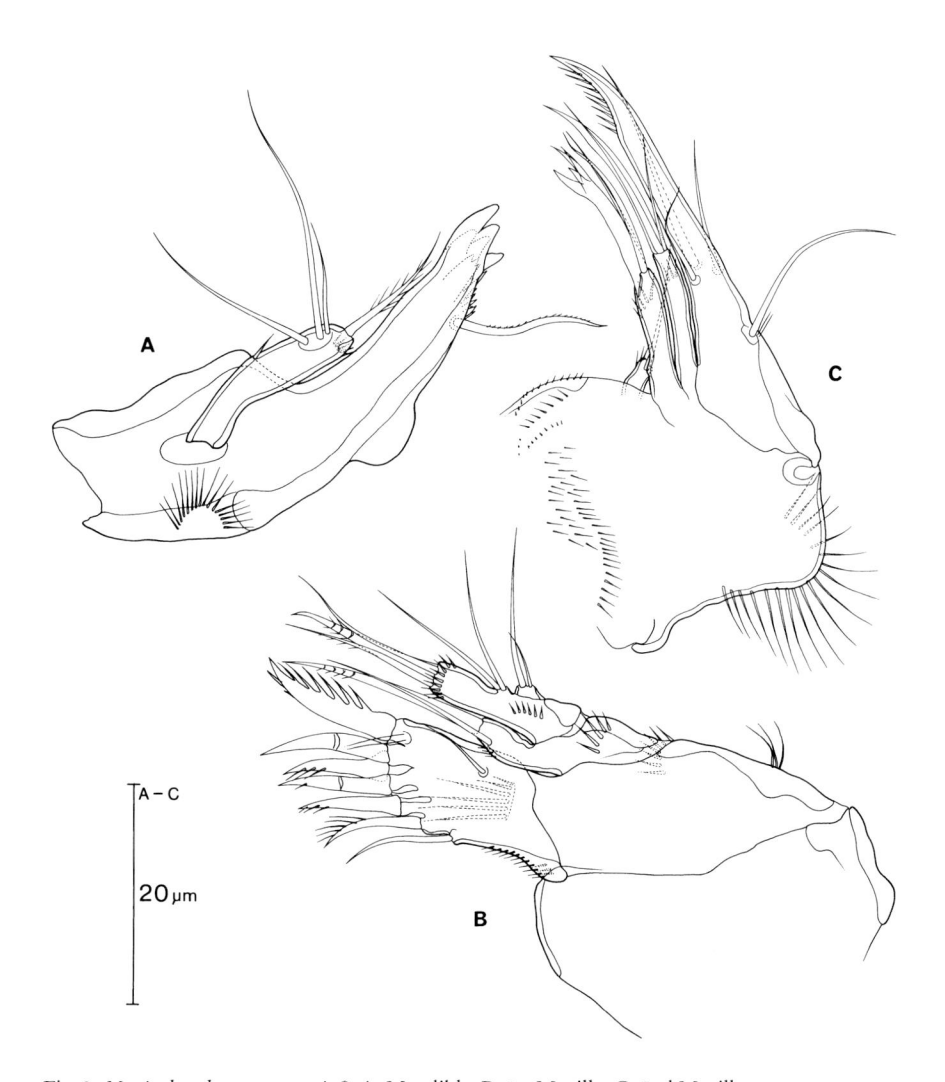

Fig. 3. *Mexicolaophonte arganoi*. ♀. A. Mandible. B. 1st Maxilla. C. 2nd Maxilla.

sites which were separated by about 100 m. Some minor differences could be ascertained (see Figs. of P.2 and endopodite P.3).

The specimens of Punta Morales are identified as *M. arganoi* though they show some morphological differences compared to COTTARELLI's description.

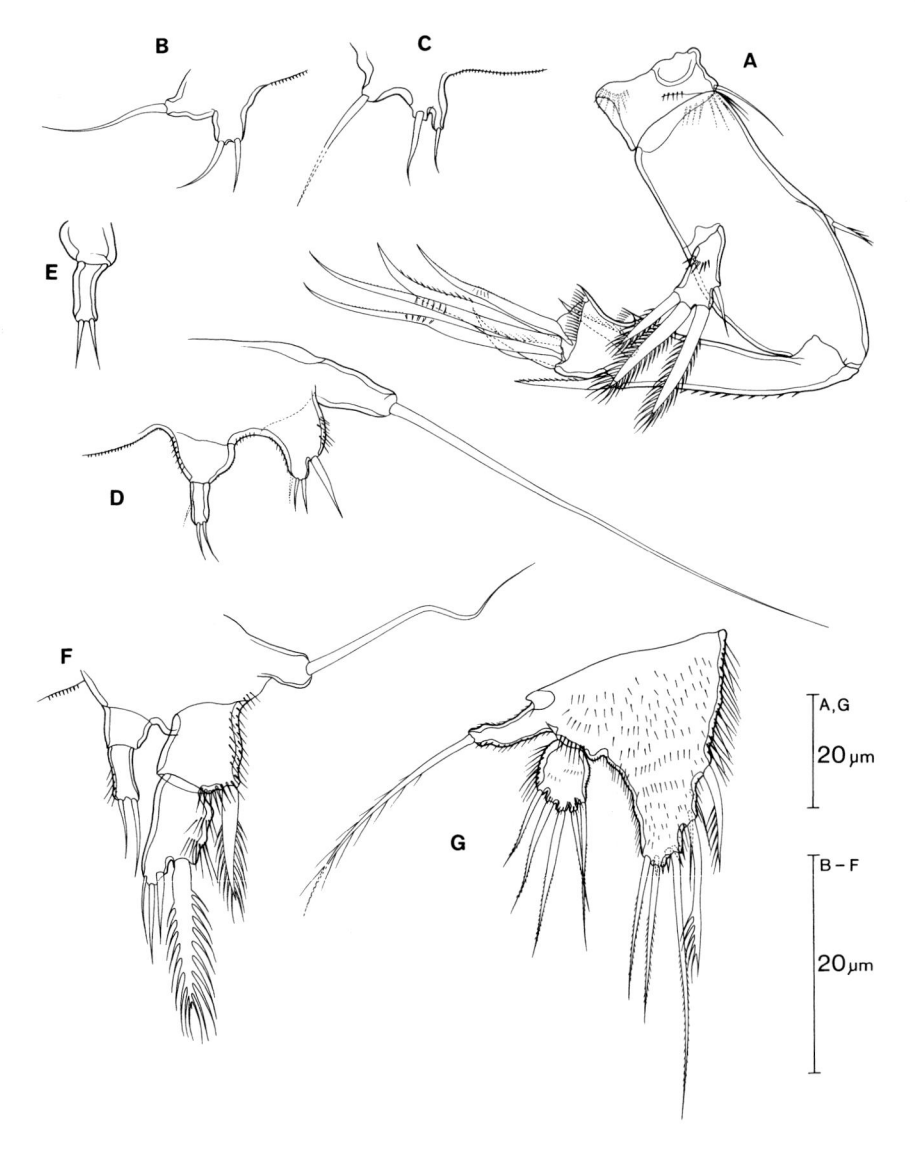

Fig. 4. *Mexicolaophonte arganoi*. ♀. A. 2nd Antenna. B. P.2. C. P.2, other specimen. D. P.3. E. Endopodite P.3, other specimen. F. P.4. G. P.5.

However, this may be due to different interpretations of the small features, e.g. palp of mandible, endites of 2nd maxilla, maxilliped, structure of pereiopods etc. One of the present ♀♀ apparently possesses an additional weak seta on the inner margin of exp. and enp. P.3 (dotted line in Fig. 4D). Both other ♀♀ obviously do not have it. More material would be necessary for clarification.

On the other hand, the similarity between the *Mexicolaophonte* ♀♀ of Costa Rica and the ♂♂ of the geographical "near" locality in Panamá seems to be more distant. However, this cannot be decided before the respective ♂♂ and ♀♀ have been found.

C. Conclusion

At present the species introduced above are only known from a few widespread localities of the Eastern Pacific Region, the Gulf of Mexico and the Caribbean Sea. Without doubt, they are to be expected in many other littoral and sublittoral sediments of that area. A detailed discussion on their problematical status – variable species or radiating species groups – should be resumed until new findings are at hand. Till then, it seems advisable to identify the present animals from Costa Rica with species that are already known rather than to interpret them to be representatives of new species.

Zusammenfassung

Im Sandstrand von Punta Morales an der Pazifikküste von Costa Rica wurden vier interstitielle Copepodenarten gesammelt. Obwohl die Tiere geringe morphologische Unterschiede im Vergleich zu den Originalbeschreibungen aufweisen, werden sie mit Arten identifiziert, die schon von anderen geographischen Regionen bekannt sind, nämlich den Küsten von Kalifornien/USA, Panamá, dem Galápagos-Archipel und Campeche/Mexiko. *Zausodes septimus* Lang, 1965 gehört zu den Harpacticidae. Die anderen Arten sind Vertreter der Laophontidae: *Klieonychocamptoides itoi* Mielke, 1981; *Afrolaophonte schmidti* Mielke, 1981 und *Mexicolaophonte arganoi* Cottarelli, 1977.

References

COTTARELLI, V. (1977): *Mexicolaophonte arganoi* n. gen. n. sp. di Laophontidae (Crustacea, Copepoda, Harpacticoida) di acque interstiziali litorali messicane. In: "Subterranean fauna of Mexico, Part III". Accad. naz. Lincei **171**, 91 – 99.

COTTARELLI, V. (1985): Laophontidae di acque interstiziali litorali dell'Indonesia (Crustacea, Copepoda, Harpacticoida). Boll. Mus. civ. St. nat. Verona **12**, 283 – 297.

COTTARELLI, V. & C. FORNIZ (1990): *Mexicolaophonte creola* n. sp. A new species of Laophontidae harpacticoid (Crustacea, Copepoda) from interstitial marine litoral waters of Sierra Leone (Western Africa). In: "Ricerche Biologiche in Sierra Leone, Part III". Accad. naz. Lincei **265**, 45 – 50.

DAHMS, H.-U. & H.K. SCHMINKE (1995): A multidisciplinary approach to the fine-systematics with *Tisbe* – an evaluation of morphological and molecular methods. Hydrobiologia **308**, 45 – 50.

FIERS, F. (1986): Harpacticoid copepods from the West Indian Islands: Laophontidae (Copepoda, Harpacticoida). Bijdr. Dierk. **56**, 132 – 164.

FIERS, F. (1991): A revision of the genus *Laophontina* Norman & T. Scott (Copepoda, Harpacticoida). Bull. Inst. Roy. Sci. Nat. Belg. **61**, 5 – 54.

GEE, J.M. & J.W. FLEEGER (1986): Two new species of harpacticoid copepod from South Orkney Islands, Antarctica, and a redescription of *Idyellopsis typica* Lang (Tisbidae). Zool. J. Linn. Soc. **88**, 143 – 165.

GOTTO, R.V. (1990): Some unsolved problems concerning copepods associated with marine invertebrates. Bijdr. Dierk. **60**, 305 – 309.

LANG, K. (1965): Copepoda Harpacticoidea from the Californian Pacific coast. Kungl. Svenska Vetenskaps. Handl. **10**, 1 – 566.

MIELKE, W. (1975): Systematik der Copepoda eines Sandstrandes der Nordseeinsel Sylt. Mikrofauna Meeresboden **52**, 1 – 134.

MIELKE, W. (1976): Ökologie der Copepoda eines Sandstrandes der Nordseeinsel Sylt. Mikrofauna Meeresboden **59**, 1 – 86.

MIELKE, W. (1981): Interstitielle Fauna von Galapagos. XXVIII. Laophontinae (Laophontidae), Ancorabolidae (Harpacticoida). Mikrofauna Meeresboden **84**, 1 – 104.

MIELKE, W. (1982): Einige Laophontidae (Copepoda, Harpacticoida) von Panama. Crustaceana **42**, 1 – 11.

MIELKE, W. (1985): Interstitielle Copepoda aus dem zentralen Landesteil von Chile: Cylindropsyllidae, Laophontidae, Ancorabolidae. Microfauna Marina **2**, 181 – 270.

MIELKE, W. (1990): *Zausodes septimus* Lang, 1965 und *Enhydrosoma pericoense* nov. spec., zwei benthische Ruderfußkrebse (Crustacea, Copepoda) aus dem Eulitoral von Panamá. Microfauna Marina **6**, 139 – 156.

MIELKE, W. (1992): Six representatives of the Tetragonicipitidae (Copepoda) from Costa Rica. Microfauna Marina **7**, 101 – 146.

NOODT, W. (1952): Marine Harpacticiden (Cop.) aus dem eulitoralen Sandstrand der Insel Sylt. Abh. math.-nat. Kl. Akad. Wiss. Lit. Mainz **3**, 103 –142.

REED, E.B. (1991): *Eudiaptomus yukonensis*, new species (Copepoda: Calanoida), from northwestern Canada. J. Crust. Biol. **11**, 647 – 652.

WELLS, J.B.J. & G.C. RAO (1987): Littoral Harpacticoida (Crustacea: Copepoda) from Andaman and Nicobar Islands. Mem. Zool. Surv. India **16**, 1 –385.

WHYBREW, D.F. (1986): Zur Systematik und Ökologie des Taxons *Paraleptastacus* Wilson, 1932 (Copepoda, Harpacticoida). Diss. Univ. Göttingen, 323 pp.

Dr. Wolfgang Mielke
II. Zoologisches Institut und Museum der Universität Göttingen
Berliner Straße 28, D-37073 Göttingen

Plasma membranes flanked by cisternae of the endoplasmic reticulum: a remarkable organization of polarized cells in small Plathelminthes

Ulrich Ehlers and Beate Sopott-Ehlers

Abstact

Polarized cells in Plathelminthes are bound together by zonulae adhaerentes and septate junctions. These two cell junctions are typical elements in the epidermis, protonephridial canals, genital ducts and epithelia lining genital organs, and of intestinal cells lining the gut lumen. Septate junctions being flanked by conspicuous narrow-spaced cisternae of the endoplasmic reticulum are described for several small free-living species belonging to different subgroups of the Plathelminthes. Such an organization is known for species of the Acoela, Catenulida and Rhabditophora and can be hypothesized for the ground pattern of the Plathelminthes.

A. Introduction

Species of the Plathelminthes possess distinct body regions that are more exposed to mechanical pressures than others. In general, the outer body covering, the outleading protonephridial canals, in- and outleading genital ducts and the intestinal cells surrounding the gut lumen, especially that of predators swallowing large prey, are such areas exposed to temporary or permanent mechanical stresses. At the EM-level, a number of different marked features can be found in the cells and tissues of such body regions.

With respect to the outer body covering, epidermal cells of free-living species display a cell web, i.e. a distinct intracellular sheet-like arrangement of microfilaments of a probable skeletal function. Such a layer of microfibrils that takes the apical epidermal cell regions is called a terminal web (see TYLER

1984). Furthermore, the epidermis of distinct taxa has become a syncytial one, for example in the ectosymbiotic species of the Temnocephalida or in many Kalyptorhynchia inhabiting high energy beaches (see EHLERS 1985). In addition, the epidermis of such Kalyptorhynchia can be strengthened by marked underlying specializations of the subepidermal ECM (see EHLERS 1989). Postlarval stages of the ecto- and endoparasitic species of the Neodermata always show a syncytial body covering, the neodermis.

Different organizations within the Plathelminthes also exist with respect to the cell junctions. The epidermis, the intestinum, the nephridia and genital structures show polarized cells, i.e. cells arranged into epithelia, that display the "typical invertebrate" junctions called zonulae adhaerentes and septate junctions. These junctions are sometimes inconspicuous but can be very pronounced in many species (see also RIEGER et al. 1991).

TYLER (1984, p. 122) reported on septate junctions in the epidermis of Acoela being flanked by cisternae of the endoplasmic reticulum. Such a situation, that is common for the species of this monophylum (see also EHLERS 1992), has not been mentioned, to our knowledge, for other representatives of the Plathelminthes. In this paper, we demonstrate that septate junctions associated with cisternae can be also found in different types of epithelial cells in other small species than acoels.

B. Material and Methods

Specimens of the species *Parotoplanina geminoducta* Ax, 1956 (Proseriata Lithophora), *Ciliopharyngiella intermedia* Ax, 1952 (the taxon *Ciliopharyngiella* is a *taxon incertae sedis* within the Neoophora; see SOPOTT-EHLERS 1997), *Proxenetes deltoides* den Hartog, 1965 (Rhabdocoela Trigonostominae), and *Jensenia angulata* (Jensen, 1878) (Rhabdocoela Dalyelliidae) were extracted from sediments taken from different localities of the Island of Sylt and the Hallig Langeneß (North Sea). For EM-processing see EHLERS (1992).

C. Results

In the following, cells joined by apicalmost zonulae adhaerentes and by septate junctions being flanked by cisternae of the endoplasmic reticulum are described from the epidermis, from the protonephridium, from epithelia lining female and male organs and genital canals and from the intestinum in a few representatives of free-living marine Plathelminthes. Such cell junctions associated with cisternae are also known from many other small species of the Catenulida, Macrostomida and several groups of the Neoophora (own unpublished observations).

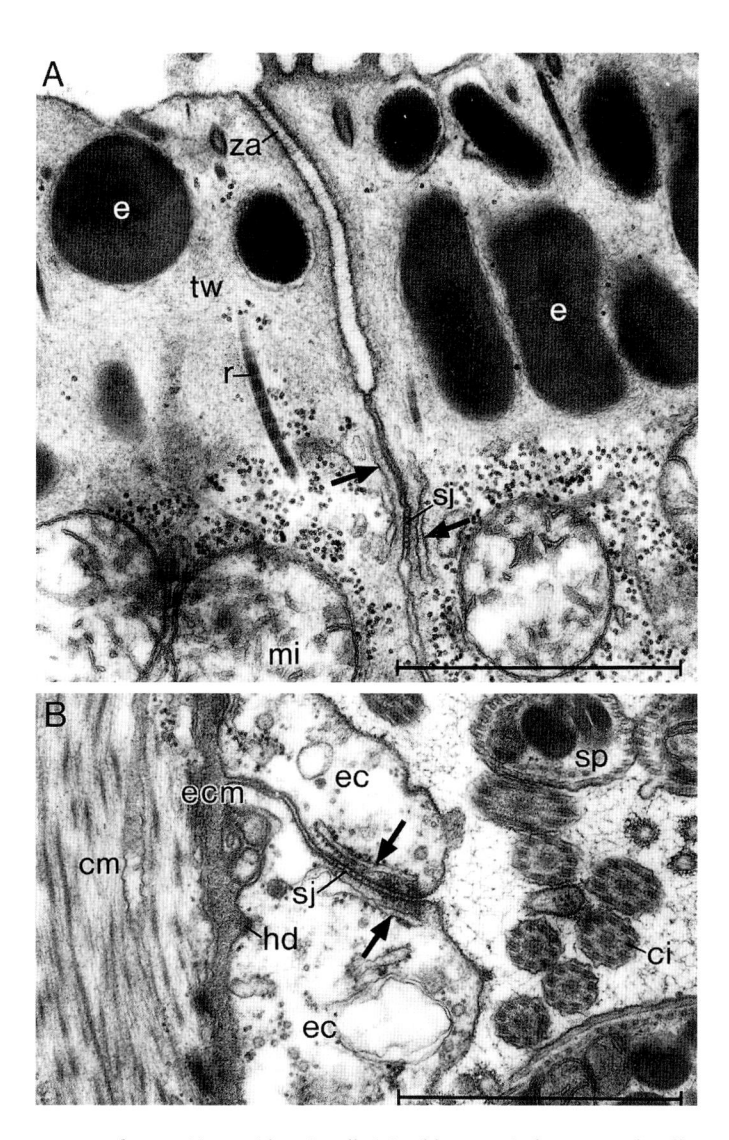

Fig. 1: *Jensenia angulata*. A. Two epidermis cells joined by an apicalmost zonula adhaerens and a septate junction being flanked by cisternae (arrows) of the endoplasmic reticulum. B. Junction of two epithelium cells lining the vesicula seminalis of the male copulatory organ. Cisternae (arrows) studied with a few ribosomes flank the septate junction. Scale bars in A and B: 1 μm.

Epidermis

Jensenia angulata displays an epidermis with a well-developed terminal web (Fig. 1 A). As in other species with such a web, this prominent microfibrillar layer is devoid of mitochondria and other cell organelles like the endoplasmic reticulum. A zonula adhaerens represents the striking intercellular junction. The septate junction is less conspicuous but flanked by cisternae present at both sides.

Proxenetes deltoides has distinct epidermal regions without strong cell webs. In such areas, more prominent septate junctions exist (Fig. 2 A). Narrow-spaced ER cisternae paralleling the joining plasma membranes are clearly visible.

Protonephridium

A protonephridial canal cell of *Ciliopharyngiella intermedia* is given as an example (Fig. 2 B). This cell is not in the form of a closed and solid tube but of a sheet-like appearance wrapped around the ciliated canal like a cuff. The longitudinal cell gap between both contiguous membranes of the canal cell is sealed by a less conspicuous zonula adhaerens and by septate junctions that extend most of the contiguous membranes. The membranes are flanked by sheet-like cisternae in nearly constant distances of about 50 nm.

Genital organs

Monolayered epithelia occur in a wide range of genital organs like the female and male genital canals or atria, bursal organs, gonoducts and copulatory organs including a vesicula seminalis and a ductus ejaculatorius. The epithelium cells of the genital atrium of *Parotoplanina geminoducta* display microvillus-like surface projections towards the lumen of the atrium (Fig. 2 C). Basally, these cells are interdigitated with each other and are joined by extensive septate junctions being flanked by cisternae of the endoplasmic reticulum. Different types of epithelial cells exist in the male genital canal of *Jensenia angulata* (Fig. 2 D). The cells are less interdigitated , but septate junctions flanked by narrow-spaced ER cisternae are obvious.

Intestinum

The intestinum of *Jensenia angulata* is a cellular one (Fig. 3). If empty, the cells with their finger-like cytoplasmic projections are clearly visible. Towards the central lumen, neighbouring cells are joined by apicalmost zonulae adhaerentes followed by extensive septate junctions. Long cisternae of the ER

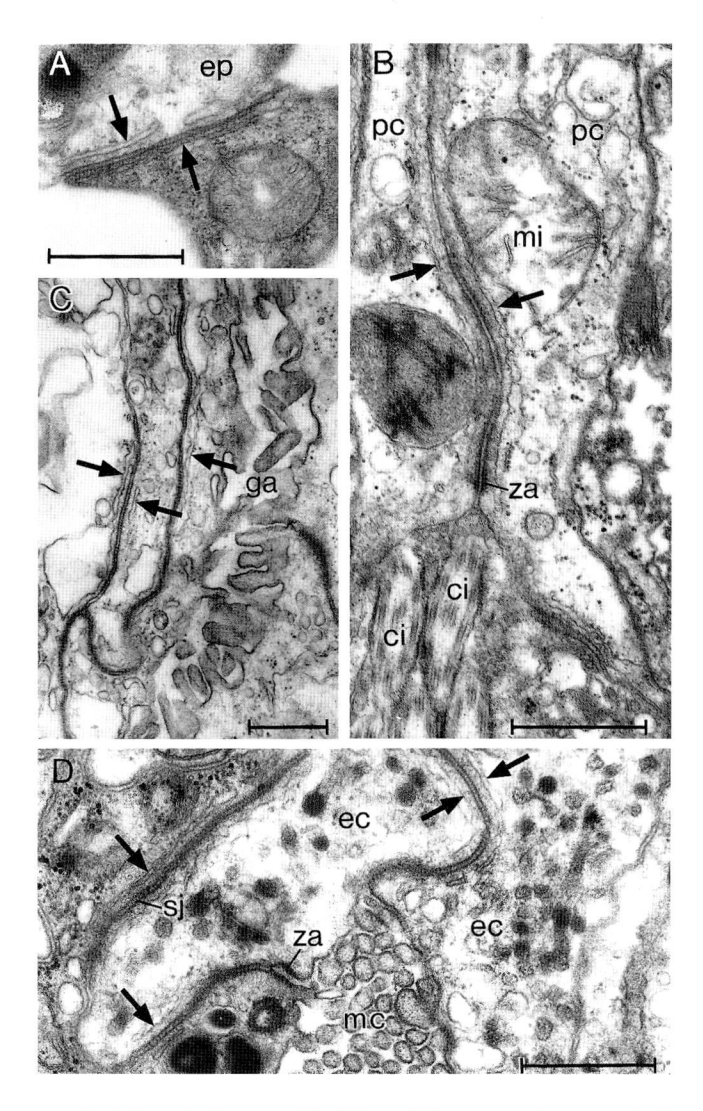

Fig. 2. The arrows in A-D mark cisternae of the ER paralleling septate junctions. A. *Proxenetes deltoides*. Junction between two epidermis cells. B. *Ciliopharyngiella intermedia*. Oblique longitudinal section of a canal cell of a protonephridium. The contiguous membranes of the cell are joined by a junction near the canal lumen (with cilia). C. *Parotoplanina geminoducta*. Part of the genital atrium and its surrounding cells. D. *Jensenia angulata*. Epithelium cells of the male genital canal. Scale bars in A-D: 0.5 µm.

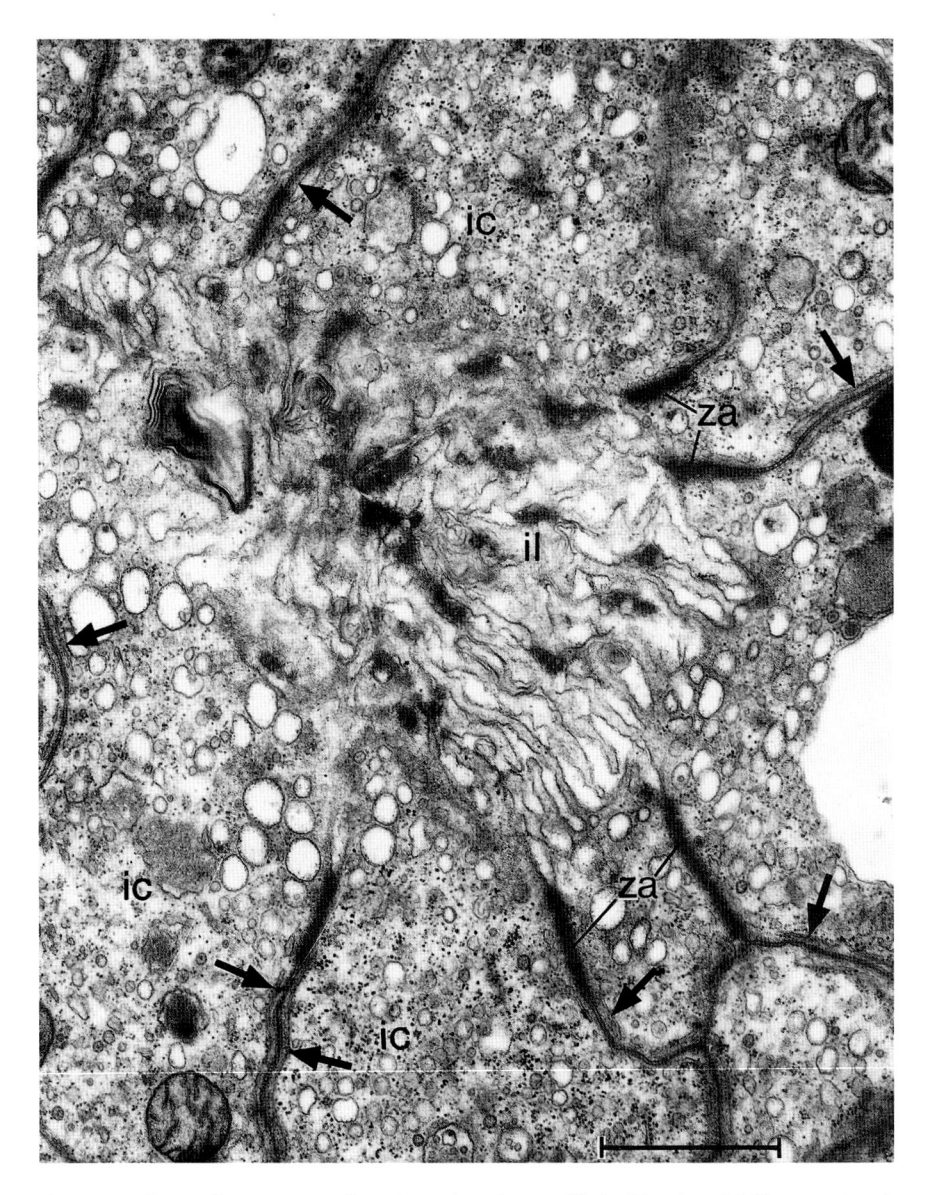

Fig. 3. Intestinum of *Jensenia angulata*. Central gut lumen filled with microvilli-like cytoplasmic projections originating from the surrounding intestinal cells joined apically by zonulae adhaerentes. Arrows point to extensive elaborations of ER cisternae. Scale bar: 1 μm.

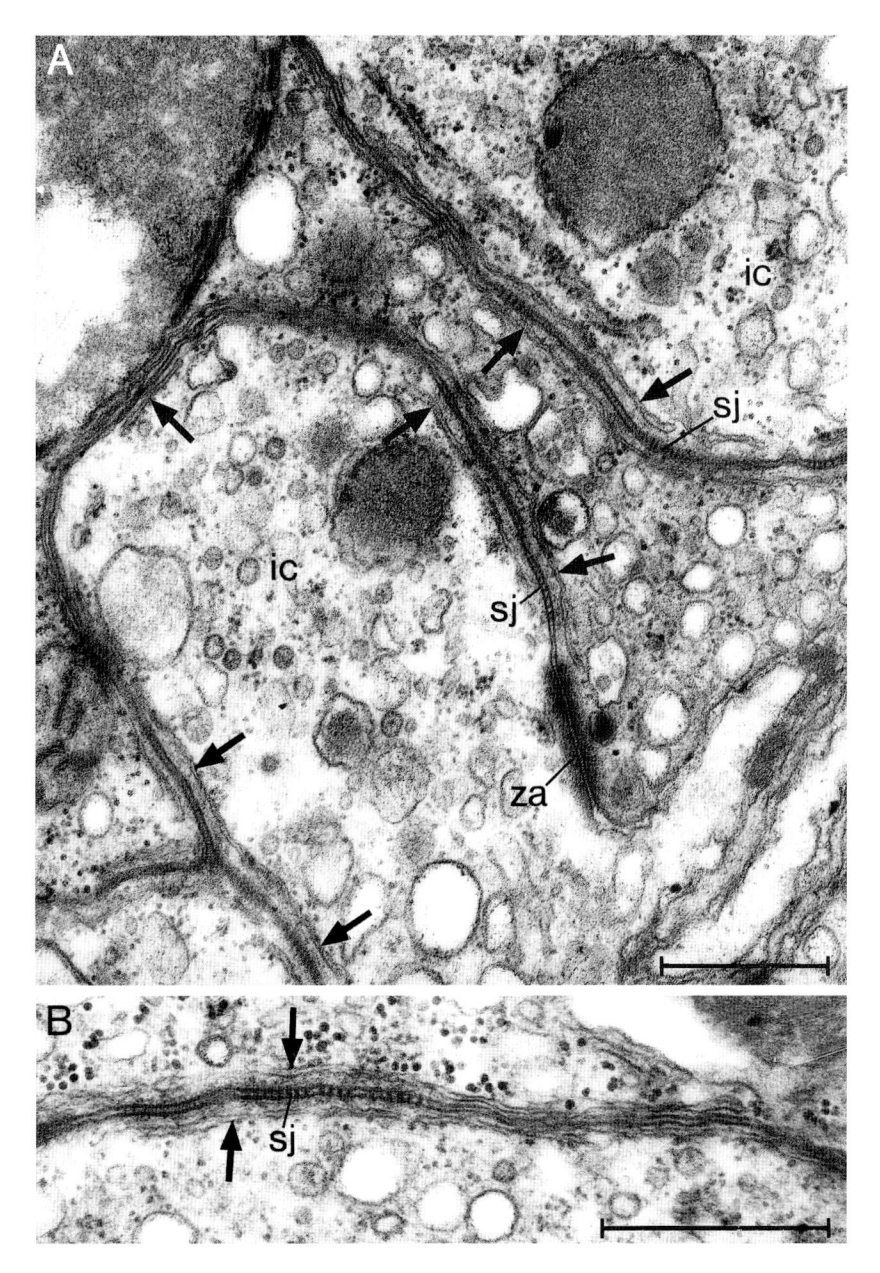

Fig. 4. *Jensenia angulata*. A and B. Intestinal cells joined by zonulae adhaerentes and septate junctions, the latter being flanked by sheet-like cisternae (arrows) of the endoplasmic reticulum. Scale bars in A and B: 0.5 μm.

flank these junctions in distances of 30–80 nm (Fig. 4 A, B). Plasma membranes without septate junctions are not flanked by cisternae. As in other epithelia, the cisternae do not extend into zonula adhaerens-regions. But in a few instances, cisternae have been observed also in these regions as is obvious in the intestinum of *Ciliopharyngiella intermedia* (Fig. 5 A). In this species, adjacent

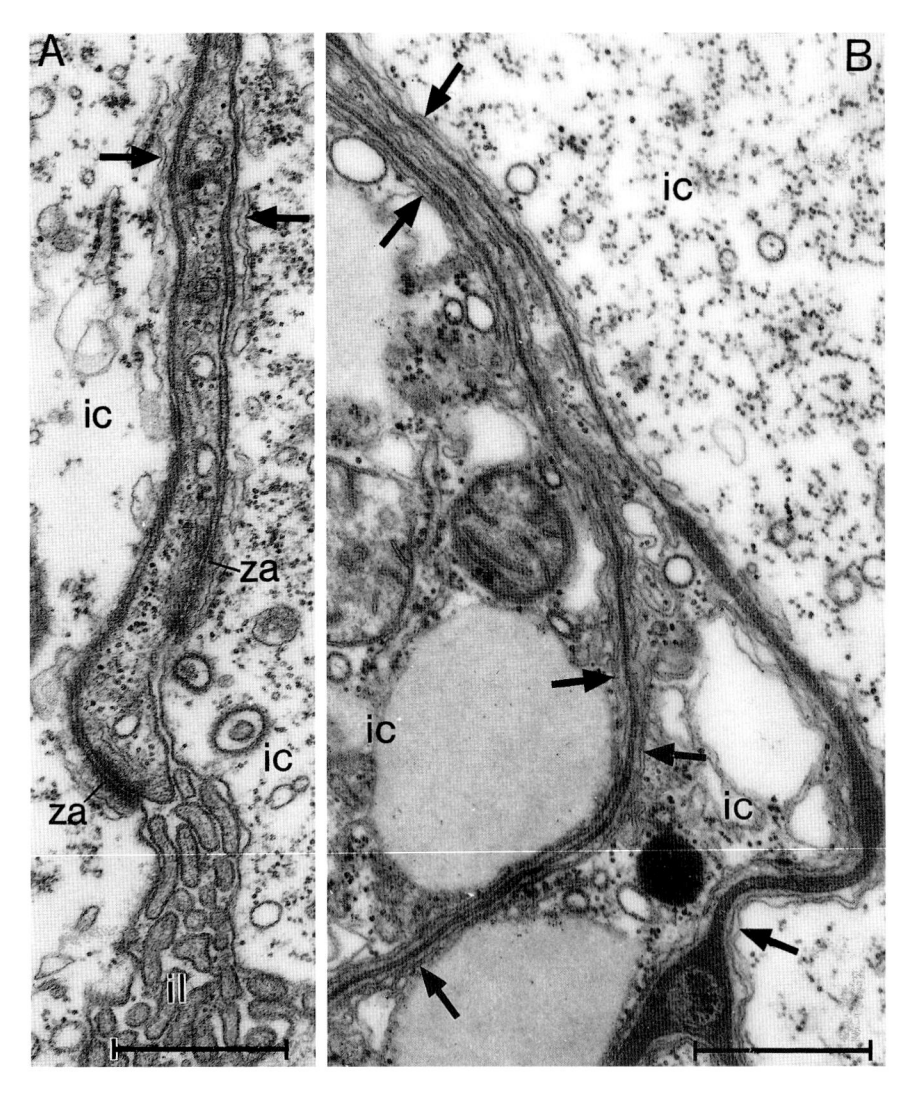

Fig. 5. *Ciliopharyngiella intermedia*. A and B. Pronounced borders of the intestinal cells with conspicuous septate junctions paralled by narrow ER cisternal spaces (arrows). Scale bars in A and B: 0.5 μm.

gut cells may be separated from each other to some extent (Fig. 5 B). Nevertheless, the plasma membranes are flanked by cisternae of the ER.

D. Discussion

The few examples given in this paper demonstrate that septate junctions being flanked by narrow-spaced cisternae of the endoplasmic reticulum are of a widespread occurrence in different epithelia of species belonging to various taxa of the Plathelminthes. That means, such an organization can be hypothesized for the ground pattern of all the Plathelminthes.

Sometimes, the cisternae may be less distinct or even absent. These differences could correspond to different physiological conditions and demonstrate, that cisternae flanking septate junctions are not permanent differentiations. As stated already by Tyler (1984) for the Acoela, the cisternae are studded with remarkable few ribosomes. Nevertheless, these cisternae paralleling the plasma membranes and septate junctions seem to be involved in the biosynthesis of the molecules required for the elaboration of the plasma membranes with the junctions. The assumption of another function is more speculative: do the cisternae contribute to the strenghthening of the cells and the epithelia like e.g. vacuoles in a variety of tissues in the Metazoa?

Finally, there is the question whether septate junctions being flanked by narrow-spaced ER cisternae are a 'normal' situation for all those Metazoa (see Ax 1995; Nielsen 1995) having this type of cell junction or whether this organization is restricted to distinct taxa and can be used for phylogenetic discussions. In this respect, it is noteworthy that Lammert (1986, p. 185 and fig. 65 B) reported on junctions flanked by cisternae in *Gnathostomula paradoxa* (Gnathostomulida).

Acknowledgements

We wish to thank Mrs. Sylvia Gubert and Mr. Bernd Baumgart for technical assistence. Financial support was provided by the Akademie der Wissenschaften und der Literatur, Mainz.

Zusammenfassung

Polarisierte Zellen von Plathelminthen werden durch Adhärenz-Verbindungen (Zonulae adhaerentes) und Septum-Verbindungen (Septate junctions) miteinander verbunden. Solche Zell-Zell-Kontakte sind charakteristische

Elemente in der Epidermis, im protonephridialen Kanalsystem, in den Epithelien von Geschlechtsgängen und Genitalorganen wie auch im Darmbereich. Bemerkenswert ist, daß die Septum-Verbindungen zwischen zwei Zellen bei vielen freilebenden Plathelminthen-Arten von flachen jedoch ausgedehnten Zisternen des Endoplasmatischen Reticulums flankiert werden. Eine solche Organisation ist von Vertretern der Acoela, Catenulida und Rhabditophora bekannt und kann für das Grundmuster der Plathelminthes hypothetisiert werden.

Abbreviations

ci	cilium	il	intestinal lumen
cm	circular muscle	mc	male canal
e	epitheliosome	mi	mitochondrion
ec	epithelium cell	pc	protonephridial cell
ecm	extracellular matrix	r	rootlet
ep	epidermis cell	sj	septate junction
ga	genital atrium	sp	spermatozoon
hd	hemidesmosome	tw	erminal web
ic	intestinal cell	za	zonula adhaerens

References

Ax, P. (1995): Das System der Metazoa I. G. Fischer, Stuttgart, Jena, New York, 226 p.
Ehlers, U. (1985): Das phylogenetische System der Plathelminthes. G. Fischer, Stuttgart, New York, 317 p.
– (1989): Duo-gland adhesive systems of *Schizochilus caecus* L'Hardy (Plathelminthes, Kalyptorhynchia). Microfauna Marina **5**, 243–260.
– (1992): Dermonephridia-modified epidermal cells with a probable excretory function in *Paratomella rubra* (Acoela, Plathelminthes). Microfauna Marina **7**, 253–264.
Lammert, V. (1986): Vergleichende Ultrastruktur-Untersuchungen an Gnathostomuliden und die phylogenetische Bewertung ihrer Merkmale. Dissertation der Math.-Nat. Fachbereiche der Universität Göttingen, 218 p.
Nielsen, C. (1995): Animal evolution. Interrelationships of the living phyla. Oxford University Press, Oxford, New York, Tokyo, 467 p.
Rieger, R.M., S. Tyler, J.P.S. Smith III & G.E. Rieger (1991): Platyhelminthes: Turbellaria. In: Microscopic anatomy of invertebrates, Vol. 3 Platyhelminthes and Nemertinea. Eds.: F.W. Harrison & B.J. Bogitsh. Wiley-Liss, New York, 7–140.
Sopott-Ehlers, B. (1997): Submicroscopic anatomy of female gonads in *Ciliopharyngiella intermedia* (Plathelminthes, Rhabdocoela, "Typhloplanoida"). Microfauna Marina **11**, 209–221.
Tyler, S. (1984): Turbellarian platyhelminths. In: Biology of the integument, Vol. 1 Invertebrates. Eds.: J. Berreiter-Hahn, A.G. Matoltsy & K.S. Richards. Springer, Berlin, 112–131.

Dr. Beate Sopott-Ehlers and Prof. Dr. Ulrich Ehlers
II. Zoologisches Institut und Museum der Universität Göttingen
Berliner Straße 28, D-37073 Göttingen

Ultrastructure of protonephridial structures within the Prolecithophora (Plathelminthes)

Ulrich Ehlers and Beate Sopott-Ehlers

Abstract

The protonephridia of *Plagiostomum lemani*, *Pseudostomum quadrioculatum*, and *Cylindrostoma monotrochum* show multiciliated terminal cells with a single filter structure. The cytoplasmic wall of the terminal cell is disintegrated into more or less irregular longitudinal projections of varying diameters. At least several of these projections have a rod-like appearance. The disintegrated wall is strengthened by accumulations of microfilaments in *P. lemani*, by vertical ciliary rootlets and microfilaments in *P. quadrioculatum*, and by individual longitudinal microtubules and microfilaments in *C. monotrochum*. In all species, the distal portions of the terminal cells form closed cytoplasmic tubes surrounded by projections of the first canal cell.

The present findings support the hypothesis that the filter regions in the ground pattern of the Neoophora were less specialized and that stengthenings by microtubules may have evolved convergently within the Prolecithophora and the Rhabdocoela.

A. Introduction

Up to now, kinship relationships of the Prolecithophora within the Neoophora or the Plathelminthes in general are a matter of controversy. For example, ROHDE (1990, 1995) discussed closer relationships with a group of flatworms comprising the Macrostomida, Polycladida, Proseriata, Tricladida, and Neodermata, but not the free-living and symbiotic Rhabdocoela ("Typhloplanoida", Kalyptorhynchia, "Dalyellioida") and the Lecithoepitheliata. On the contrary, EHLERS (1985) favoured the more traditional view (see also KAR-

LING 1974) that the Prolecithophora are part of a monophylum comprising also the Rhabdocoela (including the free-living and symbiotic taxa and the parasitic Neodermata as well) and perhaps also the Seriata. These discrepancies are mainly due to the lack of reliable characteristics: all known light-microscopical features of the Prolecithophora do not allow any precise phylogenetic hypothesis about the monophyly of this group or its sistertaxon relationship and electron microscopical investigations of species of the Prolecithophora are not very numerous.

Today, there is no doubt that the Prolecithophora constitute a monophylum; this hypothesis is substantiated by ultrastructural details of the aciliary male gametes (see EHLERS 1988; unpublished own results of many species). Furthermore, fine structural analyses of species of the Prolecithophora and of the Rhabdocoela with the "Typhloplanoida", Kalyptorhynchia, "Dalyellioida" and the Neodermata show that female gametes (germocytes and vitellocytes) of these species have common characteristics that are quite different from the situations known for female gametes in all other taxa of the Plathelminthes (see *i.a.* GREMIGNI 1997; SOPOTT-EHLERS 1995; 1997a) substantiating the view of the existence of a taxon Eulecithophora (= Prolecithophora + Rhabdocoela including the Neodermata) (see SOPOTT-EHLERS 1997b).

The ultrastructural organization of the protonephridia within the Prolecithophora is less known; EHLERS (1989) published results for the species *Archimonotresis limophila* Meixner, 1938 and WATSON & ROHDE (personal communication) for *Cylindrostoma fingalianum* (Claparède, 1861) and *Allostoma* sp. A better knowledge of nephridial organs in more species will contribute to the present discussions of the phylogenetic position of this taxon, especially to the question whether the Prolecithophora and the Rhabdocoela can be hypothesized as sistertaxa or not.

B. Material and Methods

Several specimens of three species were collected: *Plagiostomum lemani* (Du Plessis, 1874) near Tvaerminne, Baltic Sea (Finland), *Pseudostomum quadrioculatum* (Leukart, 1847) on the east side of the island of Sylt, North Sea (Germany), and *Cylindrostoma monotrochum* (v. Graff, 1882) near Roscoff, Brittany (France).

Fig. 1. *Plagiostomum lemani*. Cross-sections of a terminal cell near the origin of the tuft of cilia (in ▶ A) and the proximal region of the filter (in B). Parts of a neighboured canal cell with its nucleus are visible in A. Small arrows (in B) point to the filter diaphragms. Scale bars in all figures: 0.5 μm.

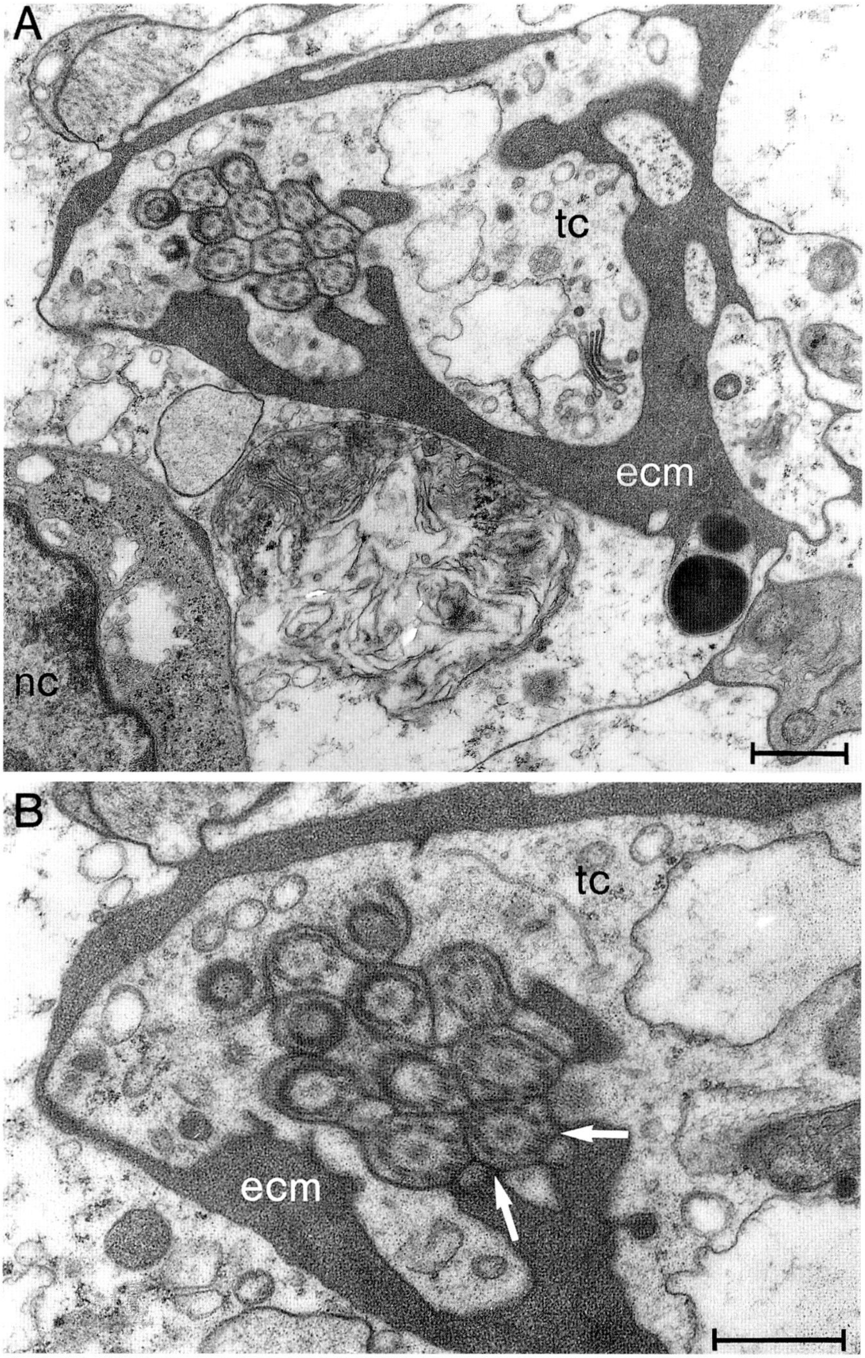
A
tc
ecm
nc
B
tc
ecm

For EM investigations, animals were fixed in 2.5% glutaraldehyde in 0.1 M sodium cacodylate buffer for 2h at 4°C, rinsed in the same buffer, postfixed in 1% OsO$_4$ for 1.5 h, and embedded in Araldite. Series of ultrathin sections stained with uranyl acetate and lead citrate were examined using Zeiss EM 900 and Zeiss EM 10B electron microscopes.

C. Results

The general organization of the protonephridial system of *Plagiostomum lemani* is described by Du Plessis (1874), Dorner (1902) and v. Hofsten (1907) whereas special reports on the systems of *Pseudostomum quadrioculatum* and *Cylindrostoma monotrochum* are lacking. But the older descriptions for other species of the Prolecithophora, e.g. by Böhmig (1890) for *Plagiostomum girardi* (O. Schmidt, 1857), are also valuable informations. In all species, the protonephridia consist of a paired system with several to many irregularly distributed fine tubules and many terminal areas.

At the EM-level, these terminal areas with the filter structures show distinct variations between the three species *P. lemani, P. quadrioculatum* and *C. monotrochum* and in comparison with *Archimonotresis limophila* as described by Ehlers (1989). Therefore, this paper focusses on the fine structures of the filter regions and the adjacent areas of the protonephridia.

I. *Plagiostomum lemani* (Figs. 1–8)

The terminal cells of this species are relatively large and ramified cells with many cytoplasmic processes. The cytoplasm is poor in organelles (Figs. 1–6), especially in comparison with the following outleading canal cells. Apparently, each terminal cell shows only one filter. The nucleus of a terminal cell takes a position proximally but distantly to the filter and mitochondria are mainly assembled in this perinuclear cell region.

At the level of the filter, Golgi complexes and vacuoles of different sizes were found and here the terminal cell starts to disintegrate into several cytoplasmic lobes (Fig. 1A). An expansive extracellular matrix (ECM) surrounds the main cytoplasmic portion of the terminal cell and its lobes of varying sizes (Figs. 1–4). Several of these lobes project deeply into the ECM and mitochondria can be seen in such cell projections (Figs. 3A, B; 4; 5). Other cytoplasmic parts are differentiated into rod-like longitudinal projections of different and varying diameters (Figs. 1B; 2A, B; 3A; 4). These rod-like structures are strengthened by accumulations of actin-like microfilaments and partly surro-

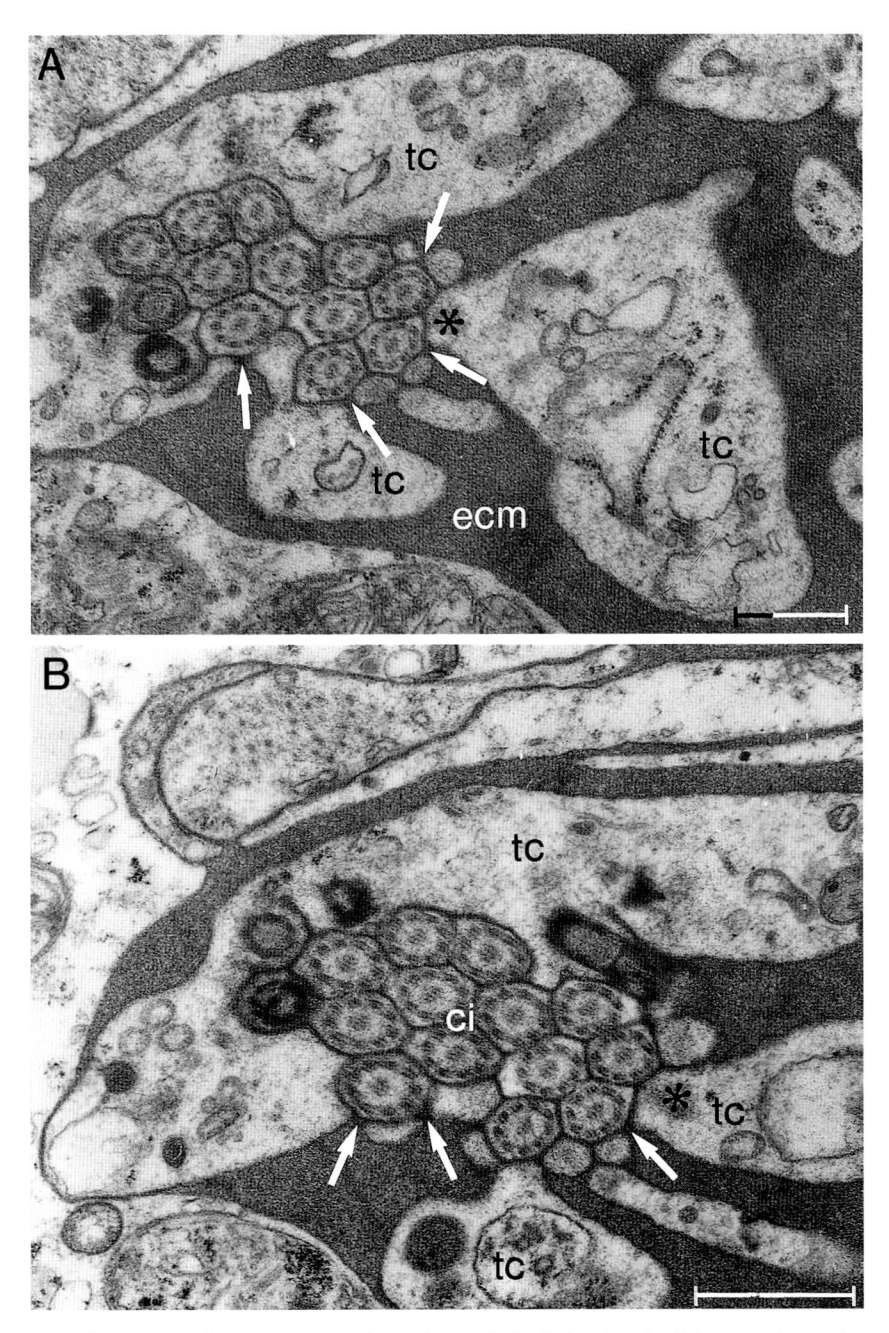

Fig. 2. *Plagiostomum lemani*. Cross-sections of a terminal cell showing the disintegrated cytoplasm at the level of the proximal filter region. Small arrows (in A and B) point to the filter diaphragms and asterisks mark accumulations of microfilaments.

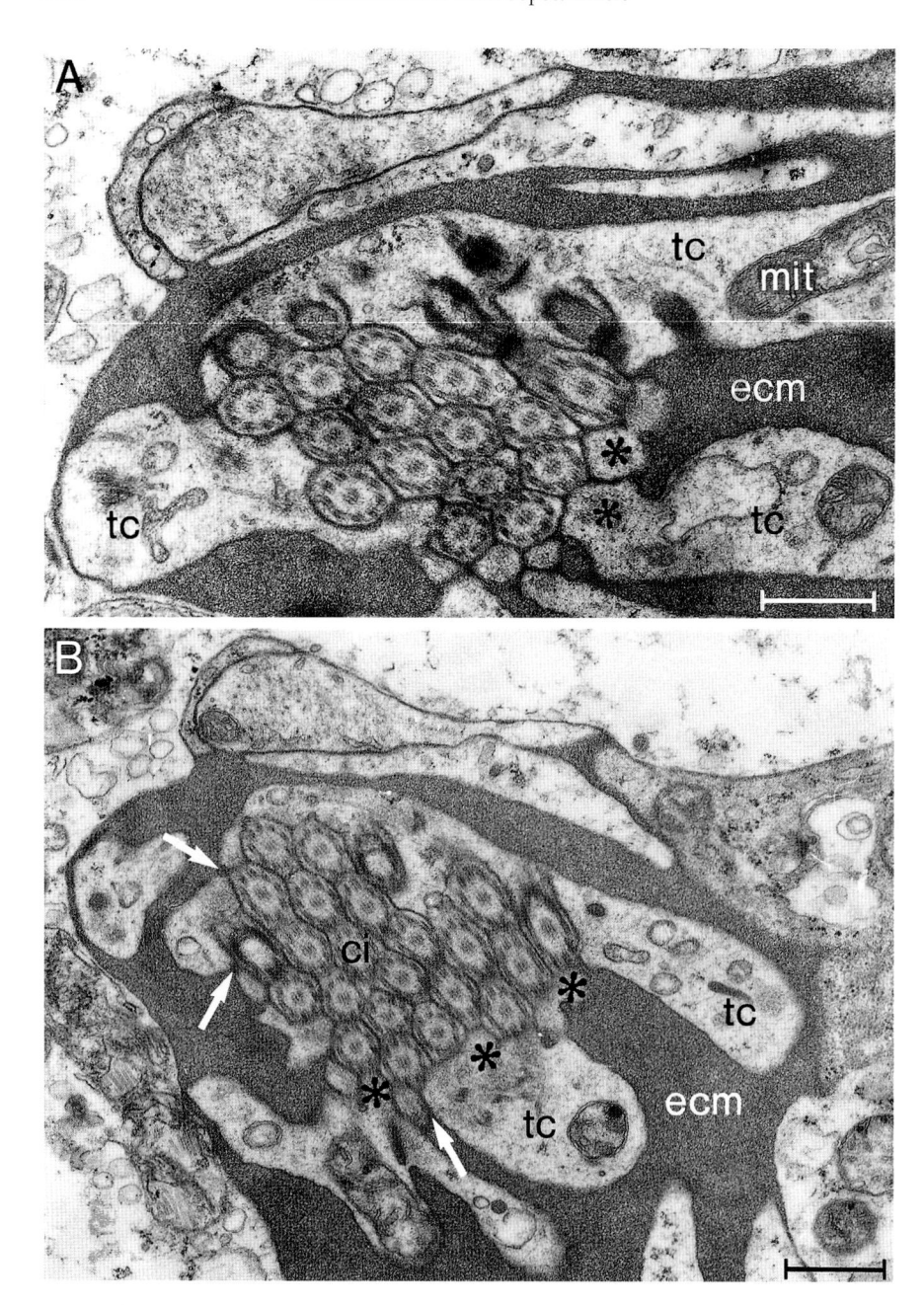

Fig. 3. *Plagiostomum lemani*. Cross-sections through the filter of a terminal cell. The cilia are surrounded by rod-like and by more irregular cytoplasmic projections strengthened by microfilaments (asterisks in A and B). Small arrows (in A and B) point to filter diaphragms.

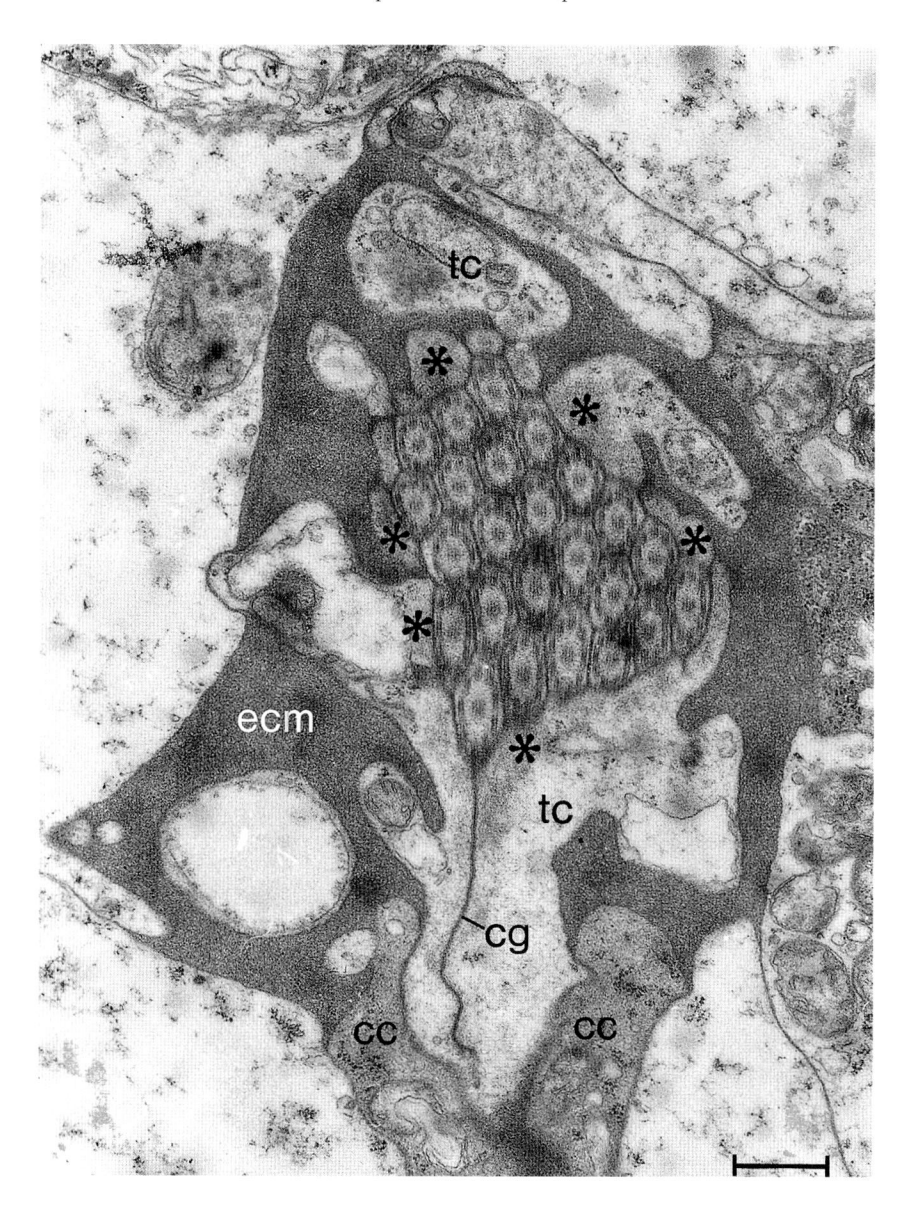

Fig. 4. *Plagiostomum lemani*. Cross-section near the distal end of the filter region showing lobed cytoplasmic areas in the upper half of the figure and fused cytoplasmic areas and a cell gap in the lower half. Asterisks mark accumulations of microfilaments within the cytoplasm of the terminal cell. Cytoplasmic projections of the first canal cell start to envelope the terminal cell.

und a tuft of cilia marking the beginning of the outleading protonephridial canal. Microfilaments are also present in the more compact periciliary cytoplasmic areas that extend between the rod-like differentiations (Figs. 3B; 4; 5). Only a single row of cytoplasmic rods exists around the tuft of cilia (Figs. 2B; 3A; 4). Any bundles of microtubules running along the length of the rods have not been observed. All the rod-like differentiations and the adjacent more compact cell areas are connected to each other by a thin layer of electron-dense extracellular material that forms the filter diaphragm around the tuft of cilia. Up to 40 cilia have been counted in a terminal cell. Each cilium shows a short vertical rootlet (Fig. 3A).

More distally, the rod-like differentiations fuse with each other and with the adjacent compact cytoplasmic areas which flank the tuft of cilia (Figs. 4; 5). A longitudinal cell gap between parts of the contiguous membranes is present (Figs. 4; 6A–C) but this gap does not reach the periphery of the cell. Thus, the distal region of the terminal cell is formed as a closed cytoplasmic tube but not as a cuff. In general, the distal region continues as long as ciliary axonemes of the tuft are present. Finally, the cell terminates as a small cytoplasmic tube (Fig. 6C).

The first canal cell can easily be distinguished from the terminal cell because of its more electron-dense cytoplasm. Cytoplasmic projections of this canal cell are already recognizable in a short distance behind the filter (Fig. 4). These projections envelope the tapering terminal cell. Sometimes, the wall of the tube-like terminal cell is interrupted (Fig. 5) so that the canal cell comes into contact with the cilia of the terminal cell. Finally, the canal cell surrounds most of the tapered region of the terminal cell (Fig. 6B, C).

The canal cells bear local tufts of cilia which display strong longitudinal rootlets (Figs. 7A; 8B). Nuclei of these cells lie in some distance to the canal lumen (Fig. 7A). The canal cells can be partly in form of closed cytoplasmic tubes and partly of a sheet-like appearance wrapped around the canal like a cuff (Figs. 7B; 8A). Often, the cytoplasm is splitted by a number of cell gaps. Such a variable and irregular arrangement is also present when different tubules of the canal system join each other (Fig. 8B).

Fig. 5. *Plagiostomum lemani*. Oblique section of the more distal region of a terminal cell still showing a bundle of cilia. The distal end of the filter in the upper part of the figure is still surrounded by rod-like cytoplasmic areas (asterisks mark microfilaments) and by the extracellular matrix. Projections of the first canal cell envelope the terminal cell in the middle and the lower part of the figure and here expansive extracellular matrices are absent.

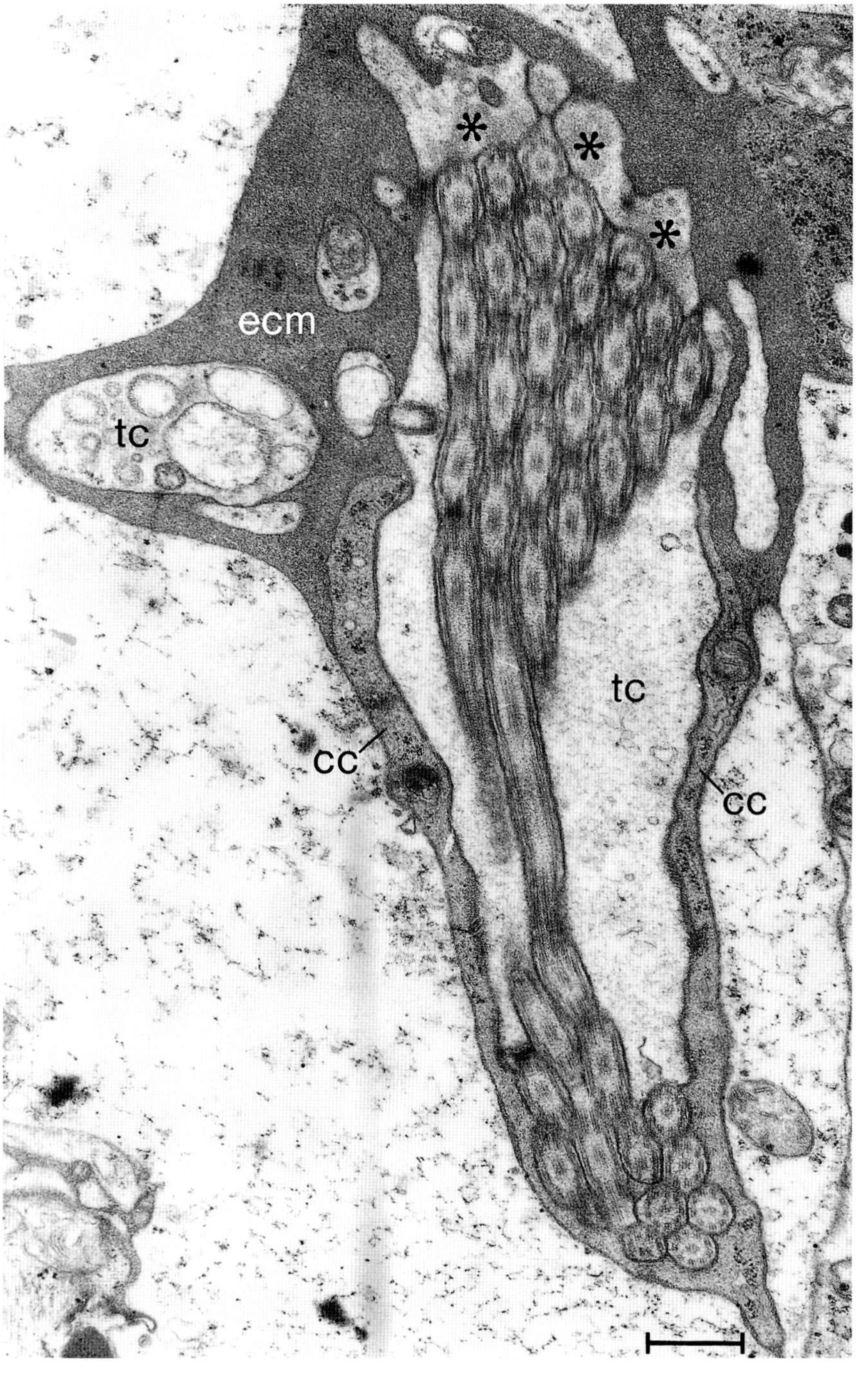
*
*
*
ecm
tc
tc
cc
cc

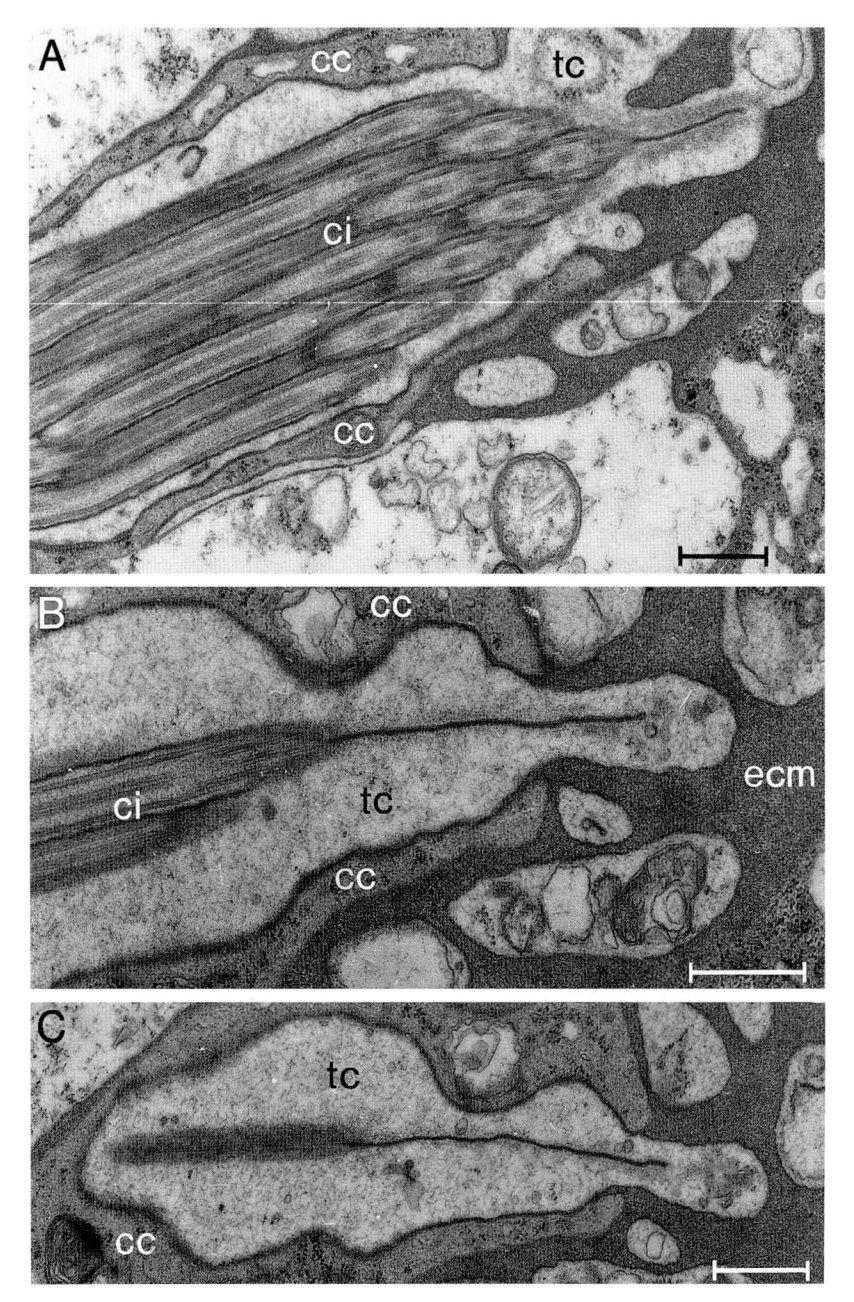

Fig. 6. *Plagiostomum lemani*. Serial oblique sections through the tube-like distal portion of a terminal cell showing the decreasing number of ciliary axonemes within the outleading canal and the increasing enveloping by the first canal cell.

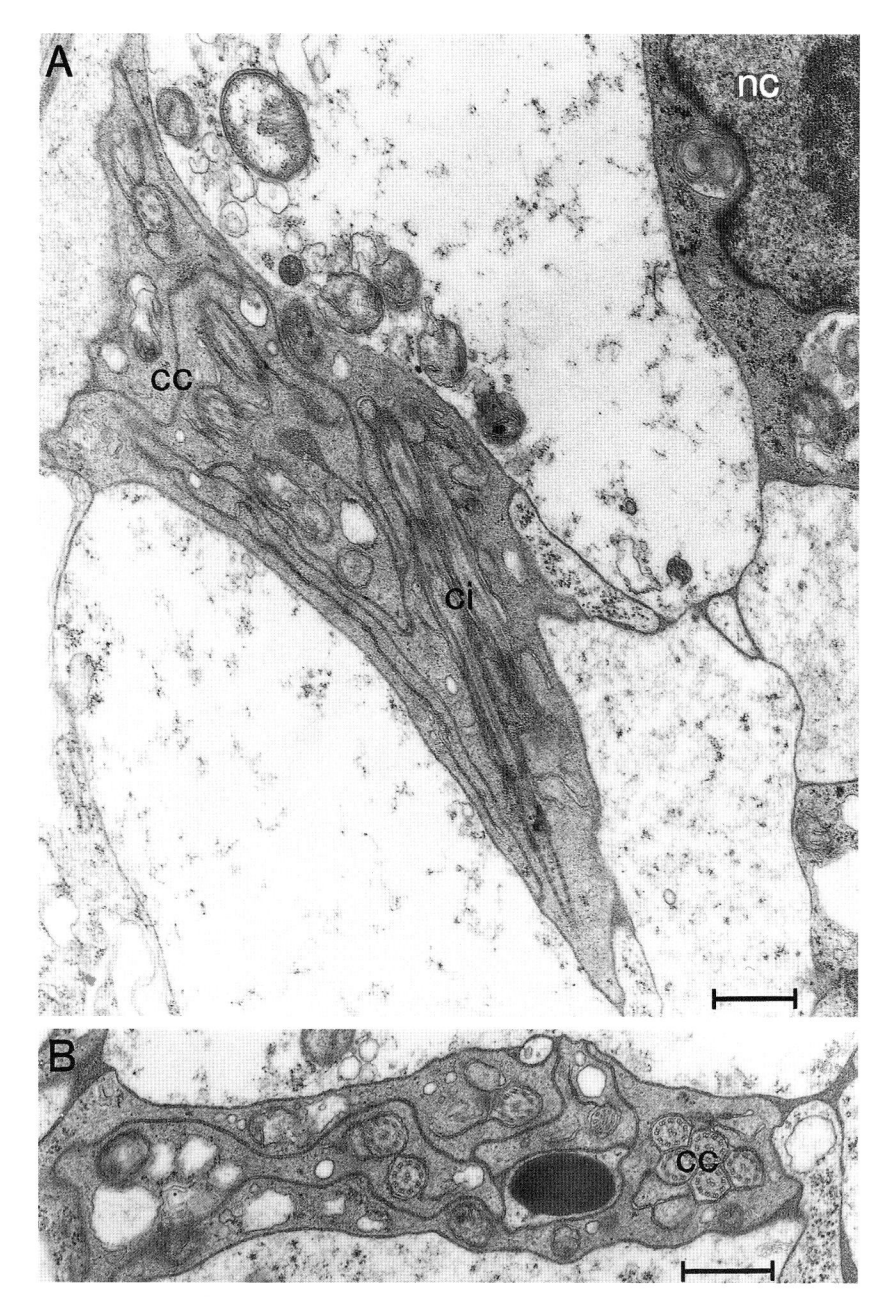

Fig. 7. *Plagiostomum lemani*. Canal cells. In A a portion of a tubule with a local tuft of few cilia and a separated nucleus of this canal cell. In B a cross-section of a tubule with cilia of different local tufts; the cytoplasm is splitted by a number of cell gaps.

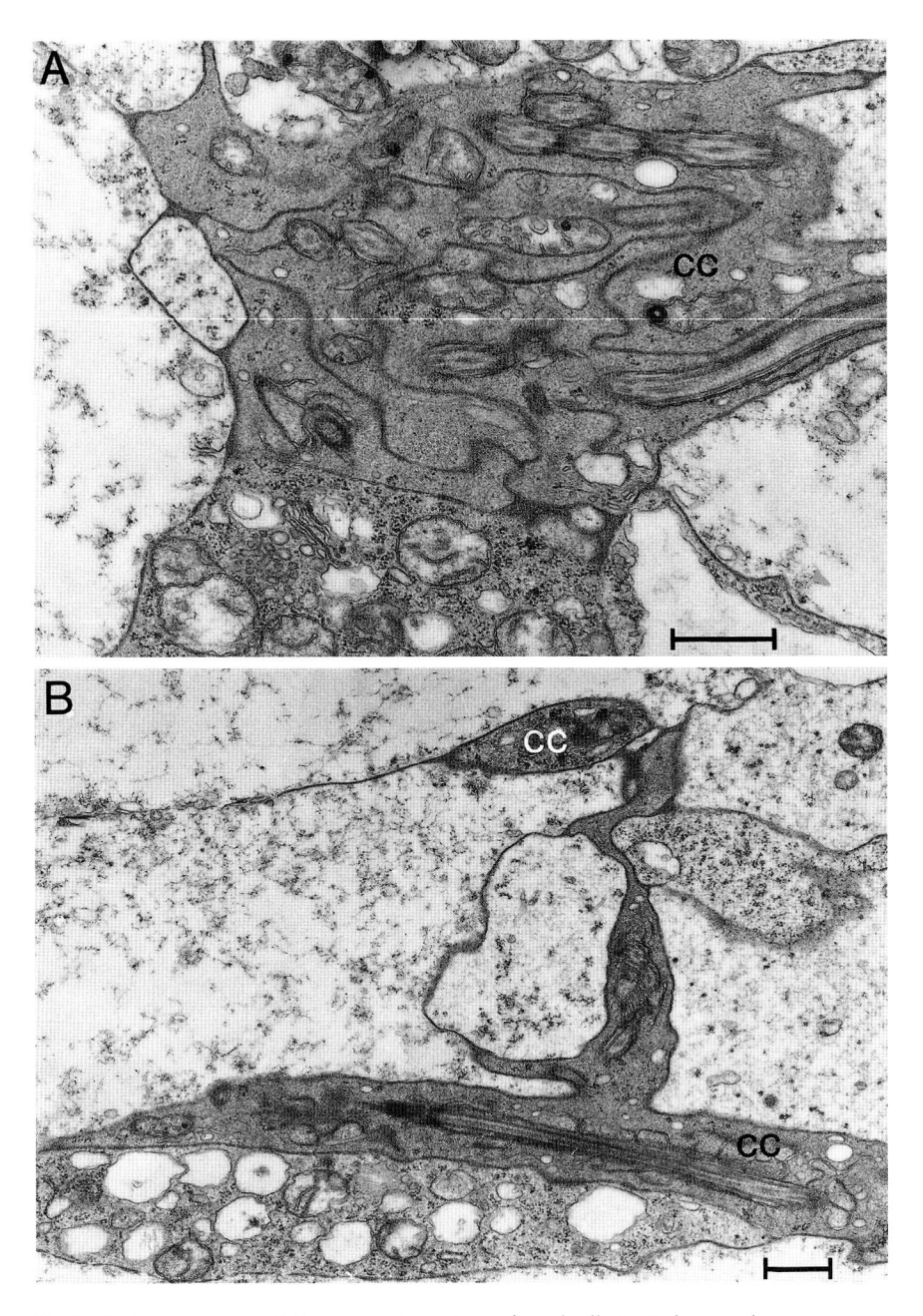

Fig. 8. *Plagiostomum lemani*. Anastomosing regions of canal cells. In A the complex arrangement of cilia and plasma membranes in such a region. In B a connection between a longitudinally sectioned tubule and a cross-sectioned tubule.

II. *Pseudostomum quadrioculatum* (Figs. 9–13)

The proximal regions of the protonephridial system of. *P. quadrioculatum* correspond mostly with those of *P. lemani* as described above. In the following, noticable differences are dealt with.

Each of the terminal cells displays only one filter per cell. In contrast to *P. lemani*, mitochondria were regularly seen in the electron-lucent cytoplasm of the cell near the filter region (Fig. 9A, B). Numerous cilia (N > 60) form a terminal tuft (Fig. 12). These cilia are arranged in regular rows, but all the cilia of a single row do not originate at the same level. In the most proximal region of a tuft, the few axonemes of the most central cilia of the middle row or rows are recognizable (Fig. 9A–C). These cilia show a strong vertical rootlet like all other cilia of a tuft (Fig. 9D). The axonemes of these first projecting cilia are surrounded by the rootlets of the cilia which will originate more distantly. This picture can be seen in all cross-sections of the proximal half of a tuft: an increasing number of ciliary axonemes surrounded by basal bodies or rootlets of the more peripherally and distally located cilia (Figs. 10; 11; 12A). In cross-sections of the distal half of a tuft, basal bodies and rootlets are no longer found, but the axonemes of all the cilia are present (Fig. 12B).

Part of the peripheral cytoplasm of the terminal cell around the tuft is disintegrated into longitudinal rod-like structures. Each of these rod-like differentiations is strengthened by a single vertical (longitudinal) rootlet (Figs. 9C, D; 10; 11A) and often by additional accumulations of actin-like microfilaments. Peripherally, these rods are surrounded by portions of the filamentous extracellular matrix (Fig. 11A). A fine but distinct layer of electron-dense extracellular material covers the small slits between adjacent membranes of the rod-like cytoplasmic projections (Figs. 10; 11) and forms the filter diaphragm.

More distally, the terminal cell is in form of a closed and solid tube (Fig. 13). This tube is present as long as any cilia of the terminal tuft are recognizable (Fig. 13 D, E). The tapering distal region of the terminal cell is enveloped by projections of the first canal cell the cytoplasm of which is more electron-dense.

III. *Cylindrostoma monotrochum* (Figs. 14–15)

As in the other two species dealt with in this paper, each terminal cell possesses only one filter. The nucleus of a terminal cell has been found close to the filter region (Fig. 14A).

Such a filter region is formed by a bundle of less than 20 cilia surrounded by the disintegrated cytoplasm of the terminal cell (Figs. 14B; 15). The cilia are not arranged to a single tuft and may not form a functional unit as can be seen

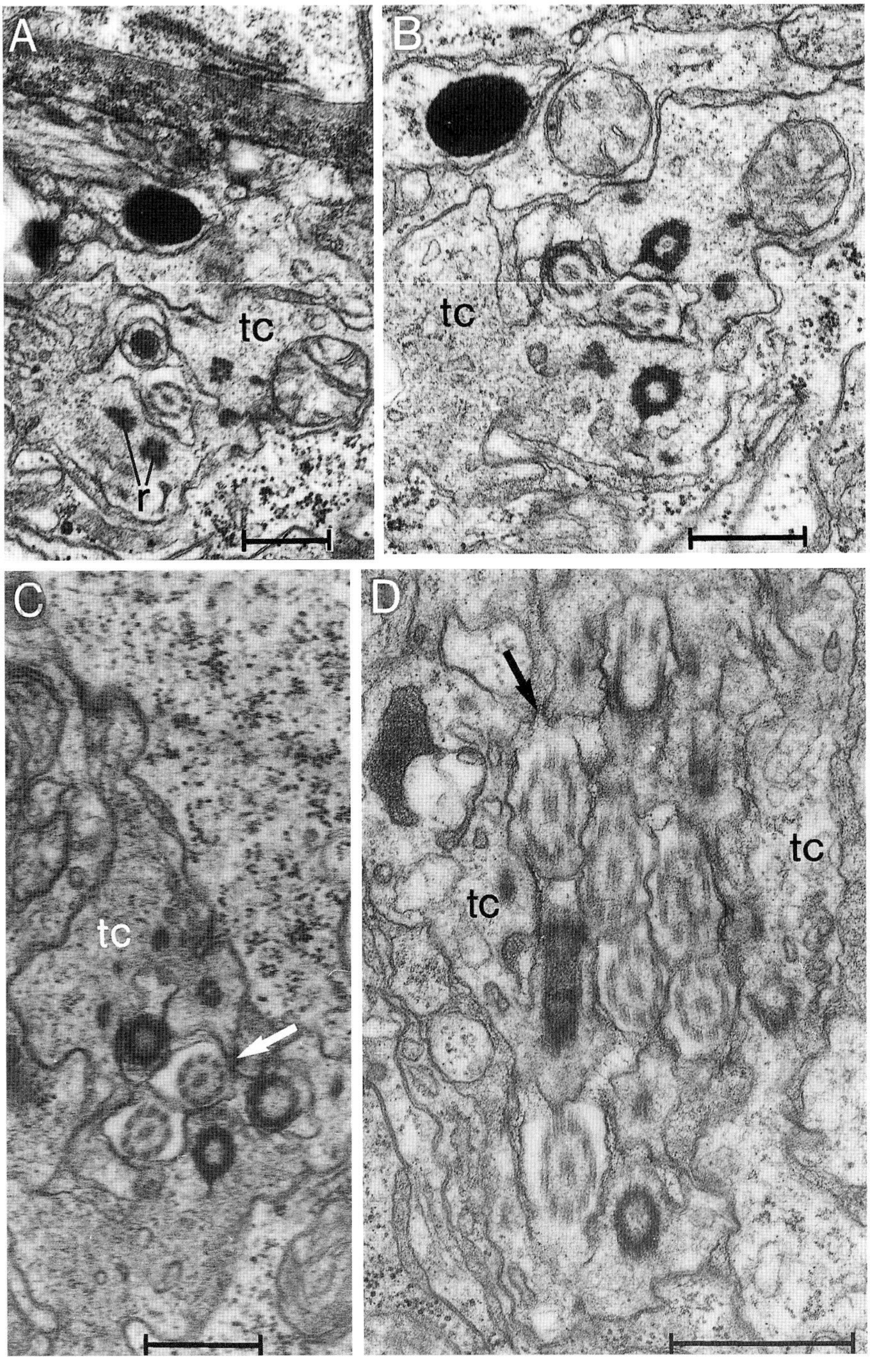
A
B
C
D
tc
tc
tc
tc
tc
r

by different positions of the microtubules in neighboured axonemes. The peripheral cytoplasmic projections of the terminal cell are strengthened by longitudinal microtubules and by accumulations of microfilaments. The microtubules are not concentrated into bundles, but are isolated from each other. There is only one row of longitudinal microtubules around the cilia of the filter. Small rod-like cytoplasmic projections show only one microtubule whereas larger cytoplasmic areas display several microtubules running in quite constant distances to each other along the length of the filter (Figs. 14B; 15). A fine layer of electron-dense extracellular materials covers the small slits between the cytoplasmic projections and forms the filter diaphragm of the protonephridium.

More distally, all the projections fuse with each other thus forming a closed cytoplasmic tube around the distal portions of the ciliary axonemes and the row of individual microtubules is no longer recognizable within the cytoplasm of this tube.

D. Discussion

The terminal areas with the filtration structures of *Plagiostomum lemani*, *Pseudostomum quadrioculatum* and *Cylindrostoma monotrochum* as described in the present paper correspond with the terminal protonephridial regions of *Archimonotresis limophila* (see EHLERS 1989) and *Cylindrostoma fingalianum* and *Allostoma* sp. (WATSON & ROHDE unpublished) in several aspects and these correspondences can be hypothesized as basic characteristics of the monophylum Prolecithophora:

(1) A terminal area is formed by a single cell, the terminal cell.

(2) Only one filter (filter region) exists in this terminal cell and the nucleus of the cell takes a position nearby or in some distance to this filter.

(3) The terminal cell is multiciliated: many (up to more than 60) cilia are anchored in the cell and form the so-called 'terminal flame'. As stressed by WATSON & ROHDE, the cilia of this 'flame' arise more or less continuously along the proximal regions of the filter region. Each cilium is provided with a vertical (longitudinal) rootlet. Prominent microvilli (so-called 'internal leptotriches') around the basal portions of the ciliary axonemes are not present.

◀ Fig. 9. *Pseudostomum quadrioculatum*. Serial cross-sections of proximal regions of a terminal cell. In A the two most proximally originating cilia surrounded by rootlets of cilia which originate more distally. The basal bodies of two (in B) and then of three (in C) of these additional cilia are visible. In C more additional cilia are sectioned. Small arrows point to the filter diaphragms.

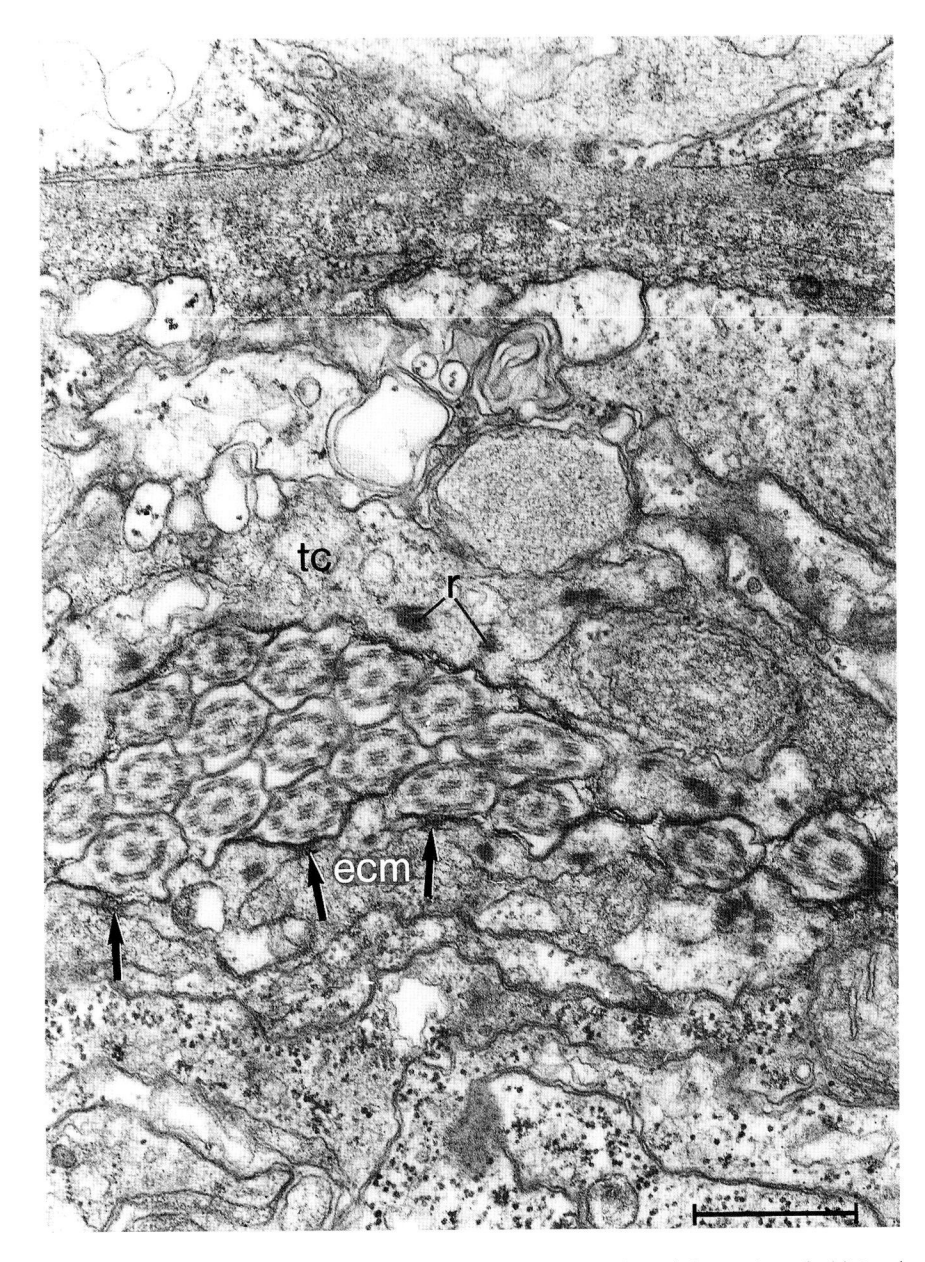

Fig. 10. *Pseudostomum quadrioculatum*. Cross-section with 18 cilia and the rootlets of additional cilia. Small arrows point to filter diaphragms.

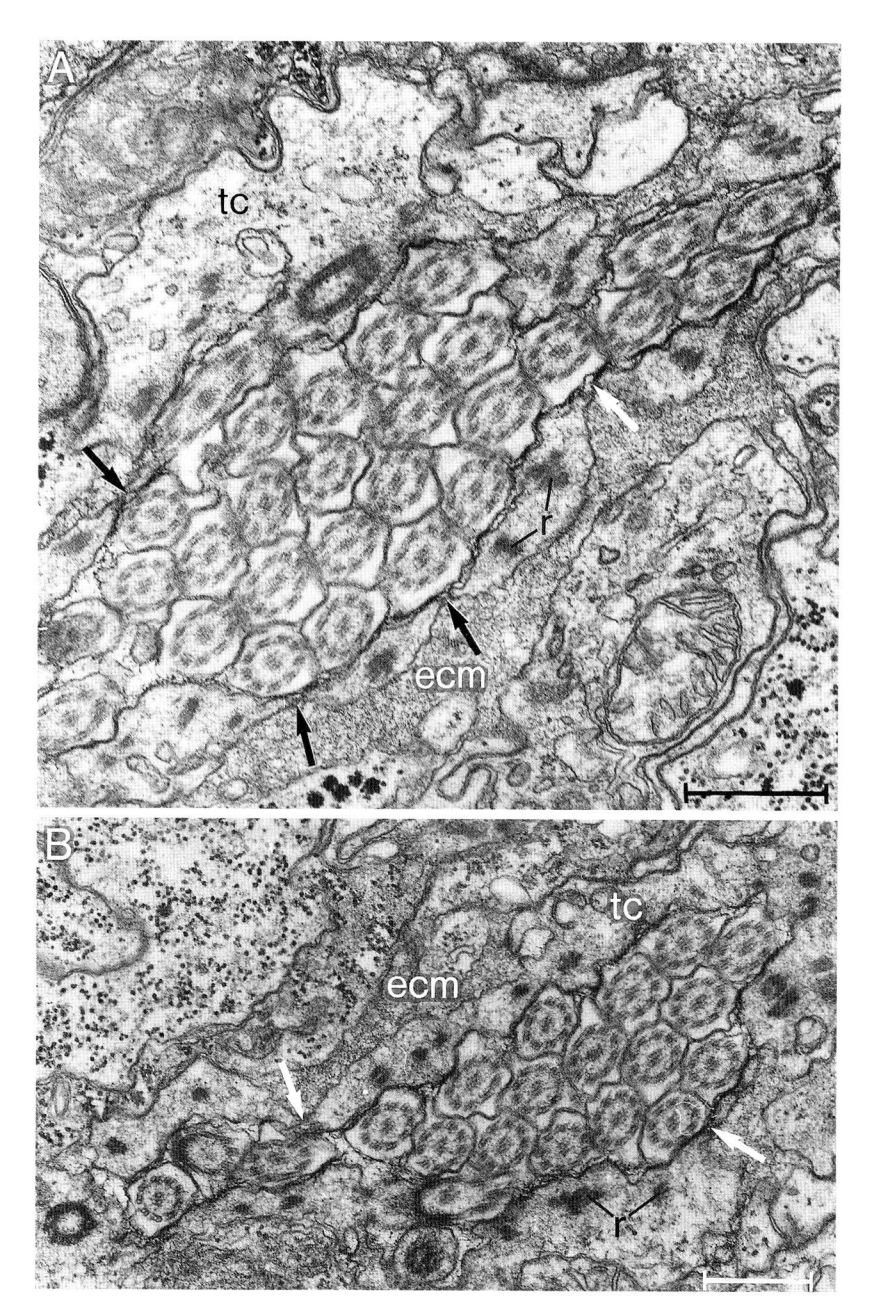

Fig. 11. *Pseudostomum quadrioculatum*. The periciliary cytoplasm of the terminal cell forms rod-like differentiations stengthened by ciliary rootlets and microfilaments. Small arrows (in A and B) point to filter diaphragms.

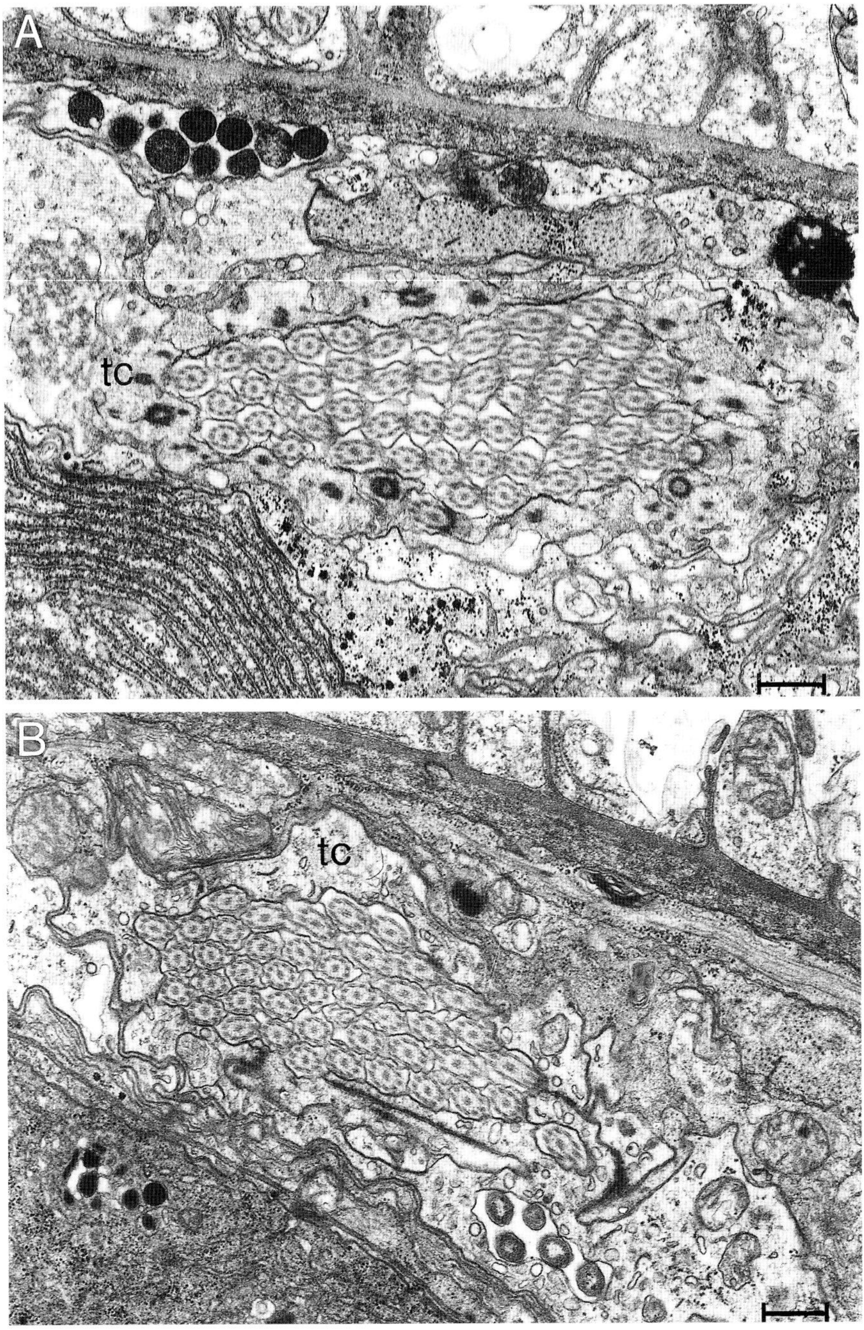
A
tc
B
tc

(4) In the filter region, the cytoplasmic wall of the terminal cell is fenestrated by many slits with variable orientations or disintegrated into more or less irregular projections of varying diameters. At least several of these predominantly longitudinal projections (or the massive parts of the wall between the fenestrations) have a rod-like appearance. Only a single row of cytoplasmic projections of the subdivided wall exists around the tuft of cilia and the slits between adjacing cytoplasmic differentiations are covered by distinct extracellular materials which form the filter diaphragms.

(5) The first canal cell is not involved in the construction of the filter. This cell envelopes the tapering distal portions of the terminal cell which is formed as a closed cytoplasmic tube but not as a cuff.

Differences between the species of the Prolecithophora investigated up to now mainly exist with respect to the organization of the cytoplasmic wall in the filter region. The longitudinal parts (projections) of the more or less fragmented wall are strengthened by accumulations of microfilaments in *P. lemani*, by vertical rootlets of cilia which originate more distally (and by additional accumulations of microfilaments) in *P. quadrioculatum* and perhaps also in *Allostoma* sp., by individual longitudinal microtubules (and additional accumulations of microfilaments) in *C. monotrochum* and probably also in *C. fingalianum* and by bundles of longitudinal microtubules in *A. limophila*. Whereas the former five species display more or less irregularly shaped but also rod-like cytoplasmic projections, *A. limophila* shows a row of rather well developed rods – a prerequisite to be strengthened by longitudinal bundles of microtubules.

Based on these differences, alternative hypotheses about the organization of the filter wall in the ground pattern of the monophylum Prolecithophora can be made.

Hypothesis A: The cytoplasmic wall of the terminal cell is fenestrated by slits orientated in various directions. This hypothesis implies that other organizations such as the existence of regular rods strengthened by bundles of microtubules as in *Archimonotresis limophila* have evolved secondarily within the Prolecithophora. Then, the organization known from many taxa of the free-living and symbiotic Rhabdocoela with rods strengthened by bundles of microtubules would be a convergence with the situation present in *A. limophila*.

◀ Fig. 12. *Pseudostomum quadrioculatum*. Cross-sections near the distal end of the filter region (in A) still showing ciliary rootlets in rod-like cytoplasmic projections of the terminal cell and (in B) of the post-filtration region lacking any ciliary rootlets.

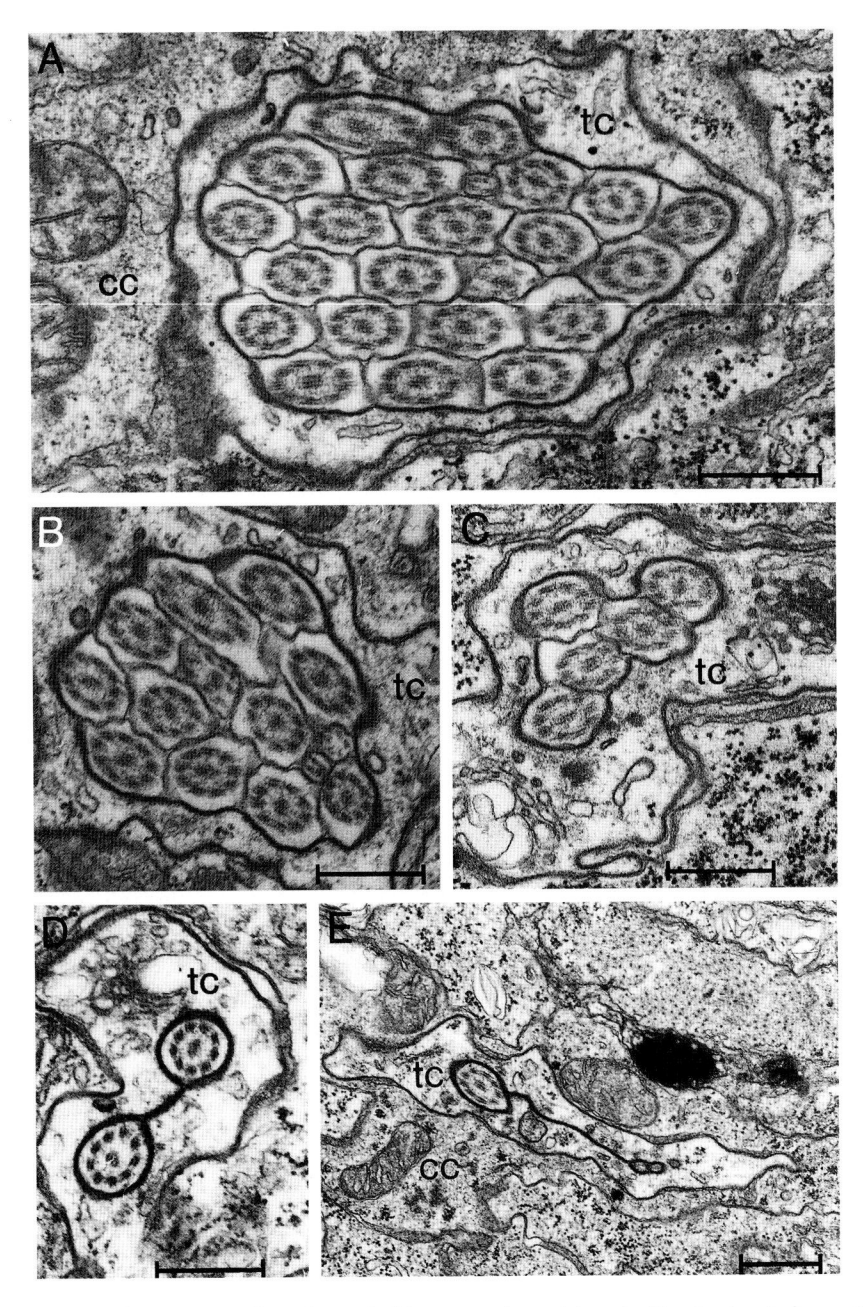

Fig. 13. *Pseudostomum quadrioculatum*. Serial cross-sections of the tube-like distal end of a terminal cell showing a decreasing number of ciliary axonemes from A to E and the proximal portion of the first canal cell enveloping the tapering area of the terminal cell.

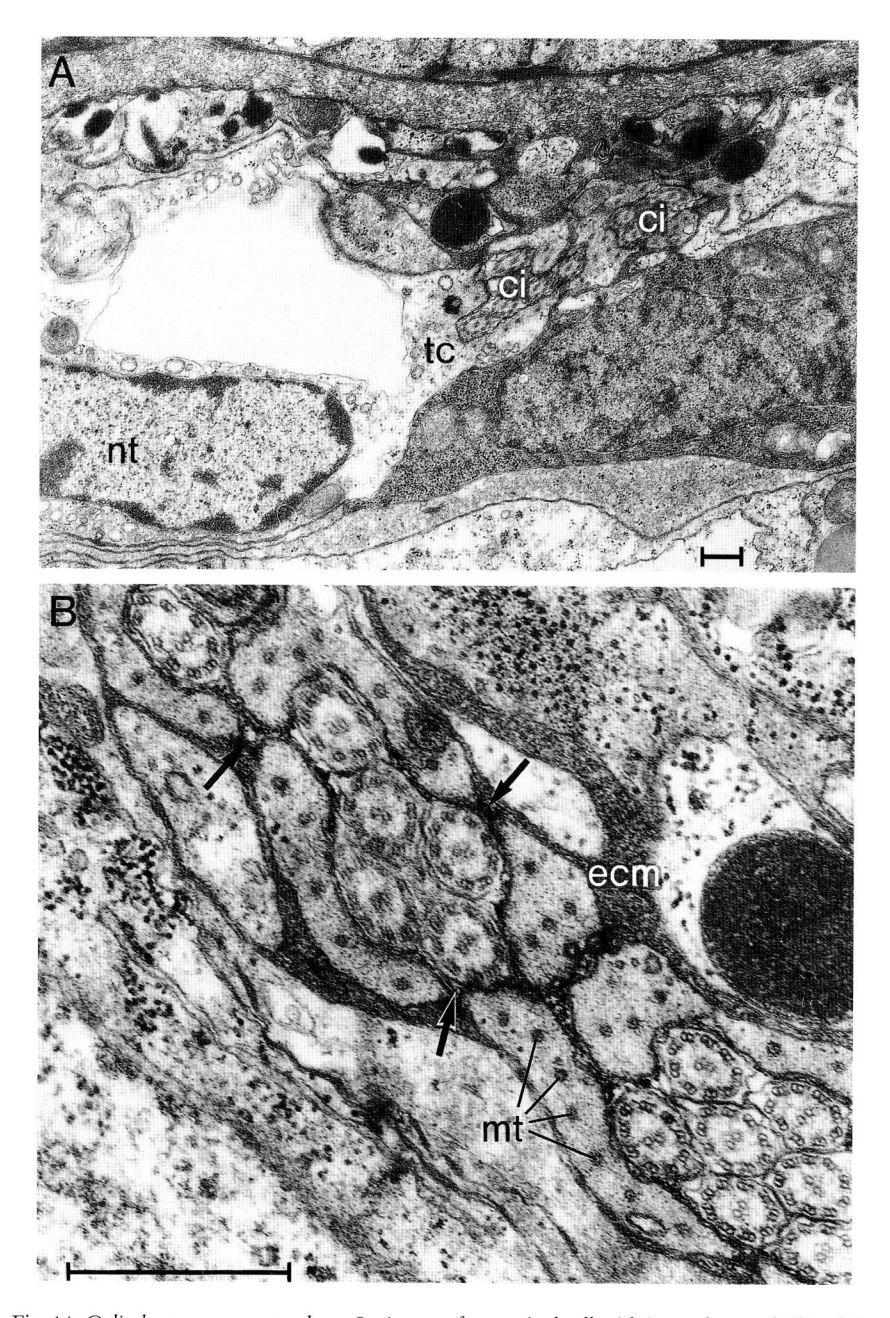

Fig. 14. *Cylindrostoma monotrochum*. In A part of a terminal cell with its nucleus and cilia of the filter region. In B a cross-section of the filter with its cilia surrounded by cytoplasmic projections which are strengthened by microtubules. Small arrows (in B) point to filter diaphragms.

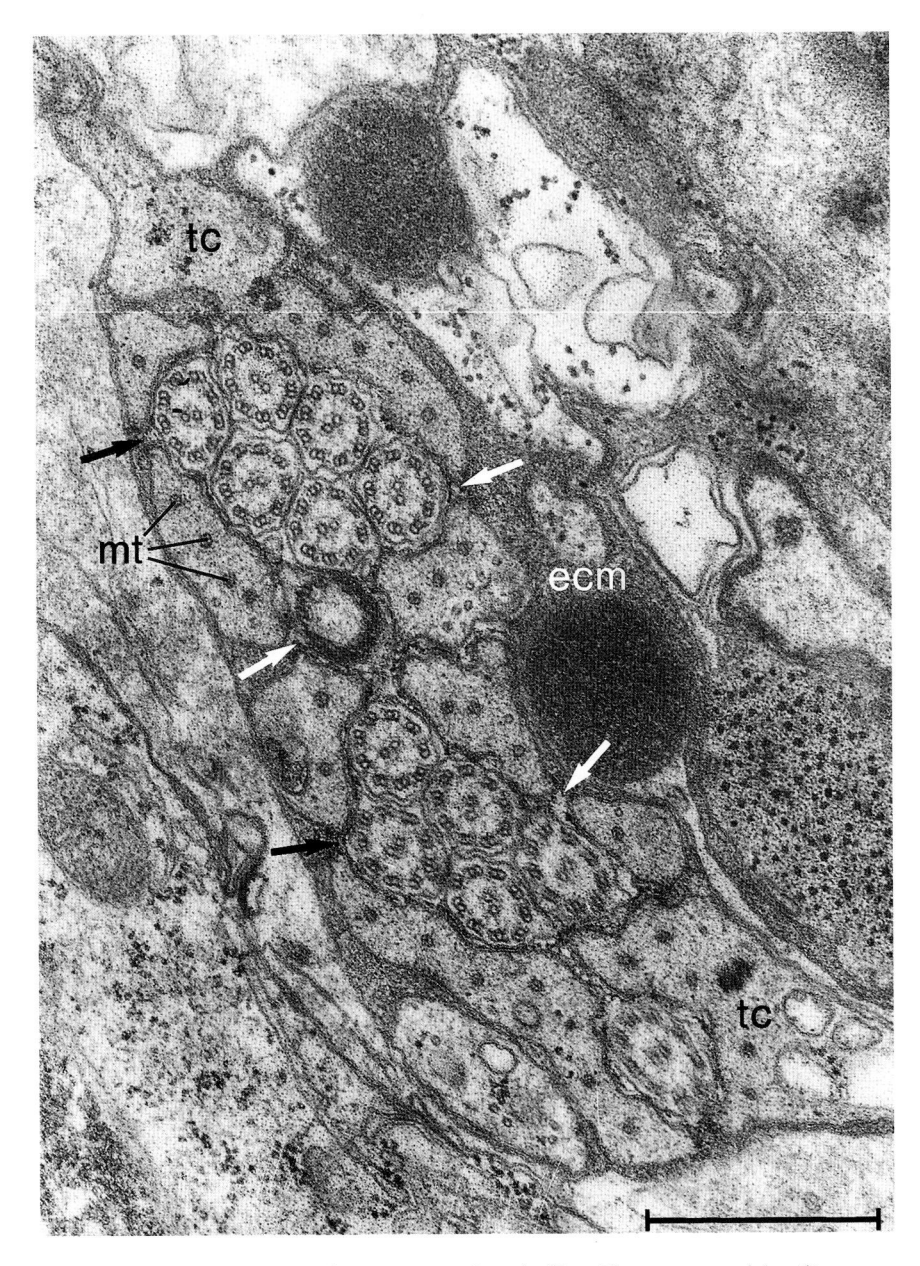

Fig. 15. *Cylindrostoma monotrochum*. Cross-section of a filter. The axonemes of the cilia are surrounded by a single layer of cytoplasmic projections (with longitudinal microtubules) of varying sizes. Small arrows point to filter diaphragms.

Hypothesis B: Rods with longitudinal bundles of microtubules mark the filter area. Then, this organization could be discussed (see ROHDE 1991) as a basic feature for a monophylum comprising more taxa like the Prorhynchidae-Lecithoepitheliata and many free-living and symbiotic species of the Rhabdocoela. All species of the Neodermata (which possess a weir and not a filter) and several other species like *Urastoma cyprinae* (see ROHDE et al. 1990) or *Kronborgia isopodicola* (see WATSON et al. 1992) would not belong to this monophylum – or would have secondarily modified filter regions in which the microtubules have been lost.

Hypothesis A is the more parsimonious one and is not contradicted by other known morphological characteristics like the fine structure of the female gametes or of the photoreceptors (see SOPOTT-EHLERS 1996). Apparently, the filter region in the ground pattern of the Neoophora was less specialized. Distinct more complicated organizations like the strengthenings of the filter walls by ciliary rootlets, by individual microtubules or by bundles of microtubules have been achieved secondarily and more than once within the Neoophora – like the formations of two-cell-weirs or weir-like organs. Such a hypothesis is comparable with that one expressed by LUMBSCH et al. (1995) on the evolution of more than one filter in a terminal cell as in many species of the free-living and symbiotic Rhabdocoela. Probably, special 'types' of terminal areas, of filter regions or of weirs are of value for understanding phylogenetic kinships at 'lower systematic levels', e.g. within the Prolecithophora and within the Rhabdocoela. But such phylogenetic hypotheses must be based on all known morphological characteristics.

The known facts on the organization of the protonephridia do not contradict the hypothesis of the existence of a taxon Eulecithophora (= Prolecithophora + Rhabdocoela including the Neodermata).

Acknowledgements

Financial support was provided by the Akademie der Wissenschaften und der Literatur Mainz. We thank Mrs. E. Hildenhagen-Brüggemann, Mrs. K. Lotz and Mrs. S. Gubert for valuable technical assistences and Mr. B. Baumgart for his help with the figures.

Zusammenfassung

Plagiostomum lemani, Pseudostomum quadrioculatum und *Cylindrostoma monotrochum* besitzen multiciliäre Terminalzellen mit einem einzigen Filterbereich. Dieser Bereich besteht aus dem zu Einzelelementen aufgelösten Cytoplasmazylinder der Terminalzelle. Die Einzelbereiche sind mehr oder weniger unregelmäßig geformt, z.T. aber auch stabförmig. Verfestigt werden sie durch Ansammlungen von Mikrofilamenten (bei *P. lemani*), durch die Vertikalwurzeln von Cilien und durch Mikrofilamente (bei *P. quadrioculatum*) oder durch einzelne Längsmikrotubuli und durch Mikrofilamente (bei *C. monotrochum*). Die distalen Bereiche der Terminalzellen aller drei Arten stellen geschlossene Cytoplasmarohre dar, umgeben von Ausläufern der angrenzenden ersten Kanalzelle.

Die derzeit bekannten Organisationsstrukturen stützen die Hypothese, daß wenig spezialisierte protonephridiale Filterbereiche in das Grundmuster der Neoophora einzusetzen sind und daß es konvergent zu Versteifungen durch Mikrotubuli innerhalb der Prolecithophora und der Rhabdocoela gekommen ist.

Abbreviations

cc	canal cell	mt	microtubules
cg	cell gap	nc	nucleus of canal cell
ci	cilium	nt	nucleus of terminal cell
ecm	extracellular matrix	r	vertical rootlet of cilium
mit	mitochondrion	tc	terminal cell

References

BÖHMIG, L. (1890): Untersuchungen über rhabdocöle Turbellarien. II. Plagiostomina und Cylindrostomina Graff. Z. wiss. Zool. **51**, 167–313.

DORNER, G. (1902): Darstellung der Turbellarienfauna der Binnengewässer Ostpreussens. Schr. phys.-ökon. Ges. Königsberg i. Pr. **63**, 1–58.

EHLERS, U. (1985): Das phylogenetische System der Plathelminthen. G. Fischer, Stuttgart, New York, 317 p.

– (1988): The Prolecithophora – a monophyletic taxon of the Plathelminthes? Fortschritte der Zoologie/Progress in Zoology **36**, 359–365.

– (1989): The protonephridium of *Archimonotresis limophila* Meixner (Plathelminthes, Prolecithophora). Microfauna Marina **5**, 261–275.

GREMIGNI, V. (1997): The evolution of the female gonad in Platyhelminthes-Turbellaria: ultrastructural investigations. Invert. Reprod. Develop. **31**, 325–330.

v. HOFSTEN, N. (1907): Zur Kenntnis des *Plagiostomum lemani* (Forel & du Plessis). Zool. stud. tillagn. T. Tullberg, Uppsala, 91–132.

KARLING, T.G. (1974): On the anatomy and affinities of the turbellarian orders. In: RISER, N.W. & M.P. MORSE (eds.): Biology of the Turbellaria. McGraw-Hill, New York, pp. 1–16.
LUMBSCH, M., U. EHLERS & B. SOPOTT-EHLERS (1995): Proximal regions of the protonephridial system in *Pseudograffilla arenicola* (Plathelminthes, Rhabdocoela): ultrastructural observations. Microfauna Marina **10**, 67–78.
du PLESSIS, G. (1874): Turbellariés limicoles. Bull. Soc. Vaudoise Sc. nat. **13**, 114–124.
ROHDE, K. (1990): Phylogeny of Platyhelminthes, with special reference to parasitic groups. Int. J. Parasitol. **20**, 979–1007.
– (1991): The evolution of protonephridia of the Platyhelminthes. Hydrobiologia **227**, 315–321.
– (1995): Aspects of the phylogeny of Platyhelminthes based on 18S ribosomal DNA and protonephridial ultrastructure. Hydrobiologia **305**, 27–35.
ROHDE, K., N. NOURY-SRAIRI, N. WATSON, J.-L. JUSTINE & L. EUZET (1990): Ultrastructure of flame bulbs of *Urastoma cyprinae* (Platyhelminthes, 'Prolecithophora', Urastomidae). Acta zool. (Stockh.) **71**, 211–216.
SOPOTT-EHLERS, B. (1995): Fine structure of vitellaria and germaria in *Polystyliphora filum* (Plathelminthes, Proseriata). Microfauna Marina **10**, 159–171.
– (1996): First evidence of mitochondrial lensing in two species of the "Typhloplanoida" (Plathelminthes, Rhabdocoela): phylogenetic implications. Zoomorphology **116**, 95–101.
– (1997a): Submicroscopic anatomy of female gonads in *Ciliopharyngiella intermedia* (Plathelminthes, Rhabdocoela, "Typhloplanoida"). Microfauna Marina **11**, 209–221.
– (1997b): Fine-structural features of male and female gonads in *Jensenia angulata* (Plathelminthes, Rhabdocoela, "Dalyellioida"). Microfauna Marina **11**, 251–270.
WATSON, N., K. ROHDE & J.B. WILLIAMS (1992): Ultrastructure of the protonephridial system of larval *Kronborgia isopodicola* (Platyhelminthes). J. Submicrosc. Cytol. Pathol. **24**, 43–49.

Prof. Dr. Ulrich Ehlers and Dr. Beate Sopott-Ehlers
II. Zoologisches Institut und Museum der Universität Göttingen
Berliner Str. 28, D-37073 Göttingen

Two *Prognathorhynchus* species (Kalyptorhynchia, Plathelminthes) from the North Inlet Salt marsh of Hobcaw Barony, South Carolina, USA

Peter Ax

In the salt marshes of New Brunswick, Canada, two species of the Kalyptorhynchia taxon *Prognathorhynchus* Meixner exist. One was described as *Prognathorhynchus eurytuba*. The other could only be named as *Prognathorhynchus* spec. (AX & ARMONIES 1987).

Populations of both species were found again in South Carolina in a comparable biotope of the intertidal (April 1995). They live in the muddy bottom with *Spartina alterniflora* in front of the Marine Field Laboratory of the Belle W. Baruch Institute for Marine Biology and Coastal Research, Georgetown.

In the case of the first species, we can confirm the identity between populations of New Brunswick and South Carolina. The second species can now be described as *Prognathorhynchus busheki* n.sp.

The North Inlet Estuary is located at 33° 20' N and 79° 20' W. The climate is considered warm temperature, with water temperatures in the major waterways ranging from 5° – 33°C. It is a high salinity system dominated by semidiurnal tides (mean range 1,4 m). Salinities in the main waterways are usually 30–35 ppt. More extreme temperatures and salinities are expected in the intertidal habitats where the Plathelminthes were collected (D. Allen, pers. communication).

Again, I am very grateful for the hospitality of my colleagues Dr. Dennis Allen and Dr. David Bushek during my stay at the Field Laboratory of the Belle W. Baruch Institute at the University of South Carolina.

Prognathorhynchus eurytuba Ax & Armonies, 1987

(Fig. 1A; 2A, B)

P. eurytuba: Ax & Armonies 1987, p. 57–60, fig. 33, 34A, B, E.

Material
Observations on several individuals (April 1995).

Fig. 1. A. *Prognathorhynchus eurytuba*. Stylet. B-D. *Prognathorhynchus busheki*. B. Proboscis hooks. C. Stylet. D. Copulatory organ with vesicula seminalis (vs), vesicula granulorum (vg) and stylet (st). North Inlet Salt Marsh, Hobcaw Barony, South Carolina.

The individuals from South Carolina are identical with the animals from New Brunswick in all details studied.

The semicircular stylet reaches to a length of 35 µm (32 µm in the Canadian population). The distal opening is strengthened at the concave side of the tube.

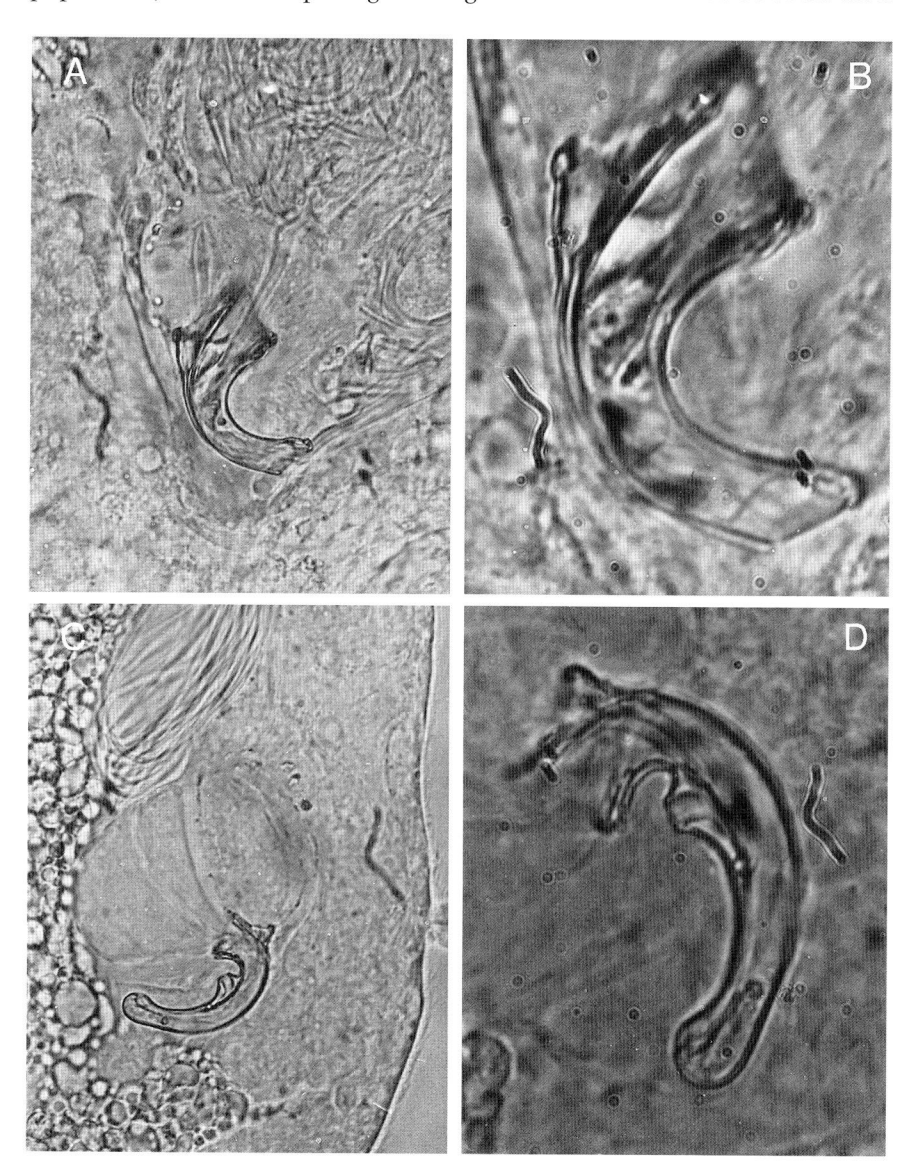

Fig. 2. A-B. *Prognathorhynchus eurytuba*. C-D. *Prognathorhynchus busheki*. Stylets in different magnifications. North Inlet Salt Marsh, Hobcaw Barony, South Carolina.

Prognathorhynchus busheki n. sp.

(Fig. 1B, C, D; 2C, D)

P. spec.: Ax & Armonies 1987, p. 60, fig. 34C, D.

Locus typicus
The North Inlet Salt Marsh of Hobcaw Barony, South Carolina.

Material
Observations on 3 specimens (April 1995).

Description
Slender body. Length about 1 mm. Uncolored. A pair of big eyes. Pharynx at the end of the first third of the body.

Simple proboscis hooks of ~ 13 µm length (18 µm in Canadian individuals).

Length of the stylet ~ 30–32 µm. The proximally curved tube has separate openings for sperm and granular secretion. Below these openings on the concave side there are two characteristic projections. The distal part of the tube terminates in a slight swelling.

No differences exist between the analysed populations of South Carolina and New Brunswick.

Discussion
The fact that populations of *P. eurytuba* and *P. busheki* live side by side in salt marsh areas of New Brunswick and of South Carolina speaks in favor of treating them as separate species.

Prognathorhynchus karlingi Ax from the Baltic is the most similar European species (Ax 1953). The three species in question have in common a small semicircular stylet with a length of only 30–35 µm. Furthermore, the length of their proboscis hooks is less than 20 µm.

References

Ax, P. (1953): *Prognathorhynchus karlingi* nov. spec., ein neues Turbellar der Familie Gnathorhynchidae aus der Kieler Bucht. Kieler Meeresforsch. **9**, 241–242.

Ax, P. & W. Armonies (1987): Amphiatlantic identities in the composition of the boreal brackish water community of Plathelminthes. A comparison between the Canadian and European Atlantic coast. Microfauna Marina **3**, 7–80.

Prof. Dr. Peter Ax
II. Zoologisches Institut und Museum der Universität Göttingen
Berliner Straße 28, D-37073 Göttingen